Nonlinear Systems and Complexity

Volume 26

Series editor

Albert C. J. Luo
Southern Illinois University
Edwardsville, IL, USA

Nonlinear Systems and Complexity provides a place to systematically summarize recent developments, applications, and overall advance in all aspects of nonlinearity, chaos, and complexity as part of the established research literature, beyond the novel and recent findings published in primary journals. The aims of the book series are to publish theories and techniques in nonlinear systems and complexity; stimulate more research interest on nonlinearity, synchronization, and complexity in nonlinear science; and fast-scatter the new knowledge to scientists, engineers, and students in the corresponding fields. Books in this series will focus on the recent developments, findings and progress on theories, principles, methodology, computational techniques in nonlinear systems and mathematics with engineering applications. The Series establishes highly relevant monographs on wide ranging topics covering fundamental advances and new applications in the field. Topical areas include, but are not limited to: Nonlinear dynamics Complexity, nonlinearity, and chaos Computational methods for nonlinear systems Stability, bifurcation, chaos and fractals in engineering Nonlinear chemical and biological phenomena Fractional dynamics and applications Discontinuity, synchronization and control.

More information about this series at http://www.springer.com/series/11433

Panayotis G. Kevrekidis ·
Jesús Cuevas-Maraver
Editors

A Dynamical Perspective on the ϕ^4 Model

Past, Present and Future

Springer

Editors
Panayotis G. Kevrekidis
Department of Mathematics and Statistics
University of Massachusetts
Amherst, MA, USA

Jesús Cuevas-Maraver
Grupo de Física No Lineal, Universidad de Sevilla
Departamento de Física Aplicada I, Escuela Politécnica Superior
Sevilla, Spain

Instituto de Matemáticas de la Universidad de Sevilla (IMUS)
Sevilla, Spain

ISSN 2195-9994 ISSN 2196-0003 (electronic)
Nonlinear Systems and Complexity
ISBN 978-3-030-11838-9 ISBN 978-3-030-11839-6 (eBook)
https://doi.org/10.1007/978-3-030-11839-6

Library of Congress Control Number: 2018968099

This Springer imprint is published by the registered company Springer Nature Switzerland AG
The registered company address is: Gewerbestrasse 11, 6330 Cham, Switzerland

Panayotis G. Kevrekidis dedicates this volume to Maria Kevrekidis, for all her sacrifices that made this book possible, to Despina and Athena Kevrekidis for all the smiles they bring to his life and to George and Despina Kevrekidis for expecting this book, before it was even conceived.

Jesús Cuevas-Maraver dedicates this volume to his beloved wife Conchi Caballero and their wonderful children Conchi and Jesús Cuevas-Caballero.

Panayotis and Jesús also want to dedicate this volume to Vassilis Tsartas for starting their collaboration on a left foot more than 15 years ago.

Preface

The study of field-theoretic models of the ϕ^4 type is of very broad appeal in Theoretical Physics and Applied Mathematics. Through its usefulness as a prototypical model in statistical mechanics and phase transitions, as well as its consideration as a hallmark of interface dynamics, nonlinear wave phenomena, and symmetry breaking instabilities, such models have enjoyed widespread attention from communities including cosmology, condensed matter physics, high-energy physics, and nonlinear partial differential equations, among many others. While different books have quantified the role of the ϕ^4 models in statistical physics, quantum mechanics, and nonlinear instability phenomena such as symmetry breaking, to the best of our knowledge, an account of some of its remarkable features in terms of nonlinear wave properties and interactions has not been previously assembled.

It turns out that since the late 70s and especially since the early 80s, some quite unexpected (especially given the "dominant", at the time, framework of integrable systems) features started to be reported in the interaction of coherent structures. The most salient manifestation of these phenomena was the existence of so-called multi-bounce windows (i.e., windows of repeated collisions before an eventual separation) in the interaction of kinks in the ϕ^4 model. Rapidly, this phenomenology spread in other non-integrable models (the double sine-Gordon equation, the parametrically modified sine-Gordon equation, later the case of coupled nonlinear Schrödinger models, etc.) These features started emerging as generic and so did their intuitive (and semi-quantitative) explanation on the basis of energy exchange between translational modes of the kink-like coherent structures and internal vibrational ones that could, in turn, decay into radiative phonon-like ones. Over time, additional surprises arose including the remarkable fractal structure of the multi-bounce windows; at the edge of two-bounce windows lied three-bounce ones, and so on and so forth, representing a self-similar structure over many scales. This, in turn, led to a systematic mathematical and computational study of this phenomenology that revealed its generic nature over multiple scales and many models. More recently, deviations from this now standard picture have began to appear. Some of these are due to models with different spectral properties than the earlier examined ones. Others are due to a careful reconsideration of some of the salient features of even some of the principal models such as the ϕ^4 one.

A parallel significant wave of activity over the past 30 years has been that of exploring nonlinear waves in discrete systems. Starting with pioneering numerical and semi-analytical observations in the late 80s, and cemented through suitable rigorous analysis over the following decade, time-periodic, spatially localized structures in the form of discrete breathers started being ubiquitously observed across models and (sub-)disciplines. From mechanical oscillators to electrical circuits, from magnetic systems to superconducting Josephson junctions, from DNA double strands to ionic or granular crystals, these patterns became a prevalent theme not only of theoretical and computational analysis but also of physical experiments. The ϕ^4 model was again a central pillar toward the consideration of such states and as a model enabling the systematic categorization of their existence, stability, and dynamics. More recently, such states have started to be argued as crucial catalysts for the dynamics and thermodynamic properties of the physical systems of interest, including, e.g., about their ability to conduct heat.

In light of these developments, we felt that this was an ideal time to summarize the current state of the art as regards the understanding of nonlinear wave phenomena in ϕ^4 models and to set an outlook toward the next decade of developments and associated challenges. In this effort, we invited some of the experts in the field, both from the early stages of study and from very recent, ongoing work on the subject to give us their perspective on the past, present, and future of this model, to summarize what it has taught us and look ahead to what more it has to offer. A brief summary of their contributions is as follows:

David Campbell, through his long-standing experience with the ϕ^4 model, contributed a historical overview of some of the early developments and offered wonderful insight into the early stages of the nonlinear model phenomenology.

Tomasz Romańczukiewicz and Yasha Shnir complemented this with a view of their recent studies on models beyond the ϕ^4, such as the ϕ^6 and the phenomenology of kink collisions in the presence, as well as absence of internal modes (including the potential presence of so-called quasi-normal modes).

Herbert Weigel shed light on the early (and quite important toward theoretical developments) models for the study of kinks through collective coordinates (CCs) enabling the characterization of their radiation and multi-bounce windows properties. Weigel raises some significant technical issues regarding such CC approaches whose resolution is a significant open challenge for future studies.

Roy Goodman presents a systematic mathematical analysis of the fractal kink–antikink collisions. This is based on qualitative ODE models related to those obtained via collective coordinate methods and uses a tractable discrete time-iterated map approximation encompassing many of the features of the dynamics.

Grant Lythe explored the nucleation of kinks; how they arise and how they may manifest, e.g., Brownian motion, in the presence of stochastic dynamics, using a combination of numerical and analytical techniques.

Dmitriev and Kevrekidis study the role of discretizations of continuum models. It turns out that numerous distinct discretizations exist, including ones that conserve momentum or energy. The properties of the kinks in those and how they approach the continuum limit, as well as how they behave in the regime of strong discreteness, are discussed in this chapter.

Cuevas-Maraver and Kevrekidis provide an overview of the relevance of discrete breather models, focusing on the contributions of Klein–Gordon field theories in these developments. Central to this chapter is the examination of not only linear but also nonlinear stability and its dynamical implications.

Another "classical" topic, namely, the continuum limit of ϕ^4 breathers, is tackled by John Boyd. Forty years later, this topic is revisited and the role of exponential asymptotics in establishing that the continuum ϕ^4 model possesses nanopteron (weakly nonlocal solitary wave) solutions is elucidated.

Igor Barashenkov explores the dynamics of a (single) wobbling kink, i.e., a kink with an excited internal mode. How the kink's wobbling fades into radiation through second harmonic generation mechanisms is elucidated, as are the dynamical implications of different types of drive such as parametric pumping or an external drive.

A somewhat unusual variant of the ϕ^4 model involving an external potential of a parabolic type is examined by Ross et al., partially motivated by atomic physics applications. Here, the existence and stability properties of the kink are dramatically modified. Nevertheless, their collisions are examined and are found to potentially feature multi-bounce windows (whose characteristics drastically depend on the trap strength).

Jean-Guy Caputo offers a glimpse into the possibilities for kinks in higher dimensions. Part of the focus is on radial structures (and the collisions of kinks between them and the domain boundaries) potentially leading to rapid breathers, while the dynamics of settings involving deformed waveguides, inhomogeneous problems, and dissipative reaction–diffusion models is also discussed.

A significant recent direction, involving the study of models such as ϕ^6, ϕ^8, ϕ^{10}, and ϕ^{12} and the $\mathcal{PT}$-symmetric ϕ^4 is explored by Saxena, Christov, and Khare. Here, there are numerous possibilities for novel phenomena, including kinks with algebraic tails, kinks with different numbers (possibly even controllable ones) of internal modes, etc.

Lastly, Xiong and Dmitriev present an intriguing set of implications of the existence of discrete breathers in models such as the ϕ^4 lattice. It is shown via systematic numerical computations (and through devising suitable diagnostics) that discrete breathers in non-translational-invariant lattices play a critical role on the thermal transport properties of the lattice leading to normal transport in consonance with Fourier's law.

We believe that these vignettes summarize the past, reflect the substantial recent developments and associated challenges of the present, and will hopefully pave the way for further surprises and lessons to be learned in the decade(s) to come in the future.

Amherst, MA, USA Panayotis G. Kevrekidis
Sevilla, Spain Jesús Cuevas-Maraver

Contents

Contributors

Igor Barashenkov University of Cape Town and National Institute for Theoretical Physics, Western Cape, South Africa;
Joint Institute for Nuclear Research, Dubna, Russia

John P. Boyd Department of Climate and Space Sciences and Engineering, University of Michigan, Ann Arbor, MI, USA

David K. Campbell Department of Physics, Boston University, Boston, MA, USA

Jean-Guy Caputo Laboratoire de Mathématiques, INSA Rouen Normandie, Saint-Etienne du Rouvray, France

Ivan C. Christov School of Mechanical Engineering, Purdue University, West Lafayette, IN, USA

Jesús Cuevas–Maraver Grupo de Física No Lineal, Universidad de Sevilla, Departamento de Física Aplicada I, Escuela Politécnica Superior, Sevilla, Spain;
Instituto de Matemáticas de la Universidad de Sevilla (IMUS), Sevilla, Spain

Robert Decker Mathematics Department, University of Hartford, West Hartford, CT, USA

Aslihan Demirkaya Mathematics Department, University of Hartford, West Hartford, CT, USA

Sergey V. Dmitriev Institute for Metals Superplasticity Problems of RAS, Ufa, Russia;
National Research Tomsk State University, Tomsk, Russia;
Institute for Metals Superplasticity Problems, Russian Academy of Sciences, Ufa, Russia

R. H. Goodman Department of Mathematical Sciences, New Jersey Institute of Technology, University Heights, Newark, NJ, USA

Panayotis G. Kevrekidis Department of Mathematics and Statistics, University of Massachusetts, Amherst, MA, USA

Avinash Khare Department of Physics, Savitribai Phule Pune University, Pune, India

Grant Lythe School of Mathematics, University of Leeds, Leeds, UK

Tomasz Romańczukiewicz Institute of Physics, Jagiellonian University, Kraków, Poland

Ryan M. Ross Department of Mathematics and Statistics, University of Massachusetts, Amherst, MA, USA

Avadh Saxena Los Alamos National Laboratory, Theoretical Division and Center for Nonlinear Studies, Los Alamos, NM, USA

Yakov Shnir Bogoliubov Laboratory of Theoretical Physics, Joint Institute for Nuclear Research, Dubna, Moscow Region, Russia

Herbert Weigel Institute for Theoretical Physics, Physics Department, Stellenbosch University, Stellenbosch, South Africa

Daxing Xiong Department of Physics, Fuzhou University, Fuzhou, Fujian, China

Acronyms

DB	Discrete breather
DNLS	Discrete Nonlinear Schrödinger
KG	Klein–Gordon
LHS	Left-hand side
NLS	Nonlinear Schrödinger
ODE	Ordinary differential equation
PDE	Partial differential equation
RHS	Right-hand side
SDE	Stochastic differential equation
sG	Sine-Gordon
SPDE	Stochastic partial differential equation

Chapter 1
Historical Overview of the ϕ^4 Model

David K. Campbell

Abstract We provide an historical overview the ϕ^4 model, placing it in the broad context of general KG theories and reviewing its applications to physics. While we will mention briefly other variants of the KG theory, we will focus chiefly on the history of the one-space, one-time dimensional [(1+1)D] degenerate minimum ϕ^4 theory that is the central topic of this book. We will also compare this theory to other nonlinear Klein–Gordon equations, in particular, to the celebrated sine-Gordon theory. We review in some detail the history of the interrelated dynamical problems of kink-antikink scattering, contrasting ϕ^4 and other non-integrable models with sine-Gordon and the search for a possible "breather" solution to ϕ^4 in the continuum limit. Our discussion is intended to set the stage for more detailed expositions in the later chapters.

1.1 Pre-history: The Klein–Gordon Equation(s) and ϕ^4 Theory

The ϕ^4 theory is one of a large class of theories based on the KG equation, first proposed by Klein [1] and Gordon [2] which was intended to describe an electron moving in a relativistic manner. In modern form, the KG equation can written as a *linear* (3+1)D relativistic equation in terms of a field ϕ that represents a particle of mass μ

$$\frac{\partial^2 \phi}{\partial t^2} - \frac{\partial^2 \phi}{\partial x^2} - \frac{\partial^2 \phi}{\partial y^2} - \frac{\partial^2 \phi}{\partial z^2} - \mu^2 \phi = 0. \tag{1.1}$$

Famously, this equation did not account for the electron's spin, which is correctly described by the Dirac equation [3]. However, for spinless scalar (or pseudo-scalar) particles, the KG theory in principle could apply and historically has been used to describe pion mesons and (perhaps) the Higgs boson [4]. However, the unstable

D. K. Campbell (✉)
Department of Physics, Boston University, 590 Commonwealth Ave, Boston, MA, USA
e-mail: dkcampbe@bu.edu

P. G. Kevrekidis and J. Cuevas-Maraver (eds.), *A Dynamical Perspective on the ϕ^4 Model*, Nonlinear Systems and Complexity 26,
https://doi.org/10.1007/978-3-030-11839-6_1

nature of these particles and the need to incorporate them into a consistent quantum field theory of elementary particles has limited the utility of the KG equation *per se* in the relativistic domain. Within the "Standard Model" of particle physics, the pion is a composite pseudo-scalar formed from an up and down quark, whereas the Higgs is an elementary particle that generates the mass of all other particles [5] (For a good recent reference to quantum field theory and the standard model, see [6]).

Our primary concern in both this chapter and the entire volume is the dynamics of a nonlinear variant of the KG theory known as the "ϕ^4 theory", and more specifically on the the one-plus-one dimensional [(1+1)D], degenerate minimum version of ϕ^4 in continuous space and time. The fundamental equation defining the theory is

$$\phi_{tt} - \phi_{xx} - \phi + \phi^3 = 0 \tag{1.2}$$

which comes from the Hamiltonian

$$H = \int \mathrm{d}x \, \frac{1}{2}\left(\frac{\partial \theta}{\partial t}\right)^2 + \frac{1}{2}\left(\frac{\partial \theta}{\partial x}\right)^2 + \frac{1}{4}(\phi^2 - 1)^2, \tag{1.3}$$

in which the quartic potential function

$$V(\theta) = \frac{1}{4}(\phi^2 - 1)^2 \tag{1.4}$$

has a double-well form with degenerate minima at $\phi = \pm 1$. The presence of two degenerate minima allows for the existence of finite energy, localized "kink" and "antikink" solutions of this equation which interpolate between these two minima as x runs from $-\infty$ to $+\infty$. Indeed, working in a co-moving frame $\xi = x - vt$, straightforward integration establishes the existence of a "kink" solution, which has the form

$$\phi_k = \tanh \frac{\gamma(v)(x - x_0 - vt)}{\sqrt{2}} \tag{1.5}$$

where $\gamma(v) = \frac{1}{\sqrt{1-v^2}}$ reflects the Lorentz invariance of the theory and the constant x_0 reflects the fact that the kink's energy is independent of its location because of the translation invariance of the theory. This solution interpolates between the minimum $\phi = -1$ as $x \to -\infty$ and $\phi = +1$ as $x \to +\infty$. Importantly, it does represent a physically localized excitation, because all the energy and momentum associated with this wave are (exponentially) localized around the kink's location, $x = x_0 + vt$. There is also an antikink solution, $\phi_{\bar{k}}$, which is just the negative of ϕ_k,

$$\phi_{\bar{k}} = -\tanh \frac{\gamma(v)(x - x_0 - vt)}{\sqrt{2}} \tag{1.6}$$

and interpolates between the minimum $\phi = +1$ as $x \to -\infty$ and $\phi = -1$ as $x \to +\infty$. These kink and antikink solutions will play a central role in our later discussion.

In an ensuing section, we shall compare the ϕ^4 theory to the celebrated "sine-Gordon" theory (defined below) which also contains kink and antikink solutions to be described later. We shall also consider a *spatially discretized* version of the ϕ^4 theory. Elsewhere in this volume, other authors will consider nonlinear KG theories involving higher order polynomials (ϕ^6, ϕ^8, etc.) (see Chaps. 2 and 12), versions of ϕ^4 in higher spatial dimensions (Chap. 11), "discrete" versions of ϕ^4 defined on a lattice (Chaps. 7 and 8), and versions of ϕ^4 in external potentials, both regular (Chap. 10) and stochastic (Chap. 5).

Before turning to these detailed discussions, let me digress to provide a brief discussion of the applications of ϕ^4 theory to physical systems.

1.2 Applications of ϕ^4 Theory in Physics

The variants of the ϕ^4 theory discussed above —in different spatial dimensions, with degenerate or non-degenerate minima, in static or dynamical contexts, in classical or quantum versions— have arisen as models for a vast array of physical phenomena, from the atomic to the cosmological scale. The following very brief introduction discusses a limited number of these applications and is intended to be indicative rather than definitive. In particular, it is focused chiefly on applications of (1+1)D, degenerate minima ϕ^4 theory (1.2) where the kink-like nonlinear excitations play a prominent role. Notwithstanding this focus, we mention also other applications, to indicate the remarkable range of applicability of this theory.

Specifically, ϕ^4 theories have permeated physics from the smallest to the largest scales. At the small scales, in elementary particle physics, the (3+1)D versions of ϕ^4 theory play a central role in the Higgs phenomenon [4] that is essential to the "standard model" of particle physics [5]. At the largest scales, we note that as early as 1975 (3+1)D ϕ^4 theories were used in cosmology to describe domain walls that can exist in models of the early universe [7].

Perhaps the first successful application of the ϕ^4 theory in physics (involving only the static version) was as a (classical) order parameter in the phenomenological theory of second-order phase transitions proposed by Ginzburg and Landau in 1950 [8]. In the ensuing decades, the one-dimensional (1+1)D ϕ^4 theory has been applied in condensed matter physics at a more microscopic level to structural phase transitions in the displacive limit [9–21] and specifically to uniaxial ferroelectrics [10, 11, 15] The kink-like excitations of ϕ^4 theory have been used to model mobile domain walls [13, 15–21] in the dynamical response functions [10, 11, 13, 15–21] leading to cluster formation [17, 18] and possibly explaining the observed "central peak" phenomenon [11, 15–21]. In quasi-one-dimensional conducting polymers, such as polyacetylene, the 1D ϕ^4 theory has served as a phenomenological model for nonlinear excitations [22–24] with exotic spin-charge relations [25–28]. Most recently, the kink excitations of ϕ^4 theory have been proposed as being relevant to dynamics in twisted graphene nanoribbons [29].

In statistical physics, ϕ^4 kinks appear as fundamental excitations in the partition function [12, 15, 19, 20] and in structural phase transitions as noted above. In quantum field theory, the (1+1)D ϕ^4 theory has been proven rigorously to exist as a well-defined quantum field theory equivalent to a 2D classical Ising model [30] and proven to have a phase transition between unbroken and broken symmetry states [31]. In the semi-classical regime, the (1+1)D ϕ^4 theory has been used to illustrate the connection between nonlinear classical excitations and quantum particles [32–35]. Further, when coupled to fermions, ϕ^4 has been used as toy model of nuclear physics [36, 37] and as an illustration of the nontrivial effects of topology on fermion number [38].

Interestingly, in condensed matter/solid state physics applications, where the field ϕ typically represents the small motion of atoms in the solid ("phonons"), the more natural formulation would seem to be the discrete, quantized version of the theory. But more often one uses the (more analytically tractable) classical continuum version, which loosely speaking amounts to making the Born–Oppenheimer approximation in which the phonon motion is treated classically. In contrast, in particle physics, the natural framework is a space-time continuum, but one often goes to discrete (lattice) theories to enable controlled (lattice) quantum Monte Carlo calculations.

Finally, we note that the static (1+1)D ϕ^4 theory, when considered in imaginary time, corresponds to the "instanton" that describes tunneling between the two degenerate classical states in the simple double well problem in quantum mechanics [39]. In sum, ϕ^4 theory is ubiquitous in physical applications.

1.3 The Saga of ϕ^4 Versus Sine-Gordon

In this section, we will compare and contrast two classical field theories defined in one spatial dimension and time by seemingly similar nonlinear PDEs: the ϕ^4 theory (1.2) and the "sine-Gordon" theory (1.7). Historically, this comparison played a crucial role in the understanding of the distinction between "solitons" and "solitary waves" and the critical differences between completely integrable and *non*-integrable classical field theories. An early and essential reference, to which I shall refer often is the article by Dashen, Hasslacher, and Neveu (DHN) [33].

We begin with a brief review of the sG equation and the analytic solutions that will be relevant to our later studies. For more detailed discussions see the reviews in [40–42]. In dimensionless units, the sG equation is

$$\frac{\partial^2\theta}{\partial t^2} - \frac{\partial^2\theta}{\partial x^2} + \sin\theta = 0 \tag{1.7}$$

and follows from the Hamiltonian

$$H = \int \mathrm{d}x \, \frac{1}{2}\left(\frac{\partial\theta}{\partial t}\right)^2 + \frac{1}{2}\left(\frac{\partial\theta}{\partial x}\right)^2 + (1 - \cos(\theta)) \tag{1.8}$$

in which the potential function

$$V(\theta) = 1 - \cos(\theta) \tag{1.9}$$

is periodic with period 2π. This means that there exists an infinite number of equivalent ground states at $\theta = \pm 2n\pi$. Thus there can exist finite energy solutions that interpolate between any of these ground states: these are of course the "N soliton" solutions described in the above mentioned reviews and elsewhere.

The name, "sine Gordon", which has been characterized by Sidney Coleman as "the sophomoric but unfortunately standard name for the theory" [43] is an obvious pun on "Klein–Gordon" and was apparently first introduced by Martin Kruskal,[1] who, given his naming (with Norman Zabusky) of the "soliton" [45], is perhaps entitled to have some fun with this name. Certainly this name does not appear in the early discussions of this equation, which initially arose in the 19th century in the context of spaces of constant negative curvature [46, 47] and later appeared in discussions of dislocations in solids [48, 49] as well as in a model field theory for elementary particle physics [50], among many other applications (again see the reviews for details). The sG equation is famously a completely integrable infinite degree of freedom Hamiltonian system. As shown by Ablowitz, Kaup, Newell and Segur [51], its initial value problem can be solved completely by an inverse scattering transform first proposed by Zakharov and Shabat [52]. In the sections below we present only the solutions that are directly relevant to our current discussion.

The single "kink" soliton solution to the SG equation is

$$\theta_k(x,t) = 4\tan^{(-1)} \mathrm{e}^{\gamma(v)(x-x_0-vt)}, \tag{1.10}$$

where $\gamma(v)$ is defined above,

Since this solution approaches 0 as $x \to -\infty$ and 2π as $x \to +\infty$, it describes a "kink" in the field θ localized around $x = x_0 + vt$. In this respect, it is similar to the "kink" soliton in the ϕ^4 theory described in (1.5).

Similarly, the single "antikink" soliton solution

$$\theta_{\bar{k}}(x,t) = 4\tan^{(-1)} \mathrm{e}^{-\gamma(v)(x-x_0-vt)}, \tag{1.11}$$

interpolates between 2π as $x \to -\infty$ and 0 as $x \to +\infty$.

The remarkable complete integrability of the sG system [51, 52] allows one to write down analytic expressions for N-soliton solutions for any N. For the case of N = 2, there are three such solutions corresponding to (1) kink-kink interactions; (2) kink-antikink interactions; and (3) a localized, oscillatory solution known as a "breather" or "bion". The solution corresponding to the interaction of a kink and

---

[1] This claim is made in a footnote on p. 259 in [44].

antikink, in which the two objects pass through each other while maintaining their structures and suffering only a phase shift, has the form (in its rest frame)

$$\theta_{k\bar{k}}(x,t) = 4\tan^{(-1)}\left(\frac{\sinh\gamma(v)vt}{v\cosh\gamma(v)x}\right). \tag{1.12}$$

The dedicated reader can verify that as $t \to -\infty$, $\theta_{k\bar{k}}$ looks like a widely separated kink and anti-kink approaching each other with velocity v. For t near 0 they interact nonlinearly, but as $t \to +\infty$, the kink and antikink emerge with their forms intact but with a "phase shift" or "time delay" which is given by

$$\delta x = 2(1-v^2)^{1/2}\ln(1/v). \tag{1.13}$$

The phase shift is such that it indicates that the kink and antikink are attracted to each other; this observation will become important in our discussion below of collective coordinate approaches to these interactions. Note that since the equation is invariant under $\theta \to \theta + 2n\pi$, a kink that interpolates between 2π and 4π is physically equivalent to one that interpolates between 0 and 2π.

The interaction of two kink solitons has the related analytic form

$$\theta_{kk}(x,t) = 4\tan^{(-1)}\left(\frac{v\sinh\gamma(v)x}{\cosh\gamma(v)vt}\right). \tag{1.14}$$

To our knowledge these interacting kink solutions were first discovered *analytically* by Seeger, Donth, and Kochendorfer [49] using Bäcklund transformation arguments based on the 19th century work on surfaces of constant curvature. But they were later independently found by Perring and Skyrme [50] via numerical simulations —"experimental mathematics" [53]— and inspired guesswork. These latter authors studied both kink-kink and kink-antikink scattering and reported their results thusly:

> The problem of two colliding particles is equivalent, by symmetry, to that of a single particle moving in the half space $x > 0$ with boundary conditions $\alpha(0) = 0$ and $\alpha(\infty) = 2\pi$. To our initial surprise the numerical integration showed that the scattering off the boundary at $x = 0$ was purely elastic, almost like hard-sphere scattering. An analytical solution was then found with these properties [...] and describes two particles centered at approximately $x = \pm vt$ and colliding elastically at $x = 0$ at $t = 0$ [...] A similar solution exists describing a particle-antiparticle collision [...]

In the sG theory, there is another nonlinear wave-like solution that can loosely be thought of as a "bound state" of a kink and antikink. Known as a "breather" or "bion", this solution is spatially-localized and periodic in time. The equation for the sG breather in its rest frame has the form

$$\theta_B(x,t) = 4\tan^{(-1)}\left(\frac{\varepsilon\sin(t/\sqrt{1+\varepsilon^2})}{\cosh(\varepsilon(x-x_0)/\sqrt{1+\varepsilon^2})}\right) \tag{1.15}$$

and a moving breather solution can be obtained by applying a Lorentz transformation. We note two key points for future reference. First, for $\varepsilon \to \infty$ the period of the breather solution goes to infinity and the solution looks like a widely separated kink-antikink pair, whose constituents "attempt" to separate to infinity but are in fact "trapped" in the oscillatory breather state. Second, and closely related, if one replaces v by $\frac{i}{\varepsilon}$ in the equation for the kink-antikink scattering (1.12), one obtains exactly the breather solution (1.15). We shall return to this final point in our discussion of the putative ϕ^4 breather.

The existence in ϕ^4 theory of kink equation (1.5) and antikink equation (1.6) solutions so similar to those of the sG equation begs the immediate question as to whether the ϕ^4 theory can contain solutions (analytic or otherwise) like the kink-antikink interaction $\theta_{\bar{k}k}$ and the breather solution θ_B in the sG system. In the 1970s and 1980s these questions were widely studied and intensely debated. Although the answer is now well known —the sG has these special solutions (and more general N-soliton solutions) precisely because it is completely integrable and the ϕ^4 theory, which is not completely integrable, does not have such solutions— in view of our goal of presenting an historical perspective on the ϕ^4 theory, we shall for the moment suspend this knowledge and trace the circuitous path by which we came to it. Precisely because the path was circuitous, we will begin here with a qualitative summary of the final results without detailing the references and then discuss the extensive quantitative details with full references in the ensuing sections. In the case of kink-antikink scattering in ϕ^4, it was found numerically that at large incoming velocities, the kink and antikink reflected off each other inelastically, losing energy to small propagating oscillations ("phonons" or "mesons", depending on the intended physical context). At lower velocities, the kink and antikink remained trapped in a long-lived oscillatory state that resembled the sG breather but clearly decayed over (long) times. In more detailed studies, it was found that regions of reflection and trapping alternated with each other in a systematic way that could be understood first heuristically and then more precisely by analytic methods. In the case of the putative ϕ^4 breather, noticing that the sG breather admitted a series expansion in terms of ε suggested attempting a similar expansion for the ϕ^4 equation, and indeed it was found that such a series expansion existed and showed that *to all orders in* ε, there was a spatially localized, time-periodic solution to ϕ^4: thus a ϕ^4 breather. However, subsequent analytic studies established that whereas this series was convergent in the case of the sG theory, for ϕ^4 it was only asymptotic, and in fact there were terms of order $e^{(-1/\varepsilon)}$ —that is, beyond all orders in perturbation theory— that destroyed the spatial localization of the putative breather and meant that only "nanopteron" solutions (infinite energy oscillating solutions with small amplitude oscillating tails extending to infinity) existed in the continuum theory. But we are getting ahead of ourselves and should turn now to the more detailed discussions.

An essential characteristic of "solitons" —as distinct from mere solitary waves— is that they persist in form despite interactions with other waves: as shown in (1.12) the sG kink and antikink pass through each other, with only a phase shift. What happens in the non-integrable ϕ^4 case? Without any obvious analytical guidance, the approach taken in the mid-seventies when the question was first studied [13, 54,

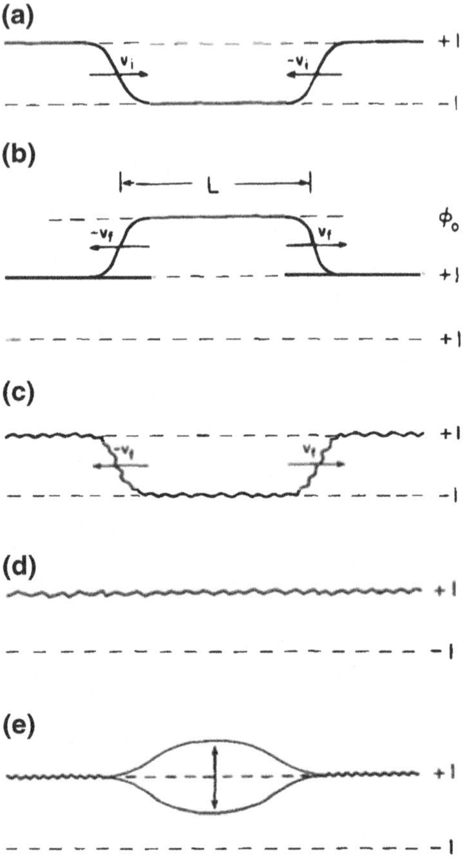

Fig. 1.1 Possible results of kink-antikink collisions in the ϕ^4 theory. The schematics of a kink-antikink ($K\bar{K}$) collision in the ϕ^4 theory, showing **a** the initial state, with the kink and antikink approaching each other with velocity $-v_i$ and v_i, respectively; **b** the energetically- forbidden state in which the kinks pass through each other; **c** a state in which the K and $\bar{K}$ are reflected, with velocities v_f and $-v_f$, respectively; **d** a state in which the K and $\bar{K}$ annihilate immediately; and **e** a trapped "breather" state, in which the excitation stays (at least for a substantial time) localized near the origin, and oscillates up and down. Adapted from [56], with permission from Elsevier

55] was to start a computer simulation with an "initial state" in which a kink and an antikink were widely separated —so that their "sum" is in fact (exponentially) close to a solution to (1.2)— and head them toward each other with initial velocities v_i and $-v_i$. In Fig. 1.1 we illustrate several possible "final" states corresponding to these initial states.

The first point to note is that for ϕ^4 it is not possible for the kink and antikink to pass through each other, for the resulting state —as indicated in Fig. 1.1b— is forbidden on energetic grounds. Since the region between the two kinks in Fig. 1.1b is at some value ϕ_0 *not* equal to a minimum of the potential, the energy of this configuration is $E \simeq V(\phi_0)L$ (where L is the separation between the kinks) and hence L can *not* approach infinity. Lest you think that this obvious "topological" distinction between the sG and ϕ^4 theories is responsible for the surprises about to be unveiled, we should stress that the ensuing results are also found in non-integrable theories where the kinks *can* pass through each other (e.g., the double sG equation [57]). In Fig. 1.1c we sketch a final state in which the kink and antikink are reflected, possibly with some additional small radiation. In Fig. 1.1d we depict a state in which the kink and antikink have annihilated rapidly, and finally, in Fig. 1.1e we show a final state consisting of a "breather"-like excitation, which stays (at least for a substantial time) localized near the origin, and oscillates up and down.

In principle, all the behaviors sketched in Fig. 1.1c–e are possible. Without analytic guidance, researchers turned to numerical simulations to study the interactions. The pioneering numerical studies of kink-antikink (K–$\bar{\text{K}}$) interactions included Kudryavtsev [55], Aubry [13], Getmanov [58], Makhankov [59], Sugiyama [60] and Ablowitz, Kruskal, and Ladik [54], this last reference published only in 1979 but based on numerical work begun in 1972. It is interesting to compare their results in the light of the potential results of the collisions mentioned above.

The study of Kudryavtsev [55] focused on a kink and antikink separated by a distance $2x_0$ and approaching each other with velocity $v = 0.1$ (in units of c). He found that rather than annihilating quickly into small oscillations, the kink and antikink "captured" each other and formed an oscillatory bound state (like the SG breather?). By calculating the effective potential energy[2] of a kink-antikink configuration as a function of x_0, he showed that the kink and antikink attracted each other and that, if one ignored "radiation", this attraction could lead to the trapped state oscillating forever. However, with the inclusion of radiation, the trapped state slowly decayed. Since he focused on only one velocity (well below the initial threshold for trapping), he did not observe reflection or the existence of alternating regions of trapping and reflection. However, he did remark on the fact that the kink and antikink cannot pass through each other but must repel (i.e., reflect from) each other, noting that:

> The growth of the potential energy at negative x_0 is the cause of soliton repulsion (note added: reflection) in this model. This fact was first noted by N. A. Voronov.

The studies of Getmanov [58] and the review of Makhankov [59] corroborated the trapping of the kink and antikink at low velocities, although their critical velocities differed somehwat from Kudryavtsev's result [55].

More interestingly, in his article, Aubry [13], studied many different velocities, and thus observed both trapping and reflection, as well as a sequence of trapping "windows" alternating with reflections. In his words:

> (a) For some values of v, usual soliton properties were found. For example, for $v = 0.5c$ walls go through each other. But after the collisions, part of the kinetic energy [...] is dissipated to small oscillations (phonons and a phase shift appears). (b) For other values of v, solitons are destroyed in the collision. For instance, for $v = 0.1c$, walls swell into each other and their total energy is slowly dissipated along the chain [...] (c) Channels of annihilation were found. For instance, walls annihilate for $v = 0.25c$ but they do not for $v = 0.22c$ and $v = 0.26c$. Also, the amplitude and sign of the collision time are velocity dependent.
>
> A more detailed analysis of these phenomena should be developed.

Since the kinks are indistinguishable, whether one says that they go through each other or reflect is to some extent semantics, but if one observes the collisions and

---

[2]This is arguably the first introduction of a collective coordinate approach in studies of the ϕ^4 theory. It was followed closely by the more detailed work of Sugiyama [60]. See the section of "Collective Coordinates" below and for a thorough discussion Chaps. 3 and 4 of this book.

compares what happens in higher order polynomial field theories (e.g. ϕ^6 or ϕ^8, see Chaps. 2 and 12) where there *are* additional degenerate minima and the kinks can move to those, the concept of reflection seems more appropriate. But notice that Aubry observed two of the resonant windows at values roughly comparable with those known accurately today [61]. Similarly, and independently, Ablowitz, Kruskal, and Ladik (AKL) [54] reporting on work begun in 1972, noted:

> Finally, we remark that (2c) (*noted added:* the ϕ^4 theory) exhibits a phenomenon not noticed by us in any other calculation nor by Kudryavasev (*sic*) [4] (*note added:* their citation number [4] is to our [55]). As v_I decreases to 0.3, v_F is seen to decrease to .135. At $v_I = .25$, an oscillatory state is observed. However, as v_I is decreased to .2, this oscillatory state changes back to emanating waves in a small region about $v_I = .2$. Indeed, at $v_I = .2$, v_F is found to be .155, which is larger than the v_F observed when $v_I = .3$! Somewhat below $v_I = .2$ ($v_I = .193$) an oscillatory state reappears. Subsequent reductions in v_I also result in oscillatory waveforms. The reason for the apparent "resonance" between these interacting aperiodic waves and the radiation is not yet fully understood.

AKL were apparently not aware of the study by Aubry and were, to our knowledge, the first to use the term "resonance" to describe the sequence of trapping and reflection in this scattering.

Sugiyama's work [60] is noteworthy in that in addition to studying the K–$\bar{\text{K}}$ collisions numerically, he introduced formal "collective coordinates" corresponding to the locations of the kink and antikink (x_0) and the amplitudes of the small oscillations ("shape modes") that are localized around the kink. In this approach, he was following the example of collective coordinates used in field theories applied to particle physics (see, for example [62]). Sugiyama calculated analytically the form of the Lagrangian for these collective coordinates, effectively reducing the ϕ^4 PDE to a two coupled ODEs and used this to discuss the "emission" of small oscillations ("bosons", in his terminology, since he was considering the theory as a model for elementary particle physics) from the kinks. He did not however use his collective coordinate formalism to study the details of the K–$\bar{\text{K}}$ collision, but merely calculated the value of the velocity at which the energy transferred to the shape modes was sufficient to permit trapping and used that to determine the critical velocity for trapping, V_c which he estimated to be $V_c \simeq 0.25$. Further, in his numerical investigations he found only reflection for $V > V_c$ and trapping below, missing the resonance windows seen by Aubry [13] and AKL [54]. In a sense he was fortunate not to attempt to use his collective coordinate formalism to calculate the full range of K–$\bar{\text{K}}$ interactions, for as pointed out by Weigel and collaborators [63, 64], there were some misprints in his evaluation of key functions in the collective coordinate model, which would have resulted in incorrect results. We will return briefly to the issue of collective coordinates in a later section, but for the full details interested readers are referred to Chap. 3 of this book. Additional early studies [65, 66] reported more than one range of initial velocities for which the collisions end in reflection and further that the reflection regions alternated with regions ending in capture.

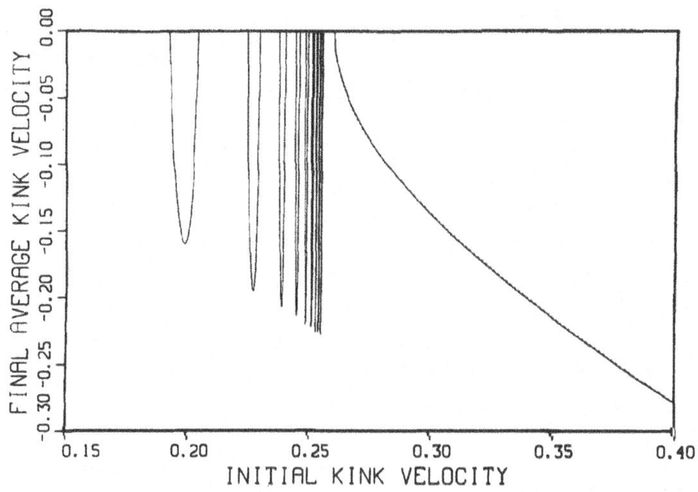

Fig. 1.2 The Kink-Antikink Collision in the ϕ^4 Theory. The final velocity as a function of the initial velocity of the kinks. A final velocity of zero means that the $K-\bar{K}$ are captured into a bound, oscillatory state. The dashed curve (which is indistinguishable from the data) is a fit to the form (constant) $\times(v_i^2 - v_{i,c}^2)^{1/2}$ for v_i above $v_{i,c}$. Adapted from [56], with permission from Elsevier

In 1983, Campbell, Schonfeld, and Wingate [61] published a detailed numerical study of the ϕ^4 kink-antikink interactions which revealed the rich resonance structure shown in Fig. 1.2 (taken from the original reference).

As seen in previous studies, at low initial velocities ($v_i < 0.193\cdots$) a trapped state is always formed whereas at high velocities ($v_i > v_{i,c} = 0.2598\cdots$) the kinks always reflect inelastically. The striking feature of Fig. 1.2 is the sequence of regions of intermediate *many* v_i in which trapping and reflection alternate. These "reflection windows" suggest an unexpected "resonance" phenomenon in the interaction of these nonlinear solitary waves, exactly as remarked in [54].

To explain this resonance phenomenon, the authors of [61] developed what they termed the "Resonant Energy Exchange Mechanism" (REEM). This mechanism explains in quantitative detail the surprising resonance structure seen in the ϕ^4 kink-antikink collisions [61] as well as similar resonances seen in other nonlinear KG theories [56, 57, 67]. First derived heuristically in those references, the REEM was later established more rigorously [68, 69] (see also Chap. 4) and recently extended and revised in important ways in [63, 64, 70] (see also Chap. 3).

The key to understanding the "resonance" mentioned by AKL and shown in detail in Fig. 1.2 was to determine what was resonating with what! This question was answered, heuristically, in [61]. The central idea involved two critical features. The first was observation, originally made by Kudryavtsev [55] (and quoted above) that the kink and antikink act like particles that are moving in an attractive potential that can lead to trapping. The second was that the kinks are individually deformable, with internal oscillations corresponding to modes localized around the separate solitary waves. In the spectrum of modes around (static) ϕ^4 kink localized at x_0, in addition to the zero-frequency Goldstone mode corresponding to the translation invariance of the kink, there is a shape mode, with frequency $\omega_s = \sqrt{3}/2$ with normalized wave function

$$\delta\phi_s = 2^{(-3/4)} \tanh \frac{(x - x_0)}{\sqrt{2}} \mathrm{sech} \frac{(x - x_0)}{\sqrt{2}} \tag{1.16}$$

which is clearly exponentially localized around the kink.[3]

When the K and $\bar{\mathrm{K}}$ first interact, the shape mode of each is excited, removing energy from the translational modes of the kinks and trapping them in their mutually attractive potential. Since the internal modes move with the kinks (essentially as adiabatic invariants), the energy transferred to them is stored, not lost. After their first collision, the kinks move apart essentially as independent particles. But since they are trapped by their mutually attractive potential, they must return to collide a second time, and when they do the energy stored in the internal nodes can be restored to the K and $\bar{\mathrm{K}}$ translational motion, leading to escape for the K and $\bar{\mathrm{K}}$, if an appropriate resonance condition is met. This condition has the form

$$\omega_s T_2 = 2n\pi + \delta \tag{1.17}$$

where ω_s is the shape mode frequency, T_2 is the time between the first and second $\mathrm{K}\bar{\mathrm{K}}$ collision, δ is an offset phase (between 0 and 2π by convention) and n is an integer, which is the "order" of the resonance. Simple particle mechanics, applied to the individual kinks [56, 61] leads to the estimate for T_2 in terms of the $\mathrm{K}\bar{\mathrm{K}}$, binding energy ϵ,[4]

$$T_2 = \frac{\pi}{\omega_0} \frac{1}{\sqrt{\epsilon}} \tag{1.18}$$

where ω_0 is the lowest frequency ($= \sqrt{2}$ for the ϕ^4 theory) in the continuum of small oscillations around the static kink. This estimate for T_2 is the leading order for $\epsilon \to 0$ [61]. As shown in detail in [61] using a combination of numerical data and heuristic arguments, one can express the binding energy of the $\mathrm{K}\bar{\mathrm{K}}$ pair as a function of their initial velocity v_i in the center of mass. For $v_i < v_c$, where v_c is the critical velocity for which the kinks first trap,

$$\epsilon(v_i) = \alpha(v_c^2 - v_i^2). \tag{1.19}$$

The constant α is determined empirically from the data for $v_i > v_c$. Combining (1.18) and (1.19) leads to an expression for T_2 in terms of v_i

[3]For purposes of our later discussion and links to other chapters in this volume, we note that the location of the kink x_0 and the amplitude of the shape mode can be taken as "collective coordinates" and used to reduce the PDE to coupled ODEs [60, 63, 64] (see also Chap. 3) or coupled maps [68, 69] (see also Chap. 4).

[4]The binding energy is amount by which the kinks are trapped in the their common potential well. See [61] for details.

$$T_2(v_\mathrm{i}) = \frac{\beta}{(v_\mathrm{c}^2 - v_\mathrm{i}^2)^{1/2}} \tag{1.20}$$

where

$$\beta = \frac{\pi}{\omega_0 \sqrt{\alpha}}. \tag{1.21}$$

Combining (1.20) with (1.17) leads to the prediction for those initial velocities, v_n, below the initial trapping velocity (v_c) which lead to resonant escape of the $\mathrm{K\bar{K}}$ pair:

$$(v_\mathrm{c}^2 - v_\mathrm{n}^2) = \frac{\beta^2 \omega_\mathrm{s}^2}{(2n\pi + \delta)^2} \tag{1.22}$$

Hence the REEM predicts a sequence of resonances, becoming increasingly dense as $v_\mathrm{i} \to v_\mathrm{c}$ from below. In addition, the width of the nth resonance is estimated to fall (for large n) as $1/n^3$ [61]. Limitations on the range of n for which the theory is expected to apply are discussed in detail in the references; in essence, one finds that n cannot be too small (otherwise the leading order estimate for T_2 in (1.18) fails), nor too large (when T_2 becomes very long, the coupling through higher order nonlinear terms to modes which remove energy from the kink becomes important).

Table 1.1 (adapted from [61]) gives a quantitative summary of the predicted versus observed structure for the first nine resonances. The fit of this theory to the resonance structure observed in ϕ^4 collisions is remarkable. Studies of $\mathrm{K\bar{K}}$ collisions in related models —the "parametrically modified sine-Gordon" [56] and the "double sine-Gordon" [57] theories— revealed equally impressive fits. Importantly, in the former case for a certain region of parameter space there are no shape modes [56] (and hence no (obvious!) internal oscillation in which energy can be stored). The REEM would thus predict that there is no resonance structure in this case, and the numerical

Table 1.1 Comparison of the numerical data on the location and widths of the various resonance windows with the predictions of the REEM. Adapted from [67], with permission from Elsevier

n	Range of v_i	Center	Width	Predicted
1	0.1926–0.2034	0.1980	0.0109	0.1990
2	0.2241–0.2288	0.2265	0.0048	0.2250
3	0.2372–0.2396	0.2384	0.0025	0.2370
4	0.2440–0.2454	0.2447	0.0015	0.2437
5	0.2481–0.2490	0.24855	0.0010	0.2478
6	0.2507–0.2513	0.2510	0.0007	0.2505
7	0.2525–0.2529	0.2527	0.0005	0.2524
8	0.2538–0.2541	0.25395	0.0004	0.2538
9	0.2548–0.2550	0.2549	0.0003	0.2548

simulations confirmed this.[5] More recently, however, studies in the ϕ^6 model [63, 64, 70] (see also Chap. 3) have shown that it is possible for a variant of the REEM to exist in cases in which there are no shape modes around the kink. The details are discussed in those references, but the key observation is that the K$\bar{\text{K}}$ stay trapped and oscillate around the origin in a manner that effectively creates a (time-dependent) "potential" that produces a bound state within the oscillating structure. This bound state plays the role of the shape modes in the original theories, and a similar "energy exchange" argument leads to the resonance windows resonance argument. As expressed by Dorey et al. [70]

> We elucidate the mechanism for this process, finding that the underlying reason is a reversible transfer of kinetic energy from the kinks to a collective bound state trapped by the $\bar{\text{K}}$K pair.

The full details of this argument can be found in the [63, 64, 70] (see also Chap. 3).

The resonances thus far described can be called "two bounce" resonances, for they arise from $\bar{\text{K}}$ interactions in which the first collision or "bounce" excites the shape mode and the second "bounce" de-excites it and liberates the kinks. This is illustrated dramatically in Fig. 1.3 where the oscillations of the field at $x = 0$ reflect the order of the resonances as shown in Table 1.1.

What about the possibility of higher "bounce" resonances, in which the second time the trapped K and $\bar{\text{K}}$ collide the resonance condition (1.17) is not fulfilled so that they remain trapped, and only after the third (or fourth, or ....) collision the appropriate resonance condition is met and the kinks escape? In the initial ϕ^4 article [61] these higher-bounce resonances were not studied in detail, although in a contemporaneous article [56] they were observed and studied in the modified sG equation. The observed "nesting" of three-bounce windows at the edges of two-bounce windows as well as the accumulation of the higher order windows shown in Fig. 1.2 strongly suggested an underlying fractal structure, which was later shown indeed to be the case in the works of Anninos et al. [72] and Goodman and Haberman [68, 69], which also placed the whole REEM on a more rigorous mathematical basis. For a thorough review of these important results, see Chap. 4 (Fig. 1.4).

One final point to close this discussion of kink-antikink interactions. In our initial short review of the early works on kink-antikink collisions, we mentioned briefly the introduction of "collective coordinates" as a possible means of reducing the nonlinear PDE to an effective ODE, a low-dimensional dynamical system. In simple terms, this approach amounts to making an Ansatz for the form of the solution that includes the motion of the center of mass of the kink (x_0) and a coordinate describing the amplitude of the shape oscillation of the kink (typically called $A(t)$ in the literature). This Ansatz is then inserted into the full Hamiltonian and integrated over all space, in the end producing an effective Hamiltonian involving x_0 and A as dynamical variables. The coefficients in the Hamiltonian are complicated functions that are typically either simplified or evaluated numerically to solve the ODEs. The

[5] There was however a subtlety in this case. Reference [56] noted that there were "quasi-resonances" in the K$\bar{\text{K}}$ related to excitation in the continuum spectrum around the kinks, and a later study by Quintero and Kevrekidis [71] showed how these "localized" phonons could in fact produce these "quasi-resonances". Interested readers should refer to these two articles for details.

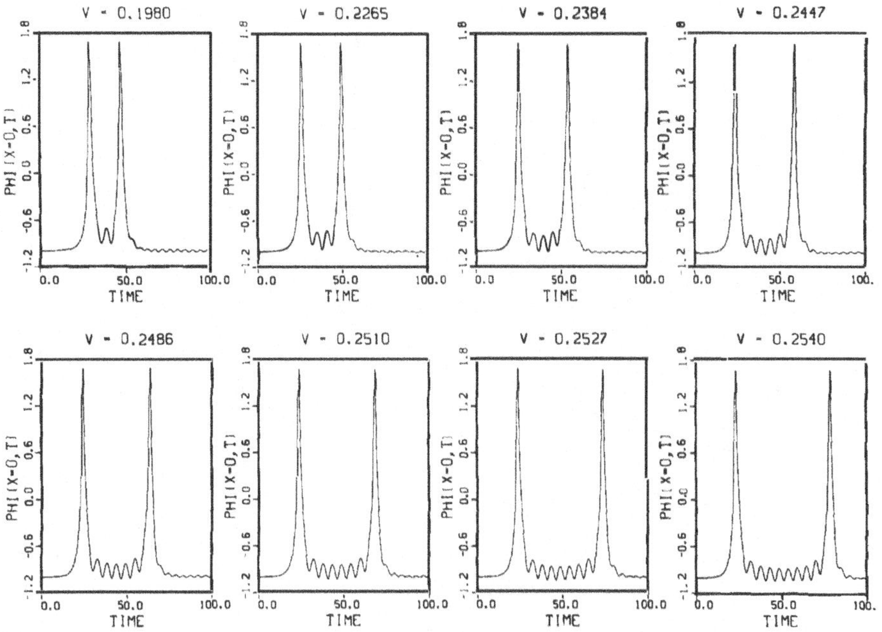

Fig. 1.3 $\phi(x = 0, t)$ versus t for incoming speeds at the center of the first eight two-bounce windows in ϕ^4 K$\bar{\text{K}}$ collisions. Adapted from [56], with permission from Elsevier

historical sequence of key references on the use of these collective coordinates starts with Kudryavtsev's 1975 paper [55], Sugiyama's 1979 paper [60], Campbell et. al's 1983 paper [61], Belova and Kudryavtsev's 1988 paper [73], Anninos et al.'s 1991 paper [72], Goodman and Haberman's 2005 and 2007 papers [68, 69] and the recent works of Weigel and Takyi [63, 64]. The details of these works are too technical to present in this historical overview, but fortunately, apart from the original references, interested readers can read thorough overviews of these results in Chaps. 3 and 4.

The results of the K–$\bar{\text{K}}$ collisions discussed above lead directly to our next topic, which is the question of the existence of a breather solution in the continuum (1+1)D ϕ^4 theory. Is the trapped oscillatory state observed in K–$\bar{\text{K}}$ collisions an indication that, for exactly the right initial conditions, a true breather solution to ϕ^4 can be found? In the late 1970s and early 1980s, there were two competing schools of thought regarding the existence of a true ϕ^4 breather. The computational physicists, who had studied numerical solutions to nonlinear PDEs similar to the ϕ^4 theory and had seen seemingly stable, localized, oscillatory "blobs" in theories like time-dependent Hartree–Fock [74] were convinced that one could "tune" parameters such that an exact breather solution could be found numerically. But the experts on integrable systems theory argued that the existence of the higher-order conservation laws in integrable systems was crucial to the existence of higher-order solitons (the sG

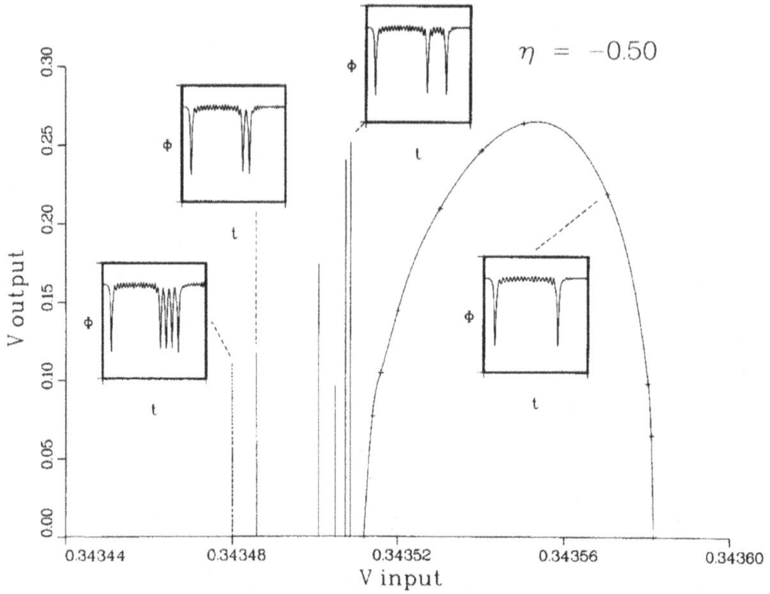

Fig. 1.4 A plot of v_{output} (=v_{f}) versus v_{input} (=v_{i}) for the double sG equation for $\eta = 0.5$ showing a sequence of three-bounce resonances near the edge of a two bounce resonance. Note the fine scale of v_{input} compared to Fig. 1.3. The insets show $\phi(x = 0, t)$. Adapted from [67], with permission from Elsevier

breather can be thought of as a two-soliton bound state), so that if a theory was *not* integrable (and ϕ^4 is not), it should not have any breathers.

The existence of a possible "breather" solution in ϕ^4 theory was confronted initially in the seminal paper of Dashen, Hasslacher, and Neveu (DHN) [32] in the context of semi-classical quantization of field theories with soliton solutions in (1+1)D. Having quantized the exact breather solution of the sG theory, DHN turned their attention to the ϕ^4 theory, with remarkably interesting results that merit revisiting.

DHN began by noting that the sG breather (whose exact form we repeat here for convenience)

$$\theta_B(x,t) = 4\tan^{(-1)}\left(\frac{\varepsilon \sin(t/\sqrt{1+\varepsilon^2})}{\cosh(\varepsilon x/\sqrt{1+\varepsilon^2})}\right) \tag{1.23}$$

could be expanded in powers of ε, with the leading order term being $\simeq 4\varepsilon \operatorname{sech} \varepsilon x$ $\sin(t(1-\varepsilon^2/2)) + \cdots$ where the approximate equality holds for small ε. We can understand the small amplitude approximation to the sG breather in terms of a "multiple-scale" asymptotic expansion [32, 75].

In this case, we start by expanding the equation itself, obtaining

$$\theta_{tt} - \theta_{xx} + \theta - \frac{1}{6}\theta^3 + \cdots = 0. \tag{1.24}$$

We seek a solution to this equation which has the spatially-localized, time-periodic form

$$\theta_B(x,t) = \varepsilon\left(A(X,T)e^{it} + c.c\right) + \varepsilon^2\theta_2 + \varepsilon^3\theta_3 + \cdots, \tag{1.25}$$

where the slow spatial scale $X = \varepsilon x$ and the slow time scale $T = \varepsilon^2 t$. Using standard multiple scale techniques [32, 75], one obtains an equation for the amplitude function A(X,T) of the form

$$2i\frac{\partial A}{\partial T} - \frac{\partial^2 A}{\partial X^2} - \frac{1}{2}|A|^2 A = 0. \tag{1.26}$$

Notice that this is the familiar NLS equation and that the signs are such that it does possess localized nonlinear solutions which are, of course, the solitons. Further, if one continues the multiple-scale expansion of (1.15) to higher orders in ε, one can argue [32] that *to all orders in* ε the expression for θ_B remains localized in space and periodic in time. Of course, this is to be expected, since we have the closed-form analytic solution for θ_B, which establishes that the multiple-scale asymptotic expansion is actually convergent.

DHN then proposed to carry out the same calculation for the ϕ^4 theory. Since ϕ^4 was already at that time known to be non-integrable, DHN did not expect to find a closed-form analytic expression for such breathers. However, the multiple-scale expansion can again be applied to provide an (asymptotic) perturbation expansion in the amplitude of the breather. Following DHN, we define $\phi_B = 1 + z_B$ and expand z_B as

$$z_B \simeq \varepsilon[A(X,T)e^{i\sqrt{2}t} + \text{c.c.}] + (\varepsilon^2)z_2 + \cdots \tag{1.27}$$

where again X and T represent scaled variables $X = \varepsilon x$ and $T = \varepsilon^2 t$. Again the resulting equation for $A(X,T)$ is the NLS equation, this time with normalizations such that

$$i2\sqrt{2}\frac{\partial A}{\partial T} - \frac{\partial^2 A}{\partial X^2} - 12|A|^2 A = 0. \tag{1.28}$$

Again the signs are such that there is an (exponentially) localized (soliton) solution for A, which has the specific form

$$A(X,T) = \sqrt{\frac{1}{6}}\text{sech}X e^{-\frac{iT}{2\sqrt{2}}}. \tag{1.29}$$

Thus to this order in ε, $z_B(x,t)$ does represent a breather solution in the sense defined above, and the ϕ^4 equation mimics exactly the behavior of sG. DHN then undertook a heroic calculation to the next leading orders, obtaining the putative ϕ^4 breather to third order in ε:

$$\phi_B^{(3)} = 1 + \frac{2\varepsilon}{\sqrt{3}}\text{sech}\xi \cos\tau - \varepsilon^2\text{sech}^2\xi + \frac{\varepsilon^2}{3}\text{sech}^2\xi \cos 2\tau + \frac{\varepsilon^3}{6\sqrt{3}}\text{sech}^3\xi \cos 3\tau, \tag{1.30}$$

where $\xi = \frac{\varepsilon\sqrt{2}x}{\sqrt{1+\varepsilon^2}}$ and $\tau = \frac{\sqrt{2}t}{\sqrt{1+\varepsilon^2}}$. DHN further argued that one could show *to all orders in* ε that this breather was exponentially localized and time-periodic. Hence at no order in ε does one see any distinction between the "breather" solutions to the integrable versus the non-integrable case. Was this the final resolution to the question of the existence of a continuum ϕ^4 breather?

Before answering this question, as an important historical aside, we note that in their work on the continuum ϕ^4 breather, DHN made a tantalizing comment about kink-antikink scattering [32]. After deriving and summing a formal infinite geometric series to obtain a closed form expression for the putative ϕ^4 breather, they proposed using the same transformation that changed the sG breather (1.15) into the sG kink-antikink interaction equation (1.12) —namely, replacing ε by $\frac{i}{v}$— to obtain a *formal* closed form expression for the kink-antikink collisions in the ϕ^4 model. Specifically, DHN obtained, for large $|t|$ (see [33]),

$$\phi = \frac{1/4\sqrt{3}v \cosh[\gamma(v)x] - \cosh[\gamma(v)vt]}{1/4\sqrt{3}v \cosh[\gamma(v)x] + \cosh[\gamma(v)vt]} \tag{1.31}$$

As DHN noted

> This looks like two well-separated kinks, moving with velocities $\pm v$. Whether or not (*text added*: this equation) could form the starting point of an approximate analytic solution for kink-antikink scattering is unknown. There are no obvious parameters in which to expand.

As we have seen above, this analytic hope was not to be, and the actual ϕ^4 kink-antikink interactions have a much richer, albeit not analytic, structure.

For several years, many authors struggled to find conclusive numerical evidence either for or against the continuum ϕ^4 breather conjecture. Our own unpublished numerical results (which finally appeared in an AIP conference proceedings [76] and hence have been generally overlooked) strongly suggested the existence of breathers in (spatially) *discretized* approximations to ϕ^4, where (in the simplest case) we approximated the continuum spatial partial derivative by

$$\frac{\partial^2\phi}{\partial x^2} \sim \frac{1}{(\Delta x)^2}\left(\phi(x+\Delta x) - 2\phi(x) + \phi(x-\Delta x)\right). \tag{1.32}$$

For finite Δx, this reduces the PDE to a set of N coupled nonlinear oscillators, where $N\Delta x = L$, the length of the spatial interval studied. For some regions of non-zero Δx (and certain values of ε), we were able to find stable localized discrete breathers and to model theoretically the regions of stability (and instability). Interested readers are referred to the original paper [76] for details of this (very) complicated structure.

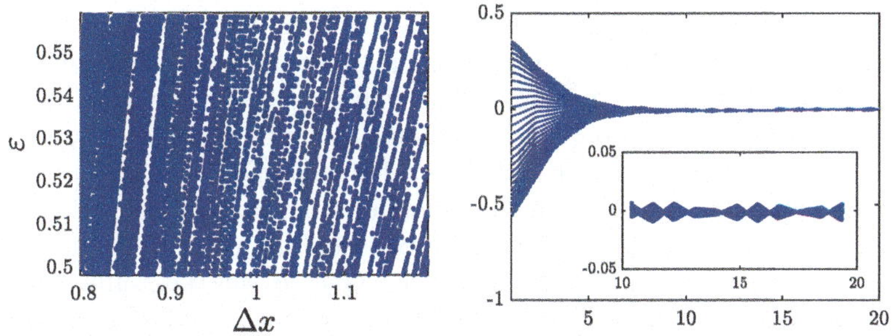

Fig. 1.5 The left panel shows results of a theoretical analysis of the stability of the discrete ϕ^4 breather. The dark regions indicate unstable breathers, the light regions stable, interwoven in a seeming fractal structure. The right panel shows an example of a discrete "breather" solution (with $\varepsilon = 0.505$ and $\Delta x = 0.9$) that is not localized in space but has small oscillatory wings extending to the end of the lattice. Such excitations have been named "nanopterons" by Boyd [78]

In essence, we found a complex "fractal"/Arnold tongue-like structure of stability regions as a function of discreteness and breather amplitude, ε. Clearly, there were regions in which stable localized breathers existed on finite discrete lattices. However, we found that as $\Delta x \to 0$, our putative breathers lost their exponential localization by coupling to the phonon mode that was nearest to twice the breather frequency, so that the solution extended over all space, with large amplitude oscillations near the origin and very small "wings" as $x \to \infty$. This was precisely the (discrete version for finite Δx) of the "nanopteron" discovered and named by Boyd [77, 78] and discussed in detail in Chap. 8.

As an indication of this subtlety, we show in Fig. 1.5 two figures whose data have been taken from those of [76]. The left panel first shows a theoretical stability analysis in the ε–Δx plane, whereas the right panel shows a delocalized nanopteron solution.

Finally, we convinced Harvey Segur to look at the problem, and together with Martin Kruskal [79], they were able to show that there were terms "beyond all orders" in ε, so that true solution was

$$\phi_{\text{true}} = \phi_b + c_1 \exp(-c_2/\varepsilon)\cos(k_c x - \omega_c t) + \cdots \tag{1.33}$$

where $\omega_c = 2\sqrt{2}(1 - \varepsilon^2/2)$, $k_c = \sqrt{\omega_c^2 - 2}$, and c_1 and c_2 are two (non-zero) constants. Hence ϕ_{true} is in fact *not* localized in space and the "breather-like" oscillations seen in K–$\bar{\text{K}}$ interactions would inevitably decay by coupling to "phonon" modes in the continuum. The "miracle" of the integrable sG equation is that when similar "beyond all orders" techniques are applied to it, the constant analogous to c_1 vanishes identically (as do all higher order non-localized terms). This remarkable result is indeed due to the infinite number of conserved quantities in that model [80].

A few closing remarks on the role of breather solutions on discrete lattices are in order. For historical reasons, we have focused on discrete breathers which were motivated by small ε expansions and hence were spread over many lattice sites, as shown by (1.30). Although these exist for certain values of Δx and ε [76], their complex regions of stability and instability render them both fragile and relatively uninteresting. A major advance in understanding "discrete breathers" came from the initial insight of Sievers and Takeno [81] on "intrinsic localized modes," which are stable, very highly localized nonlinear excitations in discrete lattices. These insights were formalized by the work of MacKay and Aubry [82] who, starting from the anti-continuum limit ($\Delta x \to \infty$, in our notation) of an isolated single-site discrete breather, were able to prove the existence of stable discrete breathers in infinite lattices for sufficiently weak coupling to adjacent lattice sites (i.e., sufficiently large Δx) and to use a constructive method of proof to continue the solutions numerically to finite coupling. This important development also provided firm analytic footing for earlier studies of ILMs [81] and led to an explosion of interest in these discrete nonlinear objects (for early reviews see [83–85]). Much more detailed discussions of these discrete breathers/ILMs can be found in Chaps. 6 and 7.

1.4 Concluding Remarks

As our presentation has been historical and pedagogical, there is no need to repeat the results in a summary or conclusion. But we should comment on topics which, in the interests of brevity, we have not covered in this chapter. Fortunately, these are covered in subsequent chapters in this volume. They include extensions beyond of the ϕ^4 theory in higher dimensions (Chap. 11), KG theories involving higher order polynomials (ϕ^6, ϕ^8, etc.) (Chaps. 2 and 12), "discrete" versions of ϕ^4 defined on a lattice (Chaps. 6 and 7), and versions of ϕ^4 in external potentials, both regular (Chap. 10) and stochastic (Chap. 5). Without question, the ϕ^4 field theory is alive and well and will remain of interest to scientists for generations to come.

Acknowledgements It is a pleasure to thank my collaborators in the work I have described here —Michel Peyrard, Jonathan Schonfeld, Pasquale Sodano, and Chuck Wingate— and to recall the great times we shared in writing the original papers. I thank Michel Peyrard for regenerating some of the old figures. I am also grateful to Panos Kevrekidis and Jesús Cuevas-Maraver for helpful comments and assistance on the manuscript.

References

1. O. Klein, Z. Phys. **37**, 895 (1926)
2. W. Gordon, Z. Phys. **40**, 117 (1926–1927)
3. P.A.M. Dirac, Proc. R. Soc. A **117**, 778 (1928)
4. P.W. Higgs, Phys. Rev. Lett. **13**, 508 (1964)
5. S. Weinberg, Phys. Rev. Lett. **19**, 1264 (1967)
6. M.D. Schwartz, *Quantum Field Theory and the Standard Model* (Cambridge University Press, Cambridge, 2014)
7. Ya.B. Zeldovich, I.Y. Kobzarev, L.B. Okun, Sov. Phys. JETP **40**, 1 (1975)
8. V.L. Ginzburg, L.D. Landau, Sov. Phys. JETP **20**, 1064 (1950)

9. For a review, R.A. Cowley, Adv. Phys. **29**, 219 (1980)
10. Y. Ishibashi, Y. Takagi, J. Phys. Soc. Jpn. **33**, 1 (1972)
11. S. Aubry, R. Pick, Ferroelectrics **8**, 471 (1974)
12. S. Aubry, J. Chem. Phys. **64**, 3217 (1975)
13. S. Aubry, J. Chem. Phys. **64**, 3392 (1976)
14. S. Aubry, Ferroelectrics **12**, 263 (1976)
15. J.A. Krumhansl, J.R. Schrieffer, Phys. Rev. B **11**, 3535 (1975)
16. C.M. Varma, Phys. Rev. B **14**, 244 (1976)
17. T. Schneider, E. Stoll, Phys. Rev. Lett. **35**, 296 (1975)
18. T.R. Koehler, A.R. Bishop, J.A. Krumhansl, J.R. Schrieffer, Solid State Commun. **17**, 1515 (1975)
19. Y. Wada, J.R. Schrieffer, Phys. Rev. B **18**, 3897 (1978)
20. G.M. Mazenko, P.S. Sahni, Phys. Rev. B **18**, 6139 (1978)
21. J.F. Currie, J.A. Krumhansl, A.R. Bishop, S.E. Trullinger, Phys. Rev. B **22**, 477 (1980)
22. M.J. Rice, Phys. Lett. A **71**, 152 (1979)
23. M.J. Rice, J. Timonen, Phys. Lett. A **73**, 368 (1979)
24. M.J. Rice, E.J. Mele, Solid State Commun. **35**, 487 (1980)
25. W.P. Su, J.R. Schrieffer, A.J. Heeger, Phys. Rev. Lett. **42**, 1698 (1979)
26. W.P. Su, J.R. Schrieffer, A.J. Heeger, Phys. Rev. B **22**, 2099 (1980)
27. H. Takayama, Y.R. Lin-liu, K. Maki, Phys. Rev. B **21**, 2388 (1980)
28. R. Jackiw, J.R. Schrieffer, Nucl. Phys. B **190**, 253 (1981)
29. R.D. Yamaletdinov, V.A. Slipko, Y.V. Pershin, Phys. Rev. B **96**, 094306 (2017)
30. R.B. Griffiths, B. Simon, Commun. Math. Phys. **33**, 145 (1972)
31. J. Glimm, A. Jaffe, T. Spencer, Commun. Math. Phys. **45**, 203–216 (1975)
32. R.F. Dashen, B. Hasslacher, A. Neveu, Phys. Rev. D **10**, 4130 (1974)
33. R.F. Dashen, B. Hasslacher, A. Neveu, Phys. Rev. D **11**, 3424 (1975)
34. J. Goldstone, R. Jackiw, Phys. Rev. D **11**, 1486 (1975)
35. R. Rajaraman, Phys. Rep. **21**, 227 (1975)
36. D.K. Campbell, Y.T. Liao, Phys. Rev. D **14**, 2093 (1976)
37. D.K. Campbell, in *Nuclear Physics with Heavy Ions and Mesons*, ed. by R. Balian, M. Rho, G. Ripka, Les Houches Summer School Session XXX, July–August 1977 (North Holland, Amsterdam, 1978), p. 674
38. R. Jackiw, C. Rebbi, Phys. Rev. D **13**, 3398 (1976)
39. S. Coleman, in *The Whys of Subnuclear Physics*, ed. by A. Zichichi (Plenum, New York, 1979), p. 845
40. A.C. Scott, F.Y.F. Chu, D.W. McLaughlin, Proc. IEEE **61**, 1443 (1973)
41. M.J. Ablowitz, H. Segur, *Solitons and the Inverse Scattering Transform* (SIAM, Philadelphia, 1981)
42. M.J. Ablowitz, P.A. Clarkson, *Solitons, Nonlinear Evolution Equations, and Inverse Scattering* (Cambridge University Press, Cambridge, 1991)
43. S. Coleman, Phys. Rev. D **11**, 2088 (1975)
44. J. Rubinstein, J. Math. Phys. **11**, 258 (1970)
45. N.J. Zabusky, M.D. Kruskal, Phys. Rev. Lett. **15**, 240 (1965)
46. R. Steuerwald, Abh. Bayer. Akad. Wiss. **40**, 1 (1936)
47. L.P. Eisenhart, *A Treatise on Differential Geometry of Curves and Surfaces* (Dover, New York, 1960)
48. A. Seeger, A. Kochendorfer, Z. Phys. **130**, 321 (1951)
49. A. Seeger, H. Donth, A. Kochendorfer, Z. Phys. **134**, 173 (1953)
50. J.K. Perring, T.H.R. Skyrme, Nucl. Phys. **31**, 550 (1962)
51. M.J. Ablowitz, D.J. Kaup, A.C. Newell, H. Segur, Phys. Rev. Lett. **30**, 1262 (1973)
52. V.E. Zakharov, A.B. Shabat, Sov. Phys. JETP **34**, 62 (1972)
53. D.K. Campbell, D. Farmer, J. Crutchfield, E. Jen, Commun. ACM **28**, 374 (1985)
54. M.J. Ablowitz, M.D. Kruskal, J.F. Ladik, SIAM J. Appl. Math. **36**, 428 (1979)
55. A.E. Kudryavtsev, JETP Lett. **22**, 82 (1975)

56. M. Peyrard, D.K. Campbell, Phys. D **9**, 33 (1983)
57. D.K. Campbell, M. Peyrard, P. Sodano, Phys. D **19**, 165 (1986)
58. B.S. Getmanov, JETP Lett. **24**, 291 (1976)
59. V.G. Makhankov, Phys. Rep. **35**, 1 (1978)
60. T. Sugiyama, Prog. Theor. Phys. **61**, 550 (1979)
61. D.K. Campbell, J.F. Schonfeld, C.A. Wingate, Phys. D **9**, 1 (1983)
62. J.L. Gervais, A. Jevicki, B. Sakita, Phys. Rev. D **12**, 1038 (1975)
63. H. Weigel, J. Phys. Conf. Ser. **482**, 012045 (2014)
64. I. Takyi, H. Weigel, Phys. Rev. D **94**, 085008 (2016)
65. R. Klein, W. Hasenfratz, N. Theodorokapoulos, W. Wünderlich, Ferroelectrics **26**, 721 (1980)
66. M. Moshir, Nucl. Phys. B **185**, 318 (1981)
67. D.K. Campbell, M. Peyrard, Phys. D **18**, 47 (1986)
68. R.H. Goodman, R. Haberman, SIAM J. App. Dyn. Sys. **4**, 119 (2005)
69. R.H. Goodman, R. Haberman, Phys. Rev. Lett. **98**, 104103 (2007)
70. P. Dorey, K. Mersh, T. Romańczukiewicz, Y. Shnir, Phys. Rev. Lett. **107**, 091602 (2011)
71. N.R. Quintero, P.G. Kevrekidis, Phys. D **170**, 31 (2002)
72. P. Anninos, S. Oliviera, R.A. Matzner, Phys. Rev. D **44**, 1147 (1991)
73. T. Belova, A. Kudryavtsev, Phys. D **32**, 18 (1988)
74. D.K. Campbell, in *Time Dependent Hartree-Fock Method*, ed. by P. Bonche, B. Giraud, Ph. Quentin (Editions de Physique, Paris, 1979), p. 218
75. D.K. Campbell, in *Complex Systems, SFI Studies in the Sciences of Complexity*, ed. by D. Stein (Addison-Wesley Longman, Boston, 1989), p. 3
76. D.K. Campbell, M. Peyrard, in *CHAOS/XAOC: Soviet-American Perspectives on Nonlinear Science*, ed. by D.K. Campbell (American Institute of Physics, New York, 1990), p. 305
77. J.P. Boyd, in *Nonlinear Topics of Ocean Physics*, ed. by A.R. Osborne, Fermi Summer School, Course LIX (North-Holland, Amsterdam, 1991), p. 527
78. J.P. Boyd, Nonlinearity **3**, 177 (1990)
79. H. Segur, M.D. Kruskal, Phys. Rev. Lett. **58**, 747 (1987)
80. B. Birnir, H. McKean, A. Weinstein, Commun. Pure Appl. Math. **47**, 1043 (1994)
81. S. Takeno, K. Kisoda, A.J. Sievers, Prog. Theor. Phys. Suppl. **94**, 242 (1988)
82. R.S. MacKay, S. Aubry, Nonlinearity **7**, 1623 (1994)
83. S. Flach, Phys. Rev. E **51**, 1503 (1995)
84. S. Flach, C.R. Willis, Phys. Rep. **295**, 181 (1998)
85. D.K. Campbell, S. Flach, Yu.S. Kivshar, Phys. Today **57**, 43 (2004)

Chapter 2
Some Recent Developments on Kink Collisions and Related Topics

Tomasz Romańczukiewicz and Yakov Shnir

Abstract We review recent works on modeling of dynamics of kinks in 1+1 dimensional ϕ^4 theory and other related models, like sine-Gordon model or ϕ^6 theory. We discuss how the spectral structure of small perturbations can affect the dynamics of non-perturbative states, such as kinks or oscillons. We describe different mechanisms, which may lead to the occurrence of the resonant structure in the kink-antikink collisions. We explain the origin of the radiation pressure mechanism, in particular the appearance of the negative radiation pressure in the ϕ^4 and ϕ^6 models. We also show that the process of production of the kink-antikink pairs, induced by radiation, is chaotic.

2.1 Introduction

Topological solitons gained increasing interest over the last decades. Many models which support classical soliton solutions, have been intensively studied in a wide variety of physical contexts, see e.g. [1, 2]. Perhaps one of the simplest examples of solitons is the class of the *kink* configurations, which appears in the (1+1) dimensional models with a potential possessing two or more degenerated minima. The sG model with infinitely degenerated vacuum is a special case of integrable theory. Other models with polynomial potentials, like for example the simple ϕ^4 model with double degenerated vacuum $U(\phi) = \frac{1}{2}(\phi^2 - 1)^2$, or the ϕ^6 model with triple degenerated vacuum, are non-integrable. The ϕ^4 model arises in many different physical situations and serves as a prototype for many non-linear systems. Indeed, this model is known in the cosmological context [1], and also has a number of applications in condensed matter physics [2]. In particular, it was applied to describe solitary waves in shape-memory alloys [3] and can also be used as as a phenomenological theory of the

T. Romańczukiewicz
Institute of Physics, Jagiellonian University, Kraków, Poland
e-mail: trom@th.if.uj.edu.pl

Y. Shnir (✉)
BLTP, JINR, Dubna 141980, Moscow Region, Russia
e-mail: shnir@theor.jinr.ru

P. G. Kevrekidis and J. Cuevas-Maraver (eds.), *A Dynamical Perspective on the ϕ^4 Model*, Nonlinear Systems and Complexity 26,
https://doi.org/10.1007/978-3-030-11839-6_2

non-perturbative transitions in polyacetylene chain [4]. Furthermore, the ϕ^4 model has been applied in biophysics to describe soliton excitations in DNA double helices [5]. The static limit of this model is known as a phenomenological theory of second order phase transitions [6]. In quantum field theory it is used as a model example to investigate transition between perturbative and non-perturbative sectors of the theory [7, 8], and also can model quantum mechanical instanton transitions in a double-well potential [9]. Some possible realization was pointed out in buckled graphene ribbons [10].

The kink solution interpolates between two different vacua of the model. This solution is topologically stable. The most interesting properties of the kinks can be observed in the processes of their scattering and collisions. Since the models with polynomial potentials are not exactly integrable, one has to encounter the energy loss to radiation in these processes. Naively, one could expect the collision between a kink and an antikink should always lead to annihilation of the solitons into large amount of radiation. However, the numerical study of the process reveals a far richer pattern [11–17]. Numerical simulations show that the processes of collisions of a kink and an antikink are chaotic, i.e., for some values of the impact velocity the solitons bounce back, while for some other impact velocities, slightly smaller or larger, they may annihilate via an intermediate oscillating bion state [11, 13].

More precisely, for initial velocities above the critical value $v_{\rm cr} = 0.2598$ the two incident kinks always escape to infinity after collision, with some energy loss due to radiation. Below $v_{\rm cr}$, the incident waves generically become trapped, but there is also a complicated pattern of narrow resonance windows, within which the kinks are again able to escape. This effect is related to reversible energy exchange between the states of perturbative and non-perturbative sectors of the model.

At the first impact, some part of the kinetic energy of the colliding solitons is transferred into excitation of the internal modes of the kinks. They then separate and propagate almost independently, however there is an attractive force between them. For initial velocities less than the critical value $v_{\rm cr}$, the kinks do not have enough energy to escape, so they turn back and collide again. At this moment the energy stored in the internal oscillating modes can be returned to the translational collective mode allowing the kinks to escape, provided that there is a resonance condition between the time interval between the collisions, and the oscillation period of the internal modes.

Another mechanism is at work in the ϕ^6 model, where the resonance windows appear due to resonant energy exchange between the internal oscillating modes trapped by the kink-antikink ($K\bar{K}$) pair, and the translational mode of the solitons [18]. We also briefly discuss the mechanism where the role of the energy storage is taken over by decaying quasi-normal modes.

The interplay between the states of the perturbative spectrum and solitons, attracted a lot of attention recently. In particular, it was pointed out that the excitations of the internal mode of the kink may produce $K\bar{K}$ pairs [16]. Another interesting observation is that the interaction between the kink and the scattering modes of the continuous spectrum results in the effect of negative radiative pressure, i.e., the ϕ^4 kink starts to accelerate towards the incoming wave [19].

A peculiar feature of many non-linear models, like the ϕ^4 theory, is that they also support time dependent non-perturbative solutions which are not captured by the linear analysis. An interesting example is the breather in the sG model. Since this model is completely integrable, the states of the continuum are completely separated from the solutions of the field equations. Thus, the breather has an infinite lifetime because it does not lose its energy into radiation. However, there are similar quasi-non-dissipative and almost periodic time dependent configurations in the ϕ^4 model: the oscillons [20]. They appear as quasi-breathers and are extremely long-lived localised field configurations, non-harmonically oscillating about the vacuum. Furthermore, the radiation energy losses of an oscillon are very small: numerical simulations show that in (1+1) dimensional ϕ^4 theory the oscillon survives even after millions of oscillations [21, 22].

The structure of this brief review is as follows. In the following section we briefly describe properties of the states of the perturbative spectrum of the ϕ^4 and ϕ^6 theories and the corresponding solitons. Then we review the effect of negative radiative pressure on the kinks in these models. We then discuss correspondence between the oscillons and the internal modes of the kinks. Production of the kink-antikink pairs from radiation is briefly discussed in Sect. 2.4. We review the resonance effects in the $K\bar{K}$ scattering in the ϕ^6 model in the Sect. 2.5. We end with a discussion of the kink boundary scattering in the ϕ^4 model on a semi-infinite line and the radiative decay of the boundary mode.

2.2 Solitons and Perturbations

2.2.1 *Spectral Structure of Small Perturbations*

Let us consider the rescaled Lagrangian density of the ϕ^4 model with two symmetric vacua $\phi_0 = \pm 1$

$$L = \frac{1}{2}(\partial_t \phi)^2 - \frac{1}{2}(\partial_x \phi)^2 - \frac{1}{2}\left(\phi^2 - 1\right)^2 . \tag{2.1}$$

The corresponding field equation is

$$\partial_t^2 \phi - \partial_x^2 \phi + 2\phi(\phi^2 - 1) = 0 . \tag{2.2}$$

Evidently, there are eigenmodes of the corresponding linearized problem, which correspond to small oscillations about one of the vacua. Suppose $\phi(x, t) = 1 + \xi(x, t)$, then the expansion in $\xi(x, t)$ yields the linearized equation

$$(\partial_t^2 - \partial_x^2 + 4)\xi(x, t) = 0 . \tag{2.3}$$

Clearly, this is the usual KG equation for scalar excitations with mass $m = 2$, there is a continuum spectrum of excitations with frequencies $\omega = \pm\sqrt{k^2+4}$.

The kink configuration is a topologically nontrivial static solution of (2.2), which interpolates between the two vacua, $\phi(-\infty) = \pm 1$, $\phi(\infty) = \mp 1$,

$$\phi_K(x) = \tanh(x - x_0)\,; \qquad \phi_{\bar{K}} = -\tanh(x - x_0)\,. \tag{2.4}$$

Here $\phi_{\bar{K}}$ is an antikink solution. The kinks are topological solitons, the field of the kink is a map $\phi : \mathbb{Z}_2 \to \mathbb{Z}_2$. Physically, they correspond to localized lumps of energy centered around $x = x_0$.

Let us consider small excitations $\xi(x,t) = \eta(x)e^{i\omega t}$ of the kink configuration (2.4) for $x_0 = 0$. The corresponding linearized equation is

$$\left(-\frac{d^2}{dx^2} + V(x)\right)\eta = \omega^2\eta\,, \qquad V(x) = 4 - \frac{6}{\cosh^2 x}\,. \tag{2.5}$$

This equation appears as a typical problem in the context of one-dimensional quantum mechanics, and describes a particle scattering from the potential $V(x)$. Such a potential has two bound modes. The lowest one

$$\eta_0 = \frac{1}{\cosh^2 x}\,, \qquad \omega_0 = 0, \tag{2.6}$$

has an interpretation of a translational mode $\phi_K(x-a) \approx \phi_K(x) - a\eta_0(x)$ and is a reflection of the translational symmetry of the model $x \to x - a$. The first mode with odd symmetry,

$$\eta_1 = \frac{\tanh x}{\cosh^2 x}\,, \qquad \omega_1 = \sqrt{3}, \tag{2.7}$$

is called the oscillatory internal mode.[1] In the linear approximation this mode oscillates with constant amplitude and frequency.

The continuum modes of the kink with eigenvalues $\omega^2 = k^2 + 4$ are

$$\eta_k(x) = \frac{3\tanh^2 x - 3ik\tanh x - 1 - k^2}{\sqrt{(k^2+1)(k^2+4)}}e^{ikx}\,. \tag{2.8}$$

It is worth mentioning that the potential generated by the sG soliton has also the form of the Pöschl–Teller potential $V_{sG}(x) = 1 - 2\,\mathrm{sech}^2 x$ and the eigenvalues for $\omega^2 = k^2 + 1$ are also known:

$$\eta_k(x) = \frac{ik - \tanh x}{\sqrt{k^2+1}}e^{ikx}\,. \tag{2.9}$$

[1]This mode is also referred to as "discrete mode" [16], or "wobbling mode" [23], or "shape mode" [11, 24].

Note that in the sG model the kink has no internal oscillating mode, and the translational mode $\eta_0 = \operatorname{sech} x$ is the only bound state of the linearized potential.

An interesting generalization of the double vacuum model (2.1) is the 1+1 dimensional ϕ^6 model, which is defined by the Lagrangian [25]

$$L = \frac{1}{2}(\partial_t\phi)^2 - \frac{1}{2}(\partial_x\phi)^2 - \frac{1}{2}\phi^2\left(\phi^2 - 1\right)^2 . \tag{2.10}$$

Clearly, there are three degenerated vacua, $\phi_0 \in \{-1, 0, 1\}$, however, like ϕ^4 theory, the model (2.10) is symmetric with respect to reflectional $\mathbb{Z}_2$ symmetry $\phi \to -\phi$ and/or $x \to -x$. The resulting equation of motion

$$\partial_t^2\phi - \partial_x^2\phi + \phi - 4\phi^3 + 3\phi^5 = 0 , \tag{2.11}$$

can be linearized about each of three vacua, the excitations $\xi(x, t)$ satisfy the linearized equations

$$\begin{aligned} (\partial_t^2 - \partial_x^2 + 1)\xi(x, t) = 0; \quad &\text{as } \phi_0 = 0 ; \\ (\partial_t^2 - \partial_x^2 + 4)\xi(x, t) = 0; \quad &\text{as } \phi_0 = \pm 1 . \end{aligned} \tag{2.12}$$

Thus, the scalar excitations about the vacua $\phi_0 = \pm 1$ have mass $m_1 = 2$ while the excitations about the symmetric vacuum $\phi_0 = \pm 0$ have mass $m_2 = 1$.

In terms of solitonic solution, the ϕ^6 model is quite richer than the ϕ^4 theory. Indeed, there are two different kinks interpolating between neighboring vacua $(\phi(-\infty) = 0, \phi(\infty) = 1)$ and $(\phi(-\infty) = -1, \phi(\infty) = 0)$:

$$\phi_{(0,\pm 1)}(x) = \pm\sqrt{\frac{1 + \tanh(x)}{2}}, \qquad \phi_{(\pm 1,0)}(x) = \phi_{(0,\pm 1)}(-x). \tag{2.13}$$

For an isolated ϕ^6 kink, however, there are no localized bound state solutions to the linearized ϕ^6 equations with the potential $V(x) = 15\phi_K^4 - 12\phi_K^2 + 1$. General continuum solutions of this equation can be written in terms of hypergeometric functions [25]. A special feature of the spectrum of linear perturbation around the ϕ^6 kinks is that, unlike the ϕ^4 model, the potential of the linearized problem V is not symmetric with respect to reflections $x \to -x$. In other words, the mass of the excitations of the continuum is different on the opposite sides of the kink [18].

2.3 Interplay Between the States of Perturbative and Non-perturbative Sectors

2.3.1 *The Effect of Negative Radiation Pressure*

Solitons in many ways are very similar to particles, yet there are some important differences. The internal structure of solitons may be revealed, for example, in the scattering spectrum of external incoming perturbations. On the other hand, the scattering on the soliton results in the radiation pressure exerted on it; this effect is similar to the familiar solar radiation pressure force on the particles in comets' tails.

Let us consider a small amplitude propagating perturbation of the scalar field, which represents a wave moving towards a soliton. In general, we can expect that the scattering spectrum contains both the reflected and transmitted waves, thus

$$\phi(x \to -\infty) = \frac{1}{2}A\left[e^{i(\omega t-kx)} + Re^{i(\omega t+kx)}\right] + \text{c.c.}\,. \tag{2.14}$$

and

$$\phi(x \to +\infty) = \frac{1}{2}ATe^{i(\omega t-kx)} + \text{c.c.}\,, \tag{2.15}$$

where A is the amplitude of the incoming wave, ω and k are the frequency and the wave number of the wave and R, T are the reflection and transmission coefficients, respectively.

The wave carries both the energy

$$\mathcal{E} = \frac{1}{2}\left(\phi_t^2 + \phi_x^2\right) + U(\phi) \tag{2.16}$$

and the momentum density

$$\mathcal{P} = \phi_x \phi_t\,. \tag{2.17}$$

The corresponding continuity equation is

$$\partial_t \mathcal{P} = -\frac{1}{2}\partial_x\left(\phi_t^2 + \phi_x^2 - 2U(\phi)\right)\,, \tag{2.18}$$

thus, there is a force acting on the soliton due to the transfer of momentum from the incoming radiation.

We can now evaluate this force averaging the flux (2.18) over the period and integrating it by parts. We obtain

$$F = \overline{\partial_t \mathcal{P}} = \frac{1}{2}k^2 A^2\left(1 + |R|^2 - |T|^2\right)\,. \tag{2.19}$$

Further, in the case of a single channel scattering we can make use of the continuity equation $|R|^2 + |T|^2 = 1$. In such a case this formula simplifies to $F = A^2k^2|R|^2$. Note that in the single channel scattering the radiative force always pushes the soliton in the direction of the incoming momentum.

One of the most surprising features of both sG and ϕ^4 model is that the solitons do not reflect any radiation in the linear approximation. The reflection coefficients read from the solutions of the linearized equations (2.8) and (2.9) are exactly 0, as there is no part proportional to e^{-ikx}. This is a general feature of the Pöschl–Teller potential $V(x) = -\frac{n(n+1)}{\cosh^2(x)}$ with $n \in \mathbb{Z}$.

The question arises of how exactly the soliton would move when it is exposed to the radiation in the full non-linearized system. As the sG model is integrable, there are analytical solutions corresponding to a static soliton with a moving cnoidal wave in its background [19, 26]. However, the sG soliton is completely transparent to the wave in all orders. As the interaction is not dynamical, there is no energy nor momentum transfer between the incoming wave and the soliton, resulting the interaction just in a phase shift. The ϕ^4 model, on the other hand, is not integrable, and some interesting effects arise therein even at the second order of the perturbation series.

We seek for the solution of the dynamical equation in the form of a perturbation series with the amplitude of incoming wave A being an expansion parameter:

$$\phi = \phi_s + \xi = \phi_s + A\xi^{(1)} + A^2\xi^{(2)} + \cdots . \tag{2.20}$$

The equation in the nth order has a general form

$$\ddot{\xi}^{(n)} + \mathbf{L}\xi^{(n)} = f^{(n)}, \tag{2.21}$$

where the linear operator is $\mathbf{L} = -\mathrm{d}^2/\mathrm{d}x^2 + U''(\phi_s(x))$, and the source term $f^{(n)}$ depends on the solutions of lower orders. In particular, $f^{(1)} = 0$. By taking the monochromatic wave $\xi^{(1)} = \frac{1}{2}e^{i\omega t}\eta_{-q}(x) + \text{c.c.}$, where $q = \sqrt{\omega^2 - 4} > 0$ is the wave number, we can evaluate the source term in the second order:

$$f^{(2)} = -6\phi_s{\xi^{(1)}}^2 = -\frac{3}{2}\left(e^{2i\omega t}\eta_{-q}^2 + 2\eta_q\eta_{-q} + e^{-2i\omega t}\eta_q^2\right). \tag{2.22}$$

The solution can be sought in the form

$$\xi^{(2)} = \xi_{+2}^{(2)}e^{2i\omega t} + \xi_0^{(2)} + \xi_{-2}^{(2)}e^{-2i\omega t}. \tag{2.23}$$

Similarly, the solution at the mth order can be, in general, expanded in a Fourier series of $\xi_n^{(m)}$ functions oscillating with the frequency $n\omega$ ($|n| \leq m$). All these solutions have an asymptotic form consisting of the inhomogeneous part, which originates from the source term, and an outgoing wave:

$$\xi_n^{(m)}(x \to \pm\infty) = \eta_{inh} + \alpha_{mn,\pm k}\eta_{\mp k}, \qquad k = \sqrt{n^2\omega^2 - 4}. \tag{2.24}$$

The coefficients $\alpha_{mn,\pm k}$ can be treated as nonlinear scattering amplitudes and can be found using Green's function technique:

$$\alpha_{mn,\pm k} = -\frac{1}{W}\int_{-\infty}^{\infty} dx'\,\eta_k(x')\,f_n^{(m)}(x')\,, \tag{2.25}$$

where $W = \eta_k'\eta_{-k} - \eta_k\eta_{-k}'$ is the Wronskian. Knowing those amplitudes we can calculate the momentum balance on both sides of the kink and hence find the force which is exerted by the wave.

Since in the ϕ^4 model the kinks are transparent at the first order, the first non-vanishing contribution to the force is proportional to the *fourth* power of the amplitude:

$$F^{(4)} = 2A^4k\left[(k+2q)|\alpha_{22,k}^2| - (k-2q)|\alpha_{22,-k}^2|\right]\,, \tag{2.26}$$

where $q = \sqrt{\omega^2-4}$ and $k = \sqrt{4\omega^2-4}$ are the wave numbers for the frequencies ω and 2ω respectively. The coefficients can be found in a closed form

$$\alpha_{22,k}(q) = -\frac{3}{2}\pi\frac{q^2+4}{q^2+1}\sqrt{\frac{q^2+4}{k^2+1}}\,\frac{1}{k\sinh\left(\frac{2q+k}{2}\pi\right)}\,, \tag{2.27}$$

thus, the force is

$$F^{(4)} = A^4 f^{(4)} = \frac{9\pi^2A^4\omega^6}{k(4\omega^2-3)(\omega^2-3)^2}\left[\frac{\omega_+}{\sinh^2\pi\omega_+} - \frac{\omega_-}{\sinh^2\pi\omega_-}\right]\,, \tag{2.28}$$

where

$$\omega_\pm := \sqrt{\omega^2-1} \pm \sqrt{\omega^2-4}\,. \tag{2.29}$$

Note that, as the force is negative the kink accelerates towards negative values of x, although the incoming wave is traveling from $-\infty$. In other words, the kink is pulled by the *negative radiation pressure* (NRP). This surprising result was confirmed by numerical simulations of the full nonlinear PDE. Both the A^4 proportionality and the frequency dependence of the force (2.28) was confirmed for the frequency range $\omega < 5$ and $A < 0.22$ with 10% accuracy. The single discrepancy, near the frequency $2\omega_d$, could be explained by the highly nonlinear resonance with the internal oscillating mode of the kink.

The effect of NRP can be surprising and counterintuitive at first. However, the physical explanation of this effect is relatively simple. In the first order the kink is transparent to the incoming radiation, and the force exerted on the kink is zero. Nonlinearities around the kink, on the other hand, produce higher frequency waves which also carry momentum. Since the second order transmission coefficient is larger than the corresponding reflection coefficient, there is a surplus of the momentum behind the kink. In order to comply with the conservation law and restore the balance, the kink starts to accelerate in opposite direction, towards the source of radiation.

The negative radiation pressure in the ϕ^4 model seems to be something exceptional. It requires that the potential, generated by the soliton configuration in a non-integrable model, is reflectionless. This is rather rare property. However, one can expect that the effect of NRP may still exist in some modifications of the ϕ^4 model with non-reflectionless perturbation potential. In such a case the force exerted on the soliton would have the form

$$F \approx A^2 f^{(2)} - A^4 f^{(4)} + O(A^6)\,. \tag{2.30}$$

This would support the effect of NRP for certain range of amplitudes, as $f^{(2)} < A^2 f^{(4)}$.

The mechanism of the NRP we described above is not unique. Indeed, the key point is to produce a surplus of momentum carried by the radiation behind the kink. This can be achieved in many different ways. For example, in more complicated multicomponent field theories, different components can have different dynamical properties. Then, even in the linear order, the scattering with increase of momentum becomes allowed. For example, a scattering with transfer of momentum from the massive field $\mathcal{P}_t \sim (\omega^2 - m^2)$ to the massless field $\mathcal{P}_t \sim \omega^2$ leads to the NRP, as long as the reflection from the scattering center remains small enough [27, 28].

A quite interesting, and even simpler example of NRP effect, can be found in the ϕ^6 model (2.10) [29]. In this model we restrict our considerations only the the first, linear order. Recall that the model (2.10) has three vacua. Small perturbations around two of them possess mass $m_{\pm 1} = 2$, but the perturbations around the third vacuum have smaller mass $m_0 = 1$. The kink interpolates between vacua of different kinds, being a sort of bridge for the waves which can travel from one vacuum to another. Waves with frequency their fulfilling $m_0 < \omega < m_{\pm 1}$ can only propagate around the vacuum $\phi = 0$ and reflect perfectly from the kink. This gives raise to a *positive radiation pressure*. But also for higher frequencies, even when they can already propagate in the second vacuum, they move more slowly and carry less momentum. This deficiency of momentum is balanced by the force pushing the kink towards $\phi = \pm 1$ vacuum.

On the contrary, the wave traveling initially through $\phi = \pm 1$ vacuum carries less momentum than after the transition through the kink. The excess of momentum behind the kink is created by pushing it again towards the $\phi = \pm 1$ vacuum. Such a wave exerts negative radiation pressure. Indeed, the solutions of the linearized equation for perturbations around the ϕ^6 kink are known in terms of hypergeometric functions [25], thus the scattering amplitudes can be found from the asymptotic form of the solution

$$\begin{cases} \eta(x \to +\infty) = e^{ikx}/B(q,k), \\ \\ \eta(x \to -\infty) = e^{iqx} + \frac{B(-q,k)}{B(q,k)} e^{-iqx} \end{cases} \tag{2.31}$$

with

$$q = \sqrt{\omega^2 - 1}, \; k = \sqrt{\omega^2 - 4},$$
$$B(q,k) = \frac{\Gamma(1 - ik)\Gamma(-iq)}{\Gamma(-\frac{1}{2}ik - \frac{1}{2}iq + \frac{5}{2})\Gamma(-\frac{1}{2}ik - \frac{1}{2}iq - \frac{3}{2})}. \tag{2.32}$$

Then the force exerted on the $\phi_{(0,1)}$ kink by the incoming wave, propagating from $+\infty$, can be evaluated as

$$F_{+\infty}(q,k) = \frac{1}{2}\frac{A^2}{|B(q,k)|^2}\left(2|B(-q,k)|^2 q^2 + qk - k^2\right). \tag{2.33}$$

Similarly, the force acting on the $\phi_{(0,1)}$ kink by a wave propagating in opposite direction, can be expressed as $F_{-\infty}(q,k) = -F_{+\infty}(k,q)$. Both forces are positive, which means that no matter from which direction the wave came, the kink would always accelerate rightwards. Moreover, as the force appears in the linearized system, in the general case of an arbitrary perturbation the contribution of all modes can be evaluated as an additive sum over all eigenfrequencies. Therefore we can say that *arbitraryly* small perturbations will always push the $\phi_{(0,1)}$ kink towards the vacuum $\phi = 1$. Considering all kinks of the ϕ^6 model we can conclude that all small perturbations will act on the solitons in such a way that the vacuum $\phi = 0$ will tend to expand.

This is a highly nontrivial conclusion with many important implications. Let us consider a system of (almost) stationary kinks and antikinks separated by a distance so large that the interactions are almost negligible. The system would last in such a state for a very long time, given the fact that the static forces decay exponentially. However, it is enough to put any radiative perturbation, which can be either a localized perturbation or even random fluctuations, filling the whole space (just like the primordial, background radiation filling the whole Universe) and the whole system would collapse. Moreover, the system would evolve in such a way, that the defects enclosing one of the vacua ± 1 would collide. Such collisions are the source of radiation which would increase the rate of further collisions. This is how an interplay between negative and positive radiation pressure can lead to a chain reaction of kink-antikink annihilations. This scenario was confirmed by numerical simulations (Fig. 2.1a–c).

In higher (D) dimensions, instead of kinks there are $D - 1$ dimensional domain walls. The walls can extend to either infinities or enclose domains of certain vacua. Large domain walls can be locally flat and the dynamics can be very similar to the dynamics of kinks. However, curved domain walls can have a surface tension which tries to close small domains [30] (Fig. 2.1d–e). The radiation pressure can increase the rate in which domains with $\phi = \pm 1$ vacua vanish and slow down the rate or even reverse for the $\phi = 0$ vacuum. For instance, random fluctuations can expand circular or spherical domains of the $\phi = 0$ vacuum above certain, critical radius (Fig. 2.1f–g).

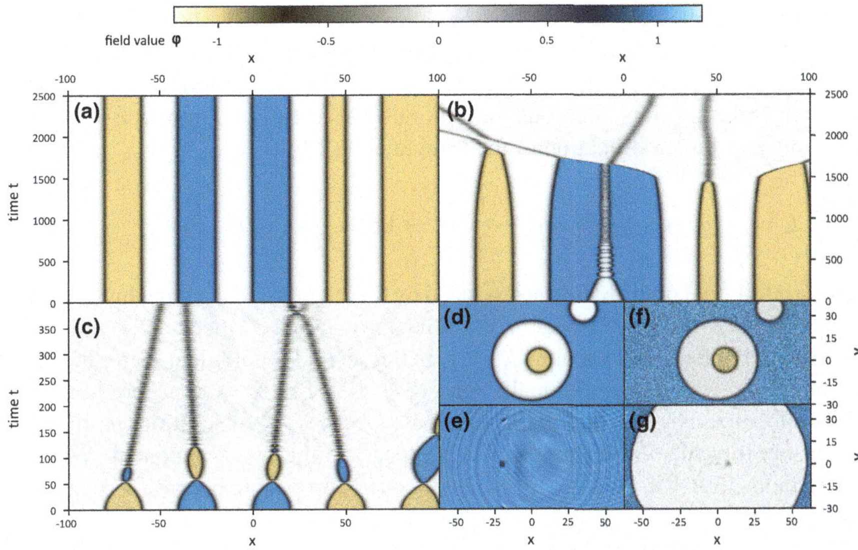

Fig. 2.1 Evolution of a system of kinks in the ϕ^6 model: **a** nearly static configuration, **b** annihilation of a close kink-antikink pair triggers the chain reaction of other annihilation, **c** with gaussian noise of amplitude $A = 0.05$. Evolution of two dimensional circular domain walls: without perturbation **d** $t = 0$ **e** $t = 75$ and with a gaussian noise of amplitude $A = 0.12$, **f** $t = 0$ **g** $t = 75$. Reprinted (without modification) from [29]. © 2017 The Author of [29], under the CC BY 4.0 license

2.3.2 *From Internal Modes to Oscillons and Back*

Apart from kinks, the simple ϕ^4 model (2.1) also possess a very interesting regular quasi-non-dissipative and almost periodic time-dependent solution, whose properties closely resemble the sG breather [31]. Such a state, observed in the process of time evolution of some initial data, is referred to as an oscillon.[2]

Since the life time of the oscillons is very large they can be well approximated by the Fourier decomposition [34]

$$\phi(x,t) = 1 + \sum_{n=0}^{N} \phi_n(x)\cos(n\omega t)\,. \tag{2.34}$$

During its evolution, the amplitude of the oscillon slowly decreases. However, the rate in which it radiates decreases even faster, being the decay rate $\frac{dE}{dt} \sim \exp(-B/E)$ beyond all orders [35–37]. The amplitude decays are synchronized with the incre-

[2]In the 3+1 dimensional theory the corresponding spherically symmetric solutions were originally discovered in 1976 by Bogolyubskii and Makhan'kov [20], who coined the term "pulson" to describe these configurations. However this observation did not attract much attention at that time, and this work remained almost forgotten until Gleiser rediscovered these solutions in 1994 [32, 33].

ments of the frequency of the oscillations. In the long time evolution, the oscillon tends to the lowest mode of the continuous spectrum with frequency $\omega = 2$.

There is an interesting correspondence between the internal modes and the oscillon, which under special conditions, can smoothly be transferred into each other [38]. Let us consider the modification of the Lagrangian (2.1)

$$\mathcal{L} = \frac{1}{2}(\partial_t \phi)^2 - \frac{1}{2}(\partial_x \phi)^2 - \frac{1}{2}\left(\phi^2 - 1\right)^2 - \frac{1}{2}V(x)(\phi - 1)^2, \tag{2.35}$$

where an additional perturbation is produced by the asymptotically vanishing Pöschl–Teller potential $V(x) = -V_0 \operatorname{sech}^2 bx$. Unlike the original ϕ^4 theory (2.1) the model (2.35) has only one trivial vacuum $\phi_0 = 1$. In the second topological sector the static configurations, which asymptotically tend to $\phi \to -1$ as $x \to \pm\infty$, are nontrivial solutions localized by the trapping potential $V(x)$. One type of solutions is similar to the non-topological soliton (lump), which appears in the two-component system of coupled fields [39, 40]. However, the lump does not propagate because it is captured by the potential.

In this sector, solutions of the second type represent a static kink-antikink pair trapped by the potential $V(x)$. This solution is unstable and a small perturbation can destabilize the configuration, which can either decay into the vacuum $\phi_0 = 1$ with two kinks escaping to infinity, or into the lump, which is a minimal energy solution in the same sector. Evidently, a particular choice of scenario depends on the explicit value of the parameter V_0 [38].

The perturbative spectrum of linearized fluctuations around the trivial vacuum $\phi_0 = 1$ of the modified model (2.35) can be found by analogy with consideration above. The corresponding eigenfunctions are solutions of the equation, which generalizes the linearized equation of the ϕ^4 model (2.5)

$$\left(-\frac{d^2}{dx^2} + 4 - \frac{V_0}{\cosh^2 bx}\right)\eta = \omega^2 \eta\,. \tag{2.36}$$

with the mass threshold $m = 2$. The ground state of the Pöschl–Teller potential is

$$\xi_0(x) = \frac{1}{\cosh^\lambda(bx)}, \quad \lambda(\lambda + 1) = \frac{V_0}{b^2}, \quad \omega^2 = m^2 - b^2\lambda^2 \tag{2.37}$$

where λ is a real parameter.

The eigenvalues ω are imaginary for $V_0 > 4 + 2b$, the corresponding modes are unstable and the system could change its ground state producing a $K\bar{K}$ pair. There is a single stable oscillating mode for $2 > V_0 > 0$, as this mode is excited to a nonlinear regime decaying via radiation that propagates through the second harmonic. The amplitude of this mode decreases according to the Manton–Merabet power law $A\,t^{-1/2}$ [16], and its frequency increases with time, as the mode evolves toward the linear regime.

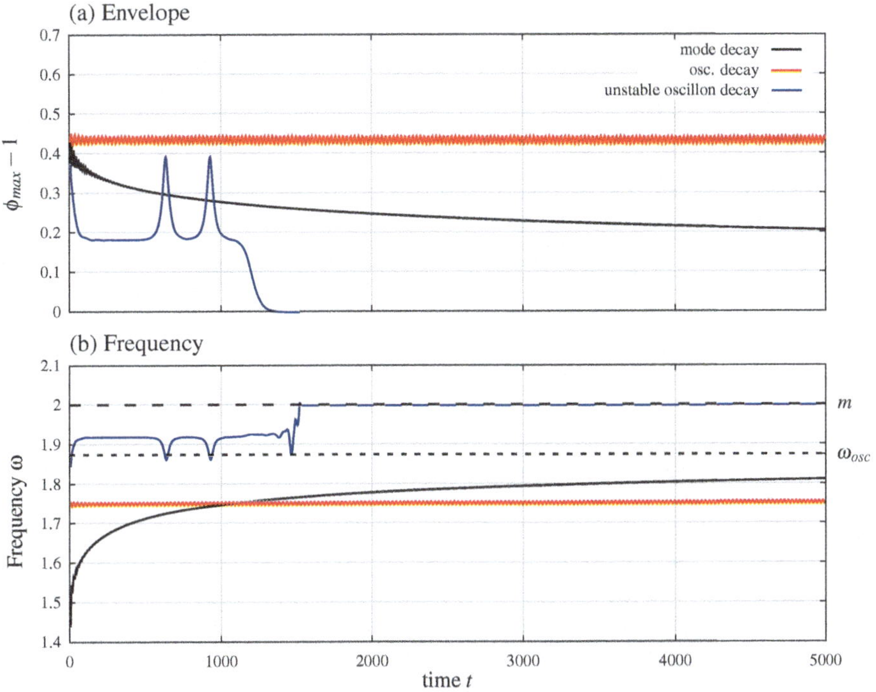

Fig. 2.2 Possible relaxation scenarios for $A_0 = 0.6$ [see (2.38)]. The plots show the maximum envelope of oscillations at $x = 0$ (upper plot) and the measured frequency ω. Black curve (smoothly decaying, middle curve in the upper plot) represents decay of the oscillating mode, red curve (almost constant curve, with small amplitude oscillations) shows the time evolution of the oscillon. Blue curve (irregular) shows decay of the unstable oscillon in the repulsive potential $V_0 < 0$. Dashed horizontal line indicates the eigenfrequency of the mode for $V_0 = 1.19$, $\omega_{\text{osc}} = 1.8735$. Reprinted (without modification) from [38]. © 2017 The Authors of [38], under the CC BY 4.0 license

Numerical analysis reveals that in the presence of the perturbation potential, the lowest bounded oscillating mode could become an effective attractor in the time evolution of some even initial data, which, in the limit $V_0 = 0$ would evolve into the oscillon-like state.

Considering the time evolution of the symmetric initial data

$$\phi(x, 0) = 1 - A_0 \operatorname{sech}^{\alpha}(x), \quad \phi_t(x, 0) = 0 \tag{2.38}$$

where α is a positive real parameter, we observe that in absence of the potential, or in case of repulsive potential, these initial data evolve into the oscillon solution.

The red curve in Fig. 2.2 illustrates the time evolution of the oscillon, which is produced in the absence of the external potential. The black curve corresponds to the evolution of the oscillating mode, which was excited up to the nonlinear regime. Due to the nonlinearity, the corresponding initial frequency is initially much lower

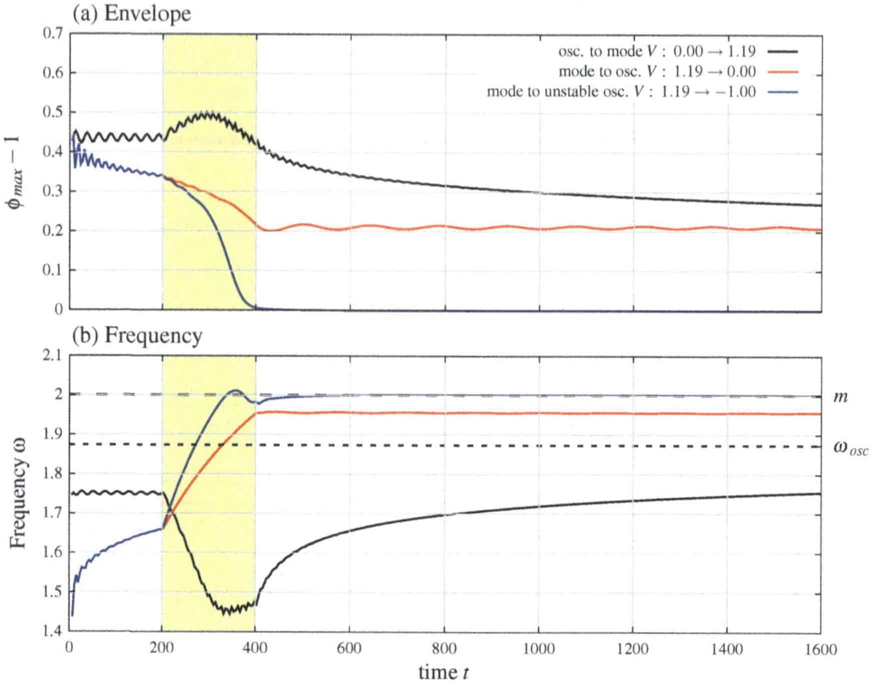

Fig. 2.3 Examples of adiabatic transformations: (i) the oscillon $A_0 = 0.6$ [see (2.38)], to the internal oscillating mode (black curve, upper in the upper plot, bottom curve in the bottom plot); (ii) the oscillating mode to the oscillon (red curve, middle); (iii) the oscillating mode to an unstable oscillon and its consequent decay (blue curve, decaying to threshold $\phi = 1$ and $\omega = m$). Reprinted (without modification) from [38]. © 2017 The Authors of [38], under the CC BY 4.0 license

than its value predicted from the linearized theory. However with time the energy is radiated away and the amplitude of oscillations decays slowly. Then the frequency becomes larger when approaching the corresponding linearized value.

An interesting scenario was observed for the situation when the initial data set rapidly converges to the oscillon configuration in the absence of the external potential, see black curves in Fig. 2.3. Within the time interval $t \in [t_1, t_2]$, we adiabatically increase the depth of the potential as $V(t) = V_0 \frac{t-t_1}{t_2-t_1}$. Then the amplitude of the oscillations increases and, when $t > t_2$, the oscillating state becomes trapped by the potential well. Consequently, the evolution pattern of this mode follows the usual $t^{-1/2}$ law of the radiative decay. In other words, the oscillon state becomes smoothly transformed into the oscillating mode.

The opposite transition is also illustrated in Fig. 2.3 (red curves). Initially, there is an excited oscillating mode trapped in the potential well with the same values of the parameters as above. When $t_1 < t < t_2$ the potential is turned off smoothly and the oscillation mode becomes transformed into an oscillon state whose amplitude is a bit above $A = 0.2$ and whose frequency is $\omega = 1.9566$. This is clearly below the mass threshold. Both the frequency and the amplitude of the configuration are modulated by small oscillations, although the corresponding average values are almost constant. Thus, we can conclude that a distinction between an oscillon and an oscillating mode is quite artificial. The only difference is that the frequency of the oscillating modes tend to the frequency calculated from the linearized model, as the amplitude decreases.

2.4 Production of the Kink-Antikink Pairs from Radiation

As we have seen, there is strong evidence of intrinsic relations between the states of the perturbative spectrum of a nonlinear non-integrable theory and the non-perturbative soliton solutions. This interplay significantly affects the process of collision of the solitons in the ϕ^4 model, as there is an intriguing pattern of scattering of the kinks [11–14] related to resonant energy exchange between the translational mode of the solitons and the excitation of the internal vibrational mode. Moreover, the opposite process of the production of the $K\bar{K}$ pairs in the collision of the states of continuum is also chaotic due to resonance effects [41, 42].

We considered two widely separated wave trains with amplitude C propagating towards the collision point which is a kink or one of the vacuum sectors of the ϕ^4 model. The initial data are (for $t < a_2/v$)

$$\phi(x,t) = \phi_0 + C[F(x+vt)\sin(\omega t + kx) + F(-x+vt)\sin(\omega t - kx)], \quad (2.39)$$

where k is the wavenumber of the incoming wave, $\omega = \sqrt{k^2+4}$ is the frequency of the continuum mode and $v = k/\omega$ is the velocity of the propagation of the wave train. The envelope of the train can be taken as $F(x) = [\tanh(x-a_1) - \tanh(x-a_2)]$, where the parameters a_1 and a_2 define the length of the train and the initial separation between the trains. ϕ_0 is the static solution, which in the case of the ϕ^4 model can be either a vacuum or a single kink.

In the case of a kink ($\phi_0 = \phi_K$) the small amplitude radiation interacts with the internal mode of the kink. The mode is excited due to a nonlinear parametric resonance (similar to the Mathieu equation) [41]. As the amplitude of the radiation grows, the response of the mode also grows. For certain amplitude the internal oscillating mode is excited so much that the energy can be released only by ejecting a $K\bar{K}$ pair. A scan through both the amplitude and frequency revealed that the creation process is chaotic, and the boundary between creation and just the excitation of the internal mode has some fractal properties. The absorption of the radiation by the oscillating mode is responsible for the aforementioned discrepancy between the second order calculations of the negative radiation pressure and the numerical results.

In the absence of the kink, i.e., for $\phi_0 = 1$, numerical simulations show that the small amplitude collision produces an oscillating lump with the frequency just a bit above the mass threshold. This lump could be identified with a non-linear excitation of the trivial vacuum, as it slowly radiates its energy away. For large amplitude

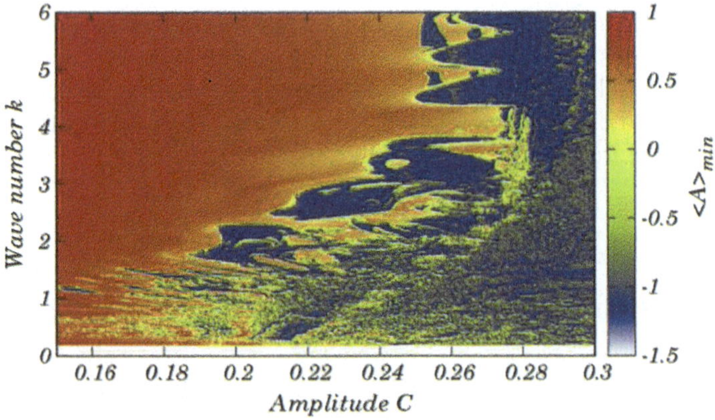

Fig. 2.4 Fractal structure in the C, k plane. Shading (or colour) represent the measured minimum of average of the field $\langle A\rangle = \frac{1}{20}\int_{-10}^{10}\mathrm{d}x\ \phi(x,t)$. The dark regions (blue in colour), where $\langle A\rangle < -1$, indicate creation of the $K\bar{K}$ pairs. Reprinted (without modification) from [42], with permission of APS

collisions, the remaining lump oscillates with frequency within the mass gap, so such a state can be identified as an oscillon.

Furthermore, for a certain range of impact parameter values, the $K\bar{K}$ pair produced in the collision leads to the emergence of an oscillon at the collision center. Notably, the regions of production of the solitons and the regions of the parametric space where this process does not take place, are separated by a fractal-like boundary (Fig. 2.4).

For certain values of impact parameters an oscillon remaining in the collision center decays into the second $K\bar{K}$ pair. Sometimes, two oscillons could also be ejected from the collision center and after a while they could decay into two pairs of $K\bar{K}$. These observations confirm the conclusion concerning the mechanism of the creation of the $K\bar{K}$ pair as a three-stage process. In the first stage, the collision of the incoming excitations in the topologically trivial sector produces an oscillon excitation. Next, the oscillon interacts with the incoming trains, and due to the parametric resonance, it may decay into the outgoing $K\bar{K}$ pair [42].

These two creation mechanisms are in some way alike to proving that the oscillon and the oscillating mode can play similar roles in the dynamics.

2.5 Kink-Antikink Scattering in the ϕ^6 Model

Since the resonance scattering of the solitons was observed in several models with different potentials [14, 24], it was suggested [14] that the existence of an internal kink mode is a necessary condition for the appearance of resonance windows. The parametrically modified sG model lent further support to this view: depending on the value of a parameter, kinks and antikinks do or do not possess an internal mode;

correspondingly, resonance windows do or do not appear [24]. Resonance windows have also been observed in vector soliton collisions [43, 44] and in the scattering of kinks on impurities [45]. Again, the mechanism always relies on the presence of a localized internal mode, either of a single kink or of an impurity, or both.

However, the mechanism of the reversible energy exchange may also work in the absence of an internal oscillatory mode. A new type of chaotic behavior, related to the interplay between the states of the perturbative spectrum and the kinks, was observed recently in the ϕ^6 model [18]. Although in this model the kink solution does not possess an internal vibrational mode, there still exist multi-bounce resonance windows in the kink-antikink collisions. Thus, the mechanism of the energy transfer should be different from the case of the ϕ^4 model discussed above.

Recall that the ϕ^6 model (2.10) has three vacua $\phi_v \in \{-1, 0, 1\}$. There are two different kink solutions (2.13), so we have to consider two types of collision of the solitons, the $K\bar{K}$ collisions in the vacuum sector $\phi_0 = 0$ and the collision in one of the symmetric sectors $\phi_0 = \pm 1$. In the former case the initial configuration of the colliding kinks, which we denote as $(0, 1) + (1, 0)$ can be taken as a superposition $\phi(x) = \phi_K(x + a) + \phi_{\bar{K}}(x - a) - 1$ where a is the separation parameter; in the latter case the initial $K\bar{K}$ configuration $(1, 0) + (0, 1)$ is $\phi(x) = \phi_K(x - a) + \phi_{\bar{K}}(x + a)$. Since both kinks do not posses an internal oscillating mode, we can naively expect that the collision will always be quasi-elastic. However, there is a wide potential well in the case of widely separated $(1, 0) + (0, 1)$ $K\bar{K}$ pair with two local minima associated with the positions of the solitons, see left column of Fig. 2.5. In a contrast, in the case of $(0, 1) + (1, 0)$ configuration these two minima are separated by a barrier. Thus, the energy of the colliding $(1, 0) + (0, 1)$ $K\bar{K}$ pair can be transferred into the excitation of the trapped oscillating states of the composite configuration. With a suitable resonance condition, this energy might be returned to the translational modes of the kinks allowing them to escape. On the contrary, there is no oscillating modes in the collective potential of the $(0, 1) + (1, 0)$ $K\bar{K}$ pair, so the collision is expected to be elastic.

Indeed, numerical simulations confirm that the pattern of collision of the $(1, 0) + (0, 1)$ $K\bar{K}$ pair is very similar to the picture observed in the ϕ^4 model, see Fig. 2.5. As we can see on the plot (a) an intricate pattern of the resonance escape windows exists up to a critical velocity $v_{\mathrm{cr}} = 0.0457$, after which the kinks always have enough energy to separate. Note that the impact velocities are very small, so the collision can be considered as an adiabatic process.

By contrast, the collision of the $(0, 1) + (1, 0)$ $K\bar{K}$ pair does not reveal any chaotic behavior with a sequences of bouncing windows; for $v < v_{\mathrm{cr}} \approx 0.289$ the pair annihilates into the vacuum $\phi_v = 0$ with a small amount of radiation emitted, while for $v > v_{\mathrm{cr}}$ the collision yields a mirror pair of solitons escaping to infinity with no bouncing: $(0, 1) + (1, 0) \to (0, -1) + (-1, 0)$.

By analogy with the $K\bar{K}$ scattering in the ϕ^4 model, we can assign a "bounce number" to each resonance window in the $(1, 0) + (0, 1)$ sector. This is the number of collisions between the solitons before their final escape to infinity. The first two-bounce windows opens at $v_{\mathrm{in}} \approx 0.0228$, it is followed by a so called "false" window at $v_{\mathrm{in}} \approx 0.0273$, the kinks become well separated after collision but they

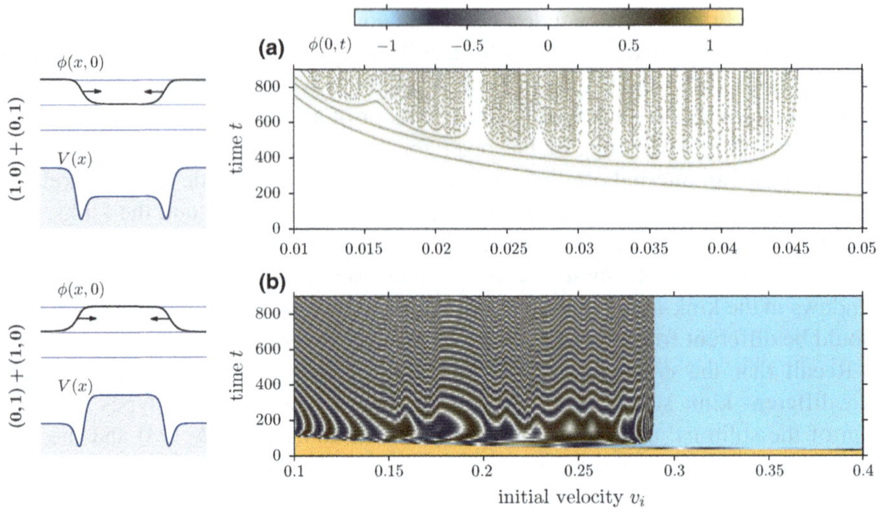

Fig. 2.5 Two types of collisions of (identical) kinks in ϕ^6 model **a** $(1, 0) + (0, 1)$ and **b** $(0, 1) + (1, 0)$. The left column shows the initial kink alignment and linearized potential, the right column the field at the center of collision $\phi(0, t)$ as a function of initial velocity and time. Note the complicated fractal structure in the first case (**a**)

cannot escape and the configuration finally collapses into the vacuum. A third "true" window opens at $v_{in} \approx 0.0303$, which is followed by a sequence of higher escape windows. Furthermore, the regions close to the edges of the windows exhibit nested structures of higher bounce windows revealing quasi-fractal structure of the interaction, which is similar to the usual ϕ^4 model. There is also the number of collective mode oscillations between two subsequent collisions, for which numerical computations show to correspond to the excitation of the lowest collective mode of the $K\bar{K}$ pair [18].

Thus, there exists a mechanism which allows resonances to occur due to the excitation of collective oscillating states in the potential well created in the space between the constituents of a suitably ordered $K\bar{K}$ pair. This does not require the existence of an internal mode localized on a single kink.

Note that the initial velocity of the kinks is very small allowing for an adiabatic approximation in the analysis of fluctuations about the kink–antikink configuration. Thus, the time between two subsequent collisions of the solitons is much larger than the corresponding period in the ϕ^4 model, and the number of internal oscillations of the kinks in the first window is quite large, $n = 12$. Further, the effect of the radiation pressure on the slow kinks is much stronger than in the ϕ^4 model, affecting the fine structure of higher order resonance windows [18].

2.6 The Role of Quasinormal Modes in Topological Defect Collisions

The internal oscillating modes, either attached to solitons or formed between them in the potential trap, play an important role during the collisions. They can live for a very long time being, consequently, the natural reservoirs storing the energy of perturbations and giving it back to the translational modes in a resonant way. The reversal energy exchange leads to the appearance of fractal-like structures in the parameter space. However, the internal oscillating modes are not the only ones where the energy can be stored [18].

Indeed, in the absence of the internal oscillating mode, perturbations spread usually rather quickly and only the modes with a slow group velocity near the mass threshold contribute to the long-time asymptotic relaxation with the usual power decay $\phi(0,t) \sim t^{-1/2}$. Among these small perturbations there often are special modes which can stay for exceptionally long times near the potential center. Such modes, which are referred to as the quasinormal modes (QNMs), decay exponentially with time. They play an important role in the relaxation processes in many physical situations, such as collisions of black holes, radioactive decay or the ringing of a bell [46].

QNMs are often referred to as resonances; however, to avoid possible confusion with the resonant structure in the collisions of the solitons we do not use this term. In a more formal description, these modes are often referred to as the poles of the Green function. They appear as a result of modifications of the boundary conditions of the corresponding linearized problem to purely outgoing wave. Such conditions break the Hermiticity of the linear operators, resulting in the appearance of the complex-valued eigenfrequencies $\omega = \Omega + i\Gamma$. The imaginary part $\Gamma > 0$ is responsible for the exponential decay of the mode, $e^{i\omega t} = e^{-\Gamma t}e^{i\Omega t}$. Note that neither ϕ^4 nor sG models support QNMs due to the reflectionless nature of the linearized potential, generated by the kinks. But this is an exception rather than the rule.

Indeed, let us consider a small perturbation of the ϕ^4 model with the deformed potential

$$U(\phi,\epsilon) = W + \frac{m^2-4}{4}\frac{\epsilon W}{W+\epsilon}. \tag{2.40}$$

where $W = \frac{1}{2}(\phi^2-1)^2$ denotes the standard ϕ^4 potential which is restored as the parameter ϵ vanishes.

In the case of small positive values of ϵ the potential $U(\phi,\epsilon)$ is almost the same as in the original ϕ^4 theory, except at the vicinity of the vacua. Near the vacua $W \approx 0$ and $U(\phi,\epsilon) \approx \frac{m^2}{4}W$, which means that the mass of the small perturbations is now equal to m rather than 2, as in the standard ϕ^4 model. The kink preserves its shape (Fig. 2.6), but it becomes deformed as approaches the vacuum, $\phi - \phi_{\text{vac}} \sim e^{-mx}$. Near the center the linearized potential

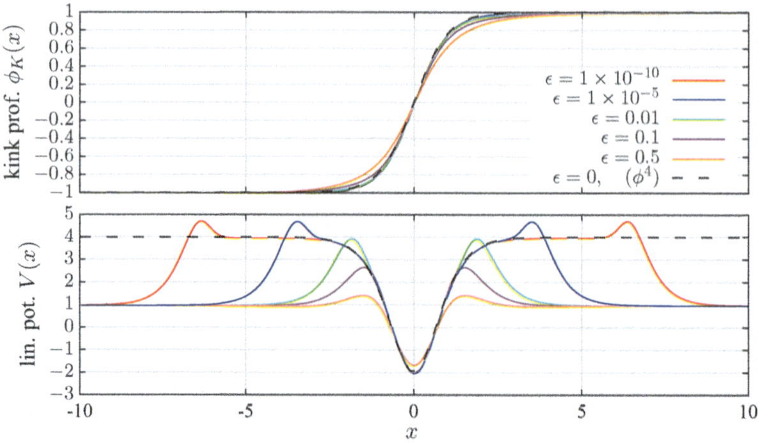

Fig. 2.6 Kink profiles and the linearized potential $V(x)$ [see (2.41)]. Reprinted (without modification) from [47]. © 2018 The Authors of [47], under the CC BY 4.0 license

$$V(x) = \frac{\partial^2 U(\phi, \epsilon)}{\partial \phi^2} \tag{2.41}$$

is similar to the Pöschl–Teller potential (2.5), but asymptotically it tends to $V \to m^2$ instead of 2, as in the original ϕ^4 theory.

If $m < 2$ the potential (2.40) corresponds to a barrier whose width is controlled by the parameter ϵ (see Fig. 2.6).

Let us consider the case $m = 1$. For small $\epsilon > 0$ the linearized equation is almost satisfied by the solution for the ϕ^4 internal oscillating mode localized by the potential (2.5). However, since the potential $U(\phi, \epsilon)$ asymptotically drops to 1, the frequency $\omega = \sqrt{3}$ is moved to the continuum part of the spectrum. The field is almost trapped near the soliton but now it can tunnel through the potential barrier. Thus, the internal oscillating mode becomes the quasi-normal mode. Note that its frequency for $\epsilon < 0.04$ can be reasonably well (2% accuracy) approximated as

$$\omega \approx 1.738 + 0.490\sqrt{\epsilon} - 2.280\epsilon + (0.325\sqrt{\epsilon} + 0.783\epsilon)i. \tag{2.42}$$

Since almost all the properties of the model are continuously controlled by the parameter ϵ, we may also expect a continuous change of the properties of the modified model. Although the QNMs lose their energy due to the radiation more effectively than the standard internal oscillating mode of the kink, numerical simulations of the $K\bar{K}$ collisions in the model with deformed potential (2.40) confirmed that the resonant structure remains visible for certain range of values of ϵ (see Fig. 2.7). The most narrow windows closes for smaller values of ϵ. The largest window close at $\epsilon \approx 0.034$. Another important result is that the critical velocity for final escape grows with ϵ. This is not a surprise since the lifetime of the QNMs decreases as ϵ increases;

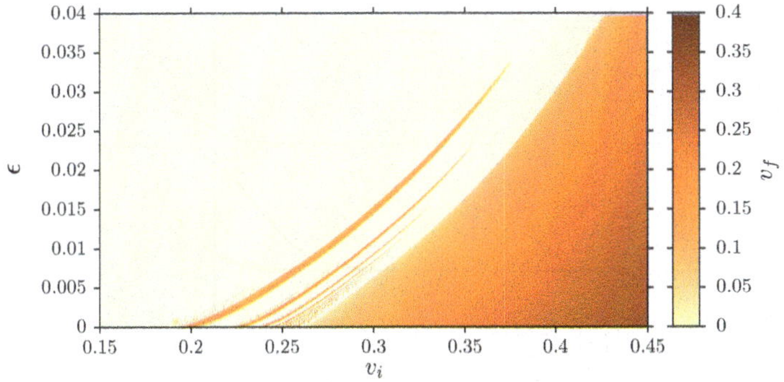

Fig. 2.7 Final velocity v_{f} of the kinks after the collision as a function of perturbation parameter ϵ and the initial velocity v_{i}. Reprinted (without modification) from [47].

i.e. the energy is stored for much shorter time. On the other hand, wide QNMs (with large Γ) can be very effective in transferring the excess of the energy away from the solitons. This may be an important general mechanism responsible for binding solitonic configurations.

2.7 Kink Boundary Scattering in the ϕ^4 Model

Yet another example of give-and-take between the perturbation spectrum and solitons is provided by the boundary ϕ^4 model (2.1) on semi-infinite line [48]. There are two vacua $\phi_0 \in \{-1, +1\}$ on the left half-line $-\infty < x < 0$. The bulk energy and Lagrangian densities are $\mathcal{E} = \mathcal{T} + \mathcal{V}$ and $\mathcal{L} = \mathcal{T} - \mathcal{V}$ respectively, where

$$\mathcal{T} = \frac{1}{2}\phi_t^2 \quad \text{and} \quad \mathcal{V} = \frac{1}{2}\phi_x^2 + \frac{1}{2}(\phi^2 - 1)^2 . \tag{2.43}$$

The boundary contribution to the energy is defined as $-H\phi_0$, where $\phi_b = \phi(0, t)$ and H can be interpreted as a boundary magnetic field. Thus, the Neumann-type boundary condition is $\phi_x(0, t) = H$ at $x = 0$.

The states of the perturbative and non-perturbative spectrum of the boundary ϕ^4 model are different from those we discussed above in Sect. 2.2.1. For $0 < H < 1$ there are four static solutions of the field equations [48], see Fig. 2.8. Two of them, $\phi_1(x) = \tanh(x - X_0)$ and $\phi_2(x) = \tanh(x + X_0)$ with $X_0 = \cosh^{-1}(1/\sqrt{|H|})$, are counterparts of the kinks on the half-line, whereas the other two solutions, $\phi_3(x) = -\coth(x - X_1)$ and $\phi_4(x) = -\coth(x + X_1)$ with $X_1 = \sinh^{-1}(1/\sqrt{|H|})$ are irregular on the full line. On the half line, ϕ_3 is non-singular and corresponds to the absolute minimum of the energy, while ϕ_1 is metastable, and ϕ_2 is the unstable saddle-point configuration between ϕ_3 and ϕ_1.

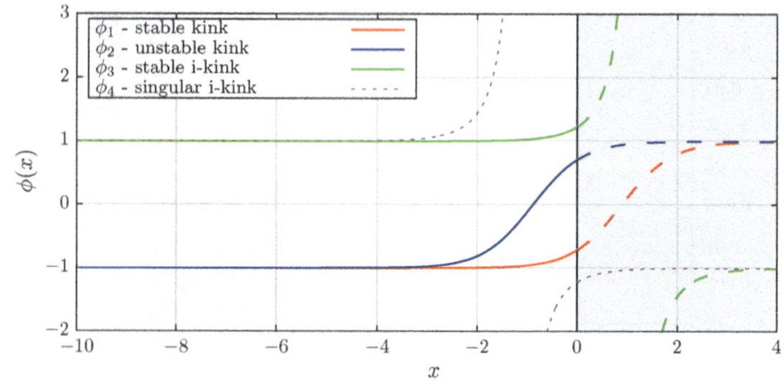

Fig. 2.8 Static solutions for $H = 1/2$ [see (2.43)]. Reprinted (without modification) from [48], © 2015 The Authors of [48], under the CC BY 4.0 license

The energies of these configurations can be found by rewriting the functionals $E[\phi] = \int_{-\infty}^{0} \mathcal{V}\,\mathrm{d}x - H\phi_0$ as

$$E[\phi] = \frac{1}{2}\int_{-\infty}^{0} \left(\phi_x \pm (\phi^2-1)\right)^2 \mathrm{d}x \mp \left[\frac{1}{3}\phi^3 - \phi\right]_{-\infty}^{0} - H\phi_0\,. \tag{2.44}$$

Since ϕ_1 and ϕ_2 satisfy $\phi_x = 1-\phi^2$ we have $\phi_1(0) = -\sqrt{1-H}, \phi_2(0) = \sqrt{1-H}$, while $(\phi_3)_x = \phi_3^2 - 1$ and so $\phi_3(0) = \sqrt{1+H}$. This yields

$$E[\phi_1] = \frac{2}{3} - \frac{2}{3}(1-H)^{3/2}, \quad E[\phi_2] = \frac{2}{3} + \frac{2}{3}(1-H)^{3/2},$$
$$E[\phi_3] = \frac{2}{3} - \frac{2}{3}(1+H)^{3/2}. \tag{2.45}$$

Then ϕ_3 remains as the only static solution for $H > 1$. For negative values of the boundary field $H < 0$, physical solutions are $\tilde{\phi}_i(x) = -\phi_i(x), i = 1\ldots 3$.

The states of the perturbative sector are also different from the corresponding excitations on the full line. Considering $\phi_1(x)$ as a static half-line solution with $0 < H < 1$ in the $\phi(-\infty) = -1$ sector, we find the equation for linear perturbations $\xi(x)e^{i\omega t}$, with boundary restriction $\partial_x\xi(x) = 0$ at $x = 0$,

$$\kappa^3 - 3\phi_0\kappa^2 + (6\phi_0^2 - 4)\kappa - 6\phi_0^3 + 6\phi_0 = 0 \tag{2.46}$$

where $\phi_0 = \phi_1(0) = -\sqrt{1-H}$, $k = i\kappa$ and now $\omega^2 = 4 - \kappa^2$. The solutions of (2.46) with $\kappa > 0$ can give rise to the localized boundary modes, and the value of κ must be smaller than the mass threshold 2 for the corresponding ω to be real and the mode stable. For $0 < H < 1$ there is just one positive solution of (2.46), which satisfies $\kappa < 2$; this is the single vibrational mode, which is localized close to the boundary. The frequency of the mode changes from the mass threshold $m = 2$ for $H = 0$ to $m = 0$ for $H = 1$. For $H < 0$ such an internal modes does not exist and the excitation spectrum is continuous. Thus, the collision of kinks with the boundary may lead to a variety of resonance phenomena, which are similar to the kink-antikink collisions in the ϕ^4 theory on a full line.

There is, however, another difference related to the dynamics of solitons near the boundary. The static force between a single antikink and the boundary can be evaluated as [48]: $F = 32\left(\frac{1}{4}H + e^{2x_0}\right)e^{2x_0}$ where $x_0 < 0$. For $H < 0$ the force is repulsive far from the boundary, and attractive in its vicinity. Therefore, if the initial velocity v_i is relatively small, the kink does not have sufficient energy to overcome the initially-repulsive force, it will be reflected without ever coming close to the boundary and without exciting any other modes.

Indeed, numerical simulations revealed that for negative values of the boundary magnetic field H and small impact velocities v_{in}, the antikink is reflected in a nearly elastic fashion, i.e., with very little radiation, see Fig. 2.9 (light smooth shading, or white and blue colors therein indicate that the state of the boundary was not changed by the collision). As v_{in} increases above some critical value, the antikink becomes trapped by the boundary, leaving only radiation in the final state (darker fast changing regions, or yellow and red in colors). Increasing v_{in} further, scattering windows begin to open, until the initial velocity exceeds an upper critical value and the antikink always escapes again. This pattern becomes more explicit for relatively small

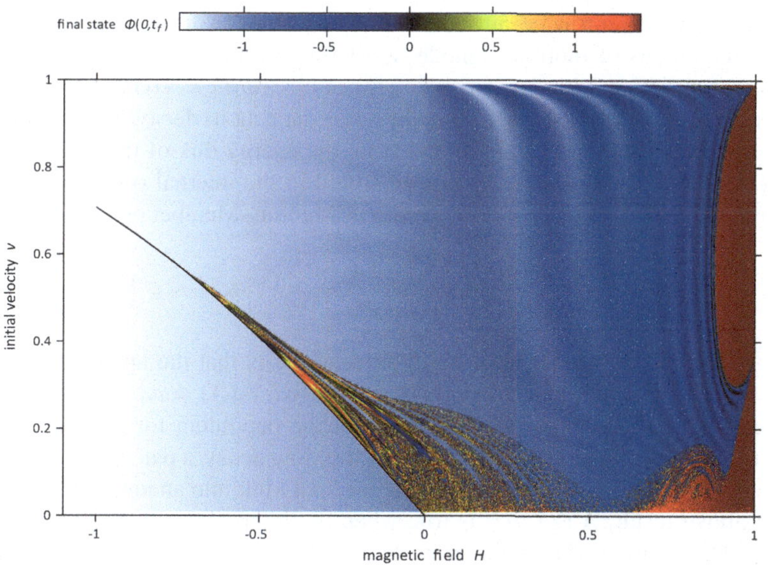

Fig. 2.9 A 'phase diagram' of antikink-boundary collisions. The plot shows the value of the field at $x = 0$, as function of the initial velocity v_{in} and the boundary magnetic field H. Reprinted (without modification) from [48]. © 2015 The Authors of [48], under the CC BY 4.0 license

positive values of H, see Fig. 2.9. Then the general structure of the resonant scattering windows remains similar to the usual case of $K\bar{K}$ scattering in the model on a full line, and it is exactly reproduced for $H = 0$. There is, however, an important difference. For positive values of H the initial translational energy of the soliton colliding with the boundary can be stored not only in the internal mode of the antikink, but also in the boundary mode. Thus, the resonance condition for the energy to be returned to the translational mode of the antikink on a subsequent impact has to be modified. It leads to the shifting of the windows we observed in numerical simulations.

For larger positive values of the magnetic field H new features appear. Firstly, the two-bounce window, which is "missing" in the full-line $K\bar{K}$ scattering [13], becomes resurrected. Secondly, as the magnitude of the oscillation of the boundary mode becomes large enough, it can be treated as an oscillon or kink-antikink bound state trapped by the boundary. As we discussed in Sect. 2.4 above, this excitation can be considered as an intermediate step in the process of production of the $K\bar{K}$ pair. Despite of all, the general pattern of non-perturbative dynamics of the solitons at $H \sim 1$ still remains obscure.

2.8 Radiative Decay of the Internal Oscillating Mode

The oscillational mode of the ϕ^4 model in the linear approximation lasts forever and oscillates with constant amplitude and frequency. However, in the full model this statement is no longer true because of the nonlinearities. In the presence of quadratic terms in equations of motion, a mode oscillating with the frequency ω_1 becomes a source of radiation with frequency $2\omega_1$ since $\cos^2(\omega_1 t) = \frac{1}{2}(\cos(2\omega_1) + 1)$. The radiation carries away the energy from the mode causing its decay. The rate in which the energy is radiated out can be found from the energy flux of the outgoing wave $\mathrm{d}E/\mathrm{d}t \sim \phi_x\phi_t \sim -A^4$. Since the energy stored in the internal oscillating mode is proportional to A^2 this leads to the relation (a Manton–Merabet power law [16])

$$\frac{\mathrm{d}E}{\mathrm{d}t} \sim -A\frac{\mathrm{d}A}{\mathrm{d}t} \sim -A^4 \Rightarrow \frac{\mathrm{d}A}{\mathrm{d}t} \sim -A^3 \Rightarrow A \sim t^{-1/2}\,. \tag{2.47}$$

An additional consequence of the nonlinearities is that the apparent frequency of the solution changes and, in general, is lowered: $\omega(A) = \omega_1 - A^2\omega^{(2)}$. This is an effect very similar to the slowing down of the pendulum for large amplitudes. When the mode is highly excited, instead of the slow decay, a pair of a kink-antikink structures can be ejected transforming the original kink into an antikink.

The above result, $A \sim t^{-1/2}$, is true when the internal oscillating mode decays through the radiation of the second harmonic. This condition is fulfilled in the ϕ^4 model although, in general, this does not need to be true. The boundary described in the previous section has an oscillating mode whose frequency can be controlled by the value of H in the range from 0 to 2. If the oscillating mode frequency is below $m/2 = 1$, the second harmonics is unable to propagate and higher orders of the

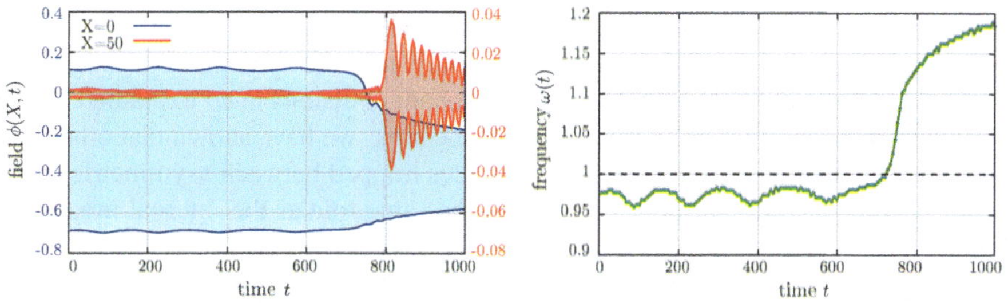

Fig. 2.10 Transition from slower to faster decay for $H = 0.8393$ and $A_0 = 0.3$. On the left panels the blue (darker) line (with initially large oscillations) shows the field at center and the red (lighter) line (with initially small oscillations), the far field. Right panel displays the frequency measured from the two subsequent maxima. Clearly, the transition occurs when the frequency crosses the value $m/2 = 1$ at $t \approx 720$

perturbation series would have to be considered. If the frequency of the oscillating mode were between $m/3$ and $m/2$ then the first propagating harmonic would be the third. For lower frequencies, even that condition is not enough and higher and higher harmonics have to be considered. In general, if the first harmonic which can propagate is the nth one, the Manton–Merabet law changes to

$$A \sim t^{-1/(2n-2)} \tag{2.48}$$

As the amplitude affects the frequency of the mode, it is possible that for certain amplitude the frequency $\omega(A) = \omega_1 - A^2\omega^{(2)}$ is below $m/2$ even for $\omega_1 > m/2$. However, as the mode decays slowly $A \sim t^{-1/4}$, its amplitude decreases and its frequency grows crossing at some point $m/2$. This event releases the second harmonics and the decay rate increases to $A \sim t^{-1/2}$. Far away from the mode this transition can be seen as a sudden burst of radiation. An example of such an evolution is presented in Fig. 2.10. For $H = 0.8393$ the eigenfrequency of the internal oscillating mode is equal to $\omega_1 = 1.2467$. A highly excited mode, with initial amplitude $A_0 = 0.3$ is slowed down by the nonlinearities and the observed frequency $\omega(A) \approx 0.98$ is below half of the threshold $m/2 = 1$. The mode decays through the third harmonic exhibiting rather slow decrease of the amplitude and small radiation in the far field zone. Around $t = 720$ the amplitude decreased to the value for which the frequency becomes $m/2 = 1$. The second harmonic is freed and the decay rate accelerates. At certain distance this event can be measured as a burst of radiation.

2.9 Conclusions

In this brief review we presented some recent results on the investigation of interactions between the kinks, related to the fine interplay between solitons and excitations of the perturbative spectrum. We demonstrate that the spectral structure of small perturbations around the kinks plays an essential role in the dynamics of the solitons. Apart from the well known mechanism of resonant collisions in ϕ^4 explained by the existence of the internal oscillating mode of the kink, we have shown that other resonant mechanisms can exist as well. The modes trapped between asymmetric kinks in ϕ^6, or the quasinormal modes, may play the same role as the internal mode of a kink.

We have also discussed a reverse process, when the kinks are created from the radiation. In this resonance process the oscillons and the internal modes of the kinks may play a role of an intermediate state, increasing the excess of energy density which is released by the emission of kink-antikink pairs.

Another interesting problem, which we briefly sketched, is the similarity between the internal oscillating modes and the oscillons. In some ways they can be transformed from one into another via adiabatic change of the model parameters.

We show that the interaction between the solitons and incoming radiation may produce some unexpected results. Indeed, in the integrable sG model, the waves pass through the kink without any exchange of energy. Similar effect may be observed in the ϕ^4 model only if we restrict ourselves to the linear approximation. The contribution of higher nonlinear terms yields both the energy and momentum transfer between the kink and the second harmonic. Furthermore, the contribution of the second harmonic produces a surplus of the momentum behind the kink pushing it towards the source of the radiation. This is an example of the negative radiation pressure effect.

In the case of the asymmetric kinks in the ϕ^6 model, the force exerted by the waves is also asymmetric. On one side of the soliton the wave causes positive radiation pressure; however, on the other side of the kink, the radiation pressure is negative. As a result, any small perturbation pushes the kink in such a way that the vacuum with smaller mass parameter always expands.

Acknowledgements We are grateful to Patrick Dorey for valuable collaboration, many results of our joint work are reviewed in this brief survey. We thank Piotr Bizoń and for inspiring and valuable discussions. Y.S. gratefully acknowledges support from the JINR Bogoljubov-Infeld Program of collaboration Kraków-Dubna. He would like to thank the Institute of Physics, Jagiellonian University Kraków, for its kind hospitality.

References

1. A. Vilenkin, E.P.S. Shellard, *Cosmic Strings and Other Topological Defects* (Cambridge University Press, Cambridge, 2000)
2. T. Bishop, A.R. Schneider, *Solitons and Condensed Matter Physics* (Springer, Berlin, 1978)
3. F. Falk, Z. Phys. B Condens. Matter **54**, 159 (1984)
4. M. Rice, J. Timonen, Phys. Lett. A **73**, 368 (1979)
5. S. Yomosa, Phys. Rev. A **27**, 2120 (1983)
6. V.L. Ginzburg, L.D. Landau, Sov. Phys. JETP **20**, 1064 (1950)
7. A. Ringwald, Nucl. Phys. B **330**, 1 (1990)

8. O. Espinosa, Nucl. Phys. B **343**, 310 (1990)
9. R.F. Dashen, B. Hasslacher, A. Neveu, Phys. Rev. D **10**, 4114 (1974)
10. R.D. Yamaletdinov, V.A. Slipko, Y.V. Pershin, Phys. Rev. B **96**, 094306 (2017)
11. P. Anninos, S. Oliveira, R.A. Matzner, Phys. Rev. D **44**, 1147 (1991)
12. T.I. Belova, A.E. Kudryavtsev, Phys. Usp. **40**, 359 (1997)
13. D.K. Campbell, J.F. Schonfeld, C.A. Wingate, Phys. D **9**, 1 (1983)
14. R.H. Goodman, R. Haberman, SIAM J. Appl. Dyn. Syst. **4**, 1195 (2005)
15. V. Makhankov, Phys. Rep. **35**, 1 (1978)
16. N.S. Manton, H. Merabet, Nonlinearity **10**, 3 (1997)
17. M. Moshir, Nucl. Phys. B **185**, 318 (1981)
18. P. Dorey, K. Mersh, T. Romańczukiewicz, Ya. Shnir, Phys. Rev. Lett. **107**, 091602 (2011)
19. P. Forgács, A. Lukács, T. Romańczukiewicz, Phys. Rev. D **77**, 125012 (2008)
20. I.L. Bogolyubsky, V.G. Makhankov, JETP Lett. **24**, 12 (1976)
21. G. Fodor, P. Forgacs, P. Grandclement, I. Racz, Phys. Rev. D **74**, 124003 (2006)
22. P. Salmi, M. Hindmarsh, Phys. Rev. D **85**, 085033 (2012)
23. H. Segur, J. Math. Phys. **24**, 1439 (1983)
24. M. Peyrard, D.K. Campbell, Phys. D **9**, 33 (1983)
25. M.A. Lohe, Phys. Rev. D **20**, 3120 (1979)
26. H. Shin, Sine-gordon soliton on a cnoidal wave background, arXiv:nlin/0410065
27. T. Romańczukiewicz, Acta Phys. Pol. B **39**, 3449 (2008)
28. P. Forgács, A. Lukács, T. Romańczukiewicz, Phys. Rev. D **88**, 125007 (2013)
29. T. Romańczukiewicz, Phys. Lett. B **773**, 295 (2017)
30. H. Arodź, R. Pełka, Phys. Rev. E **62**, 6749 (2000)
31. A.E. Kudryavtsev, Pisma Zh. Eksp. Teor. Fiz. **22**, 178 (1975)
32. E.J. Copeland, M. Gleiser, H.R. Muller, Phys. Rev. D **52**, 1920 (1995)
33. M. Gleiser, Phys. Rev. D **49**, 2978 (1994)
34. A.M. Kosevich, A.S. Kovalev, Sov. Phys. JETP **40**, 891 (1975)
35. J. Boyd, Nonlinearity **3**, 177 (1990)
36. H. Segur, M.D. Kruskal, Phys. Rev. Lett. **58**, 747 (1987)
37. G. Fodor, P. Forgacs, Z. Horvath, M. Mezei, Phys. Rev. D **79**, 065002 (2009)
38. T. Romańczukiewicz, Y. Shnir, JHEP **2018**(1), 101 (2018)
39. R. Rajaraman, Phys. Rev. Lett. **42**, 200 (1979)
40. A. Halavanau, T. Romańczukiewicz, Ya. Shnir, Phys. Rev. D **86**, 085027 (2012)
41. T. Romańczukiewicz, J. Phys. A Math. Gen. **39**, 3479 (2006)
42. T. Romańczukiewicz, Ya. Shnir, Phys. Rev. Lett. **105**, 081601 (2010)
43. R.H. Goodman, R. Haberman, Phys. Rev. E **71**, 056605 (2005)
44. J. Yang, Stud. Appl. Math. **98**, 61 (1997)
45. Y.S. Kivshar, Z. Fei, L. Vázquez, Phys. Rev. Lett. **67**, 1177 (1991)
46. I.G. Moss, Nucl. Phys. B Proc. Suppl. **104**, 181 (2002)
47. P. Dorey, T. Romańczukiewicz, Phys. Lett. B **779**, 117 (2018)
48. P. Dorey, A. Halavanau, J. Mercer, T. Romańczukiewicz, Y. Shnir, JHEP **2017**(5), 107 (2017)

Chapter 3
Collective Coordinate Methods and Their Applicability to ϕ^4 Models

Herbert Weigel

Abstract Collective coordinate methods are frequently applied to study dynamical properties of solitons. These methods simplify the field equations — typically partial differential equations — to ordinary differential equations for selected excitations. More importantly though, collective coordinates provide a practical means to focus on particular modes of otherwise complicated dynamical processes. We review the application of collective coordinate methods in the analysis of the kink-antikink interaction within the ϕ^4 soliton model and illuminate discrepancies between these methods and the exact results from the field equations.

3.1 Motivation

The ϕ^4 model represents the simplest non-linear extension of the KG theory in one time and one space dimensions that contains solutions with localized energy densities. These solutions are usually called *solitons* though more precisely they should be named solitary waves [1, 2].

Solitons have applications in almost all disciplines of physics. For example, they characterize domains that may emerge in cosmology [3, 4] or in condensed matter [5, 6]. The localized energy densities assign particle structures to solitons and picturing baryons as chiral solitons successfully describes many of their static and dynamic properties [7]. Solitons typically emerge in field theories with degenerate vacua such that the soliton configuration takes different vacuum values in distinct regimes of spatial infinity. When it takes an infinite amount of energy to (continuously) transform these vacuum configurations into another, the so-constructed solitons are named *topological* [8].

The Lagrange density of the ϕ^4 model in one space and one time dimensions may be written as compactly as

H. Weigel (✉)
Institute for Theoretical Physics, Physics Department, Stellenbosch University, Stellenbosch, South Africa
e-mail: weigel@sun.ac.za

P. G. Kevrekidis and J. Cuevas-Maraver (eds.), *A Dynamical Perspective on the ϕ^4 Model*, Nonlinear Systems and Complexity 26,
https://doi.org/10.1007/978-3-030-11839-6_3

$$\mathcal{L} = \frac{1}{2}\left[\dot{\varphi}^2 - \varphi'^2\right] - \frac{1}{2}\left(\varphi^2 - 1\right)^2 , \tag{3.1}$$

where dots and primes denote (partial) derivatives with respect to time (t) and space (x) coordinates, respectively. Fields and variables have been scaled such that the self-interaction strength occurs as an overall constant factor[1] which is not displayed. The degenerate vacua are simply $\varphi_{\text{vac}} = \pm 1$ and there are two options to connect them thereby defining kink and antikink solutions. With the above conventions these are

$$\varphi_{K,\overline{K}}(x) = \pm \tanh(x) . \tag{3.2}$$

The ϕ^4 model (anti)kink is the prototype soliton that illuminates many of the structures to be expected for the technically more challenging applications mentioned above.

Given that these are solutions within a non-linear model, any superposition of two solutions will, in general, not be a solution anymore. However, a superposition of two widely separated solitons is still approximately a solution as interference contributions to the energy vanish. As the separation is reduced practically adiabatically, the super-imposed configurations will possess some interaction energy. Within the particle interpretation of the soliton, the separation is the distance between two particles and the interaction energy becomes the inter-particle potential [9].

In general this interaction can be studied by solving the time dependent field equations as an initial value problem. Definitely, for models in one space dimension this in an option. But in higher, in particular three, dimensions such computations become increasingly demanding. In that case a possible strategy is the introduction of so-called *collective coordinates* that parameterize certain modes of the field excitations and reduce the complexity of the field equations drastically. Their introduction, to some extent, is a matter of good guess. We are in the lucky situation that for soliton models in one space dimension we can compare the solutions to the exact field equations to those from the reduced equations for the collective coordinates. This comparison will then determine the quality of the guess. Besides these technical advantages, a maybe even more important feature of collective coordinates is that they enable us to focus on special excitations of the soliton. Studying collective coordinates thus not only sheds light on their dynamical relevance but ultimately identifies modes which trigger certain processes thereby deepening our understanding of the particular soliton model. Here we will review this investigation for the kink-antikink system of the ϕ^4 model.

In the following section we will discuss the dynamics of the kink-antikink interaction resulting from the full field equations. In Sect. 3.3 we will introduce collective coordinates to (eventually) detail the dynamics of this interaction. In Sect. 3.4 we will briefly reflect on previous approximations to the collective coordinate approach and discuss one out of many possible generalizations in Sect. 3.5. Section 3.6 shows

[1] Here we restrain to the classical description, i.e. canonical commutation relations are not imposed. Then this loop-counting factor has no dynamical relevance.

how the collective coordinate approach is applied to the ϕ^6 model. We conclude in Sect. 3.7 in which we also discuss reasons why the (specific) collective coordinate approach fails to properly describe the kink-antikink interaction.

3.2 Kink-Antikink Scattering

Kink-antikink scattering can be simulated as a solution to the field equations with suitable initial conditions. Here we discuss this treatment and its results briefly.

3.2.1 *Dynamical Kink-Antikink Interaction*

The field equation emerging from the Lagrangian, (3.1) reads

$$\ddot{\varphi} - \varphi'' = 2\varphi\left(1 - \varphi^2\right) . \tag{3.3}$$

This field equation is a PDE and has a unique solution when the initial field configuration and its time derivative (velocity) are prescribed. The initial conditions suitable to describe the kink-antikink interaction read

$$\begin{aligned} \varphi(x,0) &= \varphi_{\overline{K}}\left(\frac{x}{\sqrt{1-v^2}} - X_0\right) + \varphi_K\left(\frac{x}{\sqrt{1-v^2}} + X_0\right) - 1 , \\ \dot{\varphi}(x,0) &= \frac{v}{\sqrt{1-v^2}}\left[\varphi'_{\overline{K}}\left(\frac{x}{\sqrt{1-v^2}} - X_0\right) - \varphi'_K\left(\frac{x}{\sqrt{1-v^2}} + X_0\right)\right] , \end{aligned} \tag{3.4}$$

where $\varphi'_{K,\overline{K}}(x) = \pm\,\mathrm{sech}^2(x)$. While v is (twice) the relative velocity between the approaching (anti)kinks, X_0 measures the initial separation up to a factor $\sqrt{1-v^2}$. We take X_0 large, so that initially we have a widely separated kink-antikink system and interference effects are absent. Then the actual value of X_0 is irrelevant. Initially kink and antikink approach each other and interact by energy exchange when they are close enough. Eventually energy is transferred to fluctuations about the kink-antikink system. This interaction energy is then not available as translational energy and the components do not separate. Only when the fluctuations eventually release that energy again, do the kink and antikink depart. This produces so-called *bouncing* or *resonant* solutions in which kink and antikink structures partially separate but then *turn around* and approach again. The nature of the solutions varies with the initial velocity v.

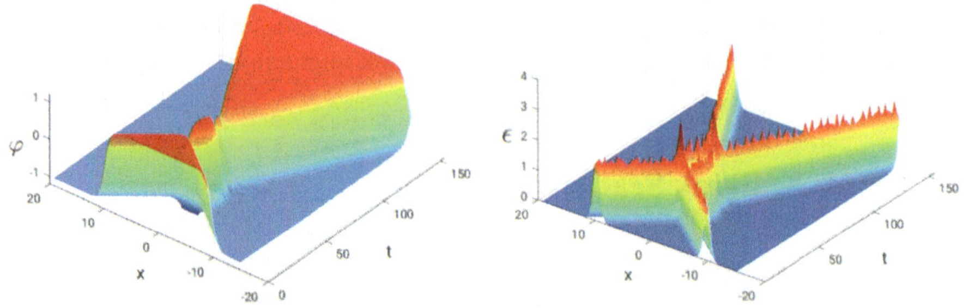

Fig. 3.1 Typical solution to the initial value problem defined through (3.4) and (3.3) for $v = 0.251$ for the field φ in the left panel and the energy density ϵ from (3.6) in the right panel

3.2.2 Bouncing Solutions and Bounce Windows

A typical solution to the above described initial value problem is shown in Fig. 3.1. This example exhibits the main features of the solutions to the full field equations. Kink and antikink approach each other, and rather than immediately departing to spatial infinity, they bounce and interact again before finally separating. Depending on the initial velocity several bounces may occur.

The bouncing solutions are most clearly discussed by defining a separation variable of the kink and antikink structures observed in the numerical simulation. We identify this to be the position of the antikink and we extract it as the expectation value of the coordinate along the positive half-line [10]

$$\langle x \rangle_t = \frac{\int_0^\infty \mathrm{d}x\, x\, \epsilon(t,x)}{\int_0^\infty \mathrm{d}x\, \epsilon(t,x)} \,. \tag{3.5}$$

where

$$\epsilon(t,x) = \frac{1}{2}\left[\ddot{\varphi} + \varphi'' + \left(\varphi^2 - 1\right)^2\right] \tag{3.6}$$

is the classical energy density from the time dependent numerical solution to (3.3) with initial conditions given in (3.4).

Typical results for $\langle x \rangle_t$ are shown in Fig. 3.2. Essentially three types of structures are observed. At low initial velocity, kink and antikink approach and bounce several times before they eventually separate. Even after separation the relative fluctuations diminish only gradually and the distinct kink and antikink configurations can be observed only for very late times. At moderate velocities, the kink-antikink system bounces a few times and then separates at slightly lower velocity. That is, at late times some of the energy remains in fluctuation modes. At large velocities the system separates without any bounce, but still some energy is contained in fluctuations about the individual (anti)kink as can be seen from the final velocity being smaller than the

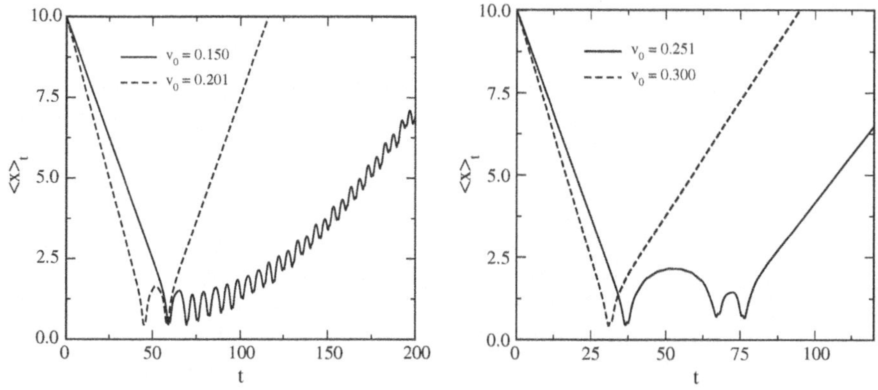

Fig. 3.2 Separation between kink and antikink extracted from the solutions to the PDE (3.3), according to (3.5) and (3.6), for different initial velocities

initial. It is obvious that at large relative velocities there is always sufficient energy available for immediate separation.

These bounce structures of the kink-antikink interaction have been exhaustively discussed in [11–15] and reviewed in [16]. In particular the critical velocity above which no bounces occur has been extracted to be $v_c = 0.260$ [11, 12]. Also certain intervals for the initial velocity below v_c, so-called bounce windows, have been identified in which zero, one, two or any larger number of bounces occurs before the final separation.

3.3 Collective Coordinates

We want to gain a deeper understanding of the rich bounce structures observed in the solutions to the PDEs. It is particularly challenging to identify those modes in which the energy is temporarily stored preventing the kink and antikink to separate immediately. For this purpose we want to model those solutions by admitting only particular modes. The introduction of collective coordinate is a very promising technique for this endeavor.

3.3.1 Large Versus Small Amplitude Fluctuations

There are two distinct types of fluctuations about classical localized configurations in a field theory. In the standard procedure small amplitude fluctuations, $\eta(x, t)$ are introduced (for quantization) [1]. In a first step the field is parameterized as

$$\varphi(x,t) = \varphi_{\rm cl}(x) + \eta(x,t)\,,$$

where $\varphi_{\rm cl}(x)$ represents the classical soliton configuration, as e.g. the kink in (3.2). In the second step, the Lagrangian is expanded in powers of η. The linear order vanishes when $\varphi_{\rm cl}(x)$ solves the field equation and contributions beyond the harmonic order are omitted. This harmonic expansion is rigorous in the $\hbar$ counting when $\eta(x,t)$ is normalized by the canonical commutation relations and produces the leading quantum corrections to properties of the classical configuration, most prominently the vacuum polarization (or Casimir) energy [17].

The spectrum of the small amplitude fluctuations is particularly interesting in the ϕ^4 model. Besides the translational zero mode (see below) there is a bound state below threshold (at frequency $\omega_{\rm th} = 2$, using dimensionless variables),

$$\eta(x,t) = {\rm e}^{-i\sqrt{3}t}\chi(x) \qquad \text{with} \qquad \chi(x) = \frac{\sinh(x)}{\cosh^2(x)}\,. \tag{3.7}$$

In the context of the kink-antikink interaction it has been conjectured that during bounces most of the energy is stored in this so-called *shape mode* [15].

However, there are also frequently non-harmonic modes, that is, modes that are not subject to a restoring force and thus may acquire large amplitudes. Those modes are typically parameterized by time dependent variables. These independent variables are called collective coordinates as they describe collective motions of the classical field configuration.

3.3.2 *Identification of Crucial Modes*

As discussed above, a number of features motivates the introduction of collective coordinates. Unfortunately, a suitable choice may be a matter of good guess. Since the collective modes are supposed not to experience any or only a small restoring force, the (would-be) zero modes of the classical configurations are good candidates. The excitation of these modes does not alter the energy. The top candidate is, of course, the translation of the localized configuration which generates its linear momentum. As a further advantage, the introduction of collective coordinates simplifies an intricate quantum field theory problem into one of quantum mechanical nature. In higher dimensions, rotations (both in coordinate and internal spaces) generate good quantum numbers that are not possessed by the classical soliton. A prominent example is the Skyrme model [18] in which canonical quantization of the collective coordinates generates baryon states [19] (for a review see [7]). This procedure can be extended to the case of approximate, or (would-be) zero modes with the symmetry violation treated in perturbation theory.

3.3.3 Separation as Collective Coordinate

For the discussion of the interaction between kink and antikink it is obvious to introduce a time dependent collective coordinate, $X(t)$, for their separation. It takes kink and antikink with opposite velocities and parameterizes the field configuration as a superposition of kink and antikink via

$$\varphi_c(x,t) = \varphi_K(\xi_+) + \varphi_{\overline{K}}(\xi_-) - 1\,. \tag{3.8}$$

where $\xi_\pm = \xi_\pm(x,t) = \frac{x}{\sqrt{1-v^2}} \pm X(t)$. Here v equals the relative velocity in the initial conditions, (3.4). Substituting this parameterization and integrating over the spatial coordinate x produces the Lagrange function

$$L(X,\dot{X}) = a_1(X)\dot{X}^2 - a_2(X)\,, \tag{3.9}$$

with mass (a_1) and potential (a_2) functions

$$\begin{aligned} a_1(X) &= \frac{1}{2}\int_{-\infty}^{\infty} \mathrm{d}x\, \left[\varphi_K'(\xi_+) - \varphi_{\overline{K}}'(\xi_-)\right]^2 \\ a_2(X) &= \frac{1}{2}\frac{1}{1-v^2}\int_{-\infty}^{\infty} \mathrm{d}x\, \left[\varphi_K'(\xi_+) \pm \varphi_{\overline{K}}'(\xi_-)\right]^2 + \frac{1}{2}\int_{-\infty}^{\infty} \mathrm{d}x\, [\varphi_c(x,t)-1]^2\,, \end{aligned} \tag{3.10}$$

that parametrically depend on the collective coordinate X via $\xi_\pm$. These integrals can be computed analytically [12, 20, 21] and an example is detailed in the appendix of [22]. It is, however, equally efficient to calculate them numerically since the equation of motion for the collective coordinate

$$\ddot{X} = -\frac{1}{2a_1(X)}\left[\frac{\mathrm{d}a_1(X)}{\mathrm{d}X}\dot{X}^2 + \frac{\mathrm{d}a_2(X)}{\mathrm{d}X}\right] \tag{3.11}$$

must also be solved numerically. Another motivation to obtain these integrals numerically is that small variations of the parameterization, (3.8), can then be easily accommodated. Since the time dependence of the field configuration only comes via $X(t)$ this is an ODE and thus technically less challenging than the full field equations (3.3).

Solutions with initial velocities analog to those in (3.4)

$$X(0) = 10 \qquad \text{and} \qquad \dot{X}(0) = \frac{-v}{\sqrt{1-v^2}} \tag{3.12}$$

are displayed in Fig. 3.3. Obviously these solutions do not produce any bounce structures. This is not surprising as there is only a single degree of freedom that cannot exchange energy with any other mode. The prime candidate mode to include for this energy exchange is the shape mode, (3.7), whose incorporation we will discuss next.

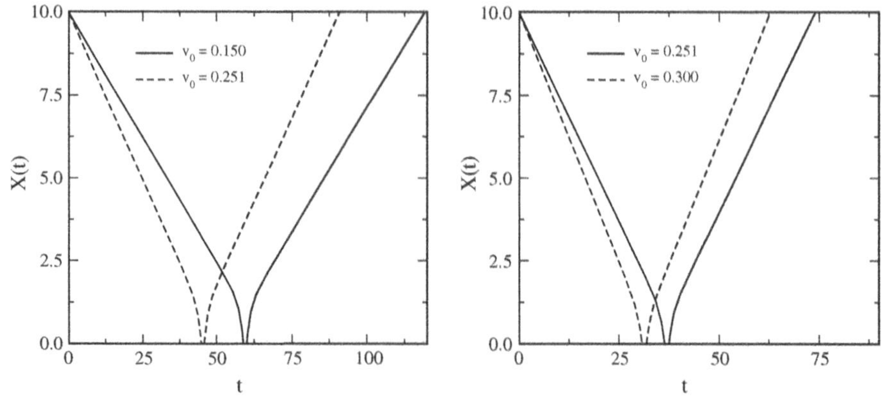

Fig. 3.3 Solutions to the equation of motion resulting from the collective coordinate Lagrangian, (3.9)

3.3.4 *Excitation of Shape Mode*

We now go one step further and include the amplitude of the shape mode as a collective coordinate [21, 22]

$$\varphi_c(x,t) = \varphi_K(\xi_+) + \varphi_{\overline{K}}(\xi_-) - 1 + \sqrt{\frac{3}{2}}\left[A(t)\chi(\xi_-) + B(t)\chi(\xi_+)\right] . \quad (3.13)$$

Note that both the kink and the antikink are accompanied by shape modes. Again this parameterization is substituted into the Lagrangian and integration over the coordinate x yields the Lagrange function

$$\begin{aligned} L(X,\dot{X},A,\dot{A},B,\dot{B}) = a_1(X)\dot{X}^2 - a_2(X) + a_3\left[\dot{A}^2+\dot{B}^2\right] - a_4\left[A^2+B^2\right] \\ + \overline{a}_3\dot{A}\dot{B} - \overline{a}_4 AB + a_5(X)\left[A-B\right] + \cdots . \end{aligned} \quad (3.14)$$

Only a few of the many terms that exhibit the essential structures have been displayed. The full Lagrange function as well as explicit expressions for the coefficient functions are listed in [23].

Complicated Euler Lagrange equations that generalize (3.11) are derived from (3.14). At $t=0$ kink and antikink approach each other while the shape modes are not excited initially, i.e. $A(0)=B(0)=0$ and $\dot{A}(0)=\dot{B}(0)=0$. In general the amplitudes A and B are independent. However, the above listed terms, in particular the linear term involving $a_5(X)$, suggest (and that is found for the full Lagrange function) that if $A(t)$ is a solution, so is $B(t)\equiv -A(t)$. That is, initial conditions with $B(0)=-A(0)$ and $\dot{B}(0)=-\dot{A}(0)$ produce $B(t)\equiv -A(t)$. In case there is no shape mode initially it is thus sufficient to only consider $A(t)$. From now on we will therefore adopt that specific collective coordinate configuration. Then the shape mode part in the collective coordinate parameterization becomes [21]

$$\sqrt{\frac{3}{2}}\left[A(t)\chi(\xi_-) + B(t)\chi(\xi_+)\right] \quad \longrightarrow \quad \sqrt{\frac{3}{2}}\left[\chi(\xi_-) - \chi(\xi_+)\right] A(t)\,. \qquad (3.15)$$

As $X \to 0$ we have $\xi_- \to \xi_+$ and the coefficient of $A(t)$ vanishes. This leads to a null-vector singularity [24, 25] as the amplitude is not well defined in that limit and causes a major obstacle to introducing the amplitude of the shape mode as a collective coordinate. This obstacle can only be circumvented by particular approximations or modifications.

The comparison with the (integrable) sG soliton further motivates to consider the shape mode as the prime candidate for intermediate energy storage. That model actually lacks a shape mode type solution as a small amplitude fluctuation off the soliton. Neither do bounces occur in the soliton-antisoliton interaction as the known analytic expression [1] for the time-dependent system reveals; only some time delay or phase shift is observed in the soliton-antisoliton interaction [12]. With generalizations or additions of impurities, bounces emerge [26]. These bounces have been analyzed using collective coordinate methods in [27, 28].

3.3.5 Approximations in Collective Coordinate Calculations

Early approaches to solve the equations of motion for the collective coordinates $X(t)$ and $A(t)$ circumvented the null-vector problem by omitting the X dependence in a_3 and a_4, equating these coefficient functions to their constant values from the respective $X \to \infty$ limits. Essentially, this omits the interference between the two shape modes at $\xi_\pm$. Furthermore non-harmonic terms in A, i.e. those beyond quadratic order in the Lagrangian, were discarded [14, 15, 25, 29]. Calculations (denoted 'lit' in Fig. 3.4) based on these approximations indeed showed surprising agreement with the results from the full field equation (3.3). Unfortunately those calculations took $a_5(X)$ from the original paper [21] which suffers from a misprint that made its way through the literature. Correcting[2] the misprint [22] and re-analyzing the equations of motion using the same approximations removes any agreement with the results from (3.3) as displayed in Fig. 3.4. The separation coordinate turns negative with a significant modulus and wild oscillations. Only occasionally it returns to the positive regime but essentially it is trapped below zero. Obviously these approximations, in particular omitting higher powers in A, are not consistent with the results from numerically simulating the full field equations.

[2]Remarkably, the fitted function $a_5(X) \propto \mathrm{e}^{-X}$ studied in [25, 29] reasonably resembles the actual behavior.

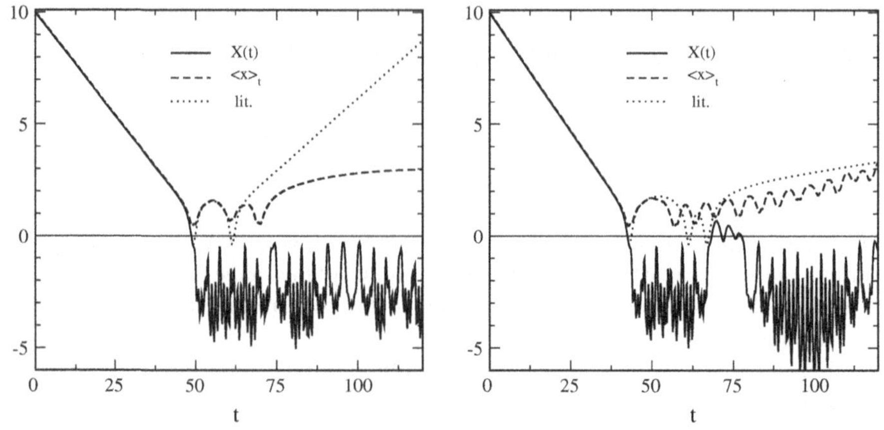

Fig. 3.4 Kink-antikink separation as function of time. Collective coordinate calculation of (3.13) and (3.15): full lines; field equations: dashed lines; collective coordinate approach from [14, 15]: dotted lines. Left panel: $v = 0.184$; right panel $v = 0.212$

3.3.6 Orbits of Collective Coordinates

Comparing the numerical solutions with (Fig. 3.4) and without (Fig. 3.3) a collectively excited shape mode with that of the full solution in Fig. 3.2 suggests that this mode may indeed be significant, but that its incorporation as the only collective coordinate on top of the separation exaggerates its role. To further investigate this conjecture we scale its source by a constant γ:

$$a_5(X) \longrightarrow \gamma a_5(X) \tag{3.16}$$

and again solve the equations of motions within the above discussed approximations. The resulting time dependences for both the separation X and the shape mode amplitude A are displayed in Fig. 3.5. We see that a moderate rescaling produces trajectories for $X(t)$ that compare reasonably well with $\langle x \rangle_t$ extracted from the full field equations. For small γ the bounce structure disappears (cfr. Sect. 3.3.3) while taking the source with full strength produces too large amplitudes for the shape mode and ties the separation coordinate into the negative domain. However, even for $\gamma = 0.5$ we find the critical velocity to be about $v_c = 0.33$ which is significantly larger than the value extracted from the exact solution. This suggests that, though the shape plays a significant role, other modes seem equally relevant.

Note, however, that these are intermediate conclusions, as the collective coordinate approach has been furnished by a number of approximations. We will elaborate on these approximations in the following section.

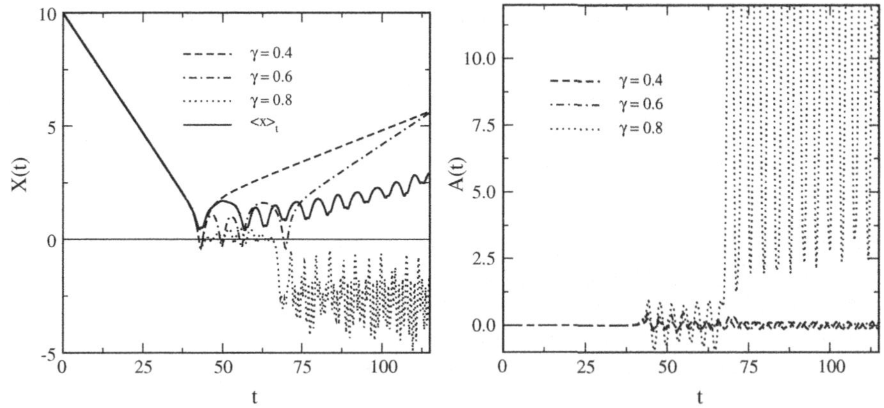

Fig. 3.5 Collective coordinates from scaled source, (3.16), for $v = 0.212$ in the harmonic approximation

3.4 Comments on Approximations

In the previous section we have seen that, in the context of kink-antikink scattering, frequently adopted approximations within the collective coordinate formulation do not reproduce the results from the field equations. It is therefore suggestive that a better agreement will be obtained when these approximations will be abandoned.

Most of the approximations for the collective coordinate approach have been implemented to gain simplifications that lead to analytic results or to avoid technical problems. With a fully numerical approach, some of these approximations can easily be abandoned.

3.4.1 Non-harmonic Contributions

When omitting the non-harmonic contributions of the shape mode amplitude, the collective coordinate equations of motion can be utilized to analytically predict the critical velocity [15, 21]. One may also argue for their omission on the basis that the shape mode as a bound state in the background of a single kink results from exactly that approximation. However, the results shown in Fig. 3.5 are actually inconsistent with the harmonic approximation. It is obvious that the amplitude of the shape mode may be large and therefore $O(A^2)$ terms may not be omitted from the equations of motion. Inclusion of these higher order terms may hamper the amplitude of the shape mode from becoming large by absorbing a significant amount of energy. In turn this might improve the quality of the collective coordinate approach. Stated otherwise, the non-harmonic terms are hoped to assist towards reducing the amplitude A such that the small amplitude approximation, which introduced the shape mode in the first place, is indeed justified.

3.4.2 Null-Vector Singularity

We have already mentioned that the collective coordinate parameterization of (3.15) produces an ill-defined amplitude of the shape mode when $X \to 0$ for the solution with $A(t) = -B(t)$, which is realized when the initial conditions obey this relation. To show that this holds true even without the identification $A \equiv -B$ we consider the kinetic terms involving the shape mode. The associated coefficients are

$$\begin{aligned} a_3(X) &= \frac{3}{4}\int \mathrm{d}x\, \chi^2(\xi_\pm) = \frac{3}{4}\sqrt{1-v^2}\int \mathrm{d}x\, \chi^2(x) \\ \overline{a}_3(X) &= \frac{3}{2}\int \mathrm{d}x\, \chi(\xi_+)\chi(\xi_-)\,. \end{aligned} \tag{3.17}$$

When $X = 0$ we have that $\xi_+ = \xi_- = \frac{x}{\sqrt{1-v^2}}$ so that $\overline{a}_3(0) \to 2a_3(0)$. Then the coefficient matrix in the equations of motion becomes

$$\begin{pmatrix} 2a_3(X) & \overline{a}_3(X) \\ \overline{a}_3(X) & 2a_3(X) \end{pmatrix} \sim 2a_3(0)\begin{pmatrix} 1 & 1 \\ 1 & 1 \end{pmatrix},$$

whose null-vector $\begin{pmatrix} 1 \\ -1 \end{pmatrix}$ is indeed the solution $A(t) = -B(t)$ [24]. The above matrix cannot be inverted and it becomes impossible to formulate the equations of motion as $\ddot{A} = \cdots$ and $\ddot{B} = \cdots$. The potential part containing $a_4(0)$ and $\overline{a}_4(0)$ has the same null-vector. In the numerical approach this singularity allows arbitrarily large amplitudes of the shape mode when $X \sim 0$. This suggests to modify the collective coordinate parameterization such that $X \to 0$ is energetically disfavored. We will get back to this in the next section.

3.4.3 Disagreement with Solution to Field Equations

In view of the above discussion, it is therefore suggestive to solve the collective coordinate equations as they emerge from the Lagrange function (3.14) for $X(t)$ and $A(t)$ $(= -B(t))$. At this point the sole approximation is to omit $\overline{a}_3(X)$ and $\overline{a}_4(X)$ (that is, keep the coefficients of the harmonic terms at their $X \to \infty$ values) to avoid the technical null-vector problem discussed above. The results of these calculations are shown in Fig. 3.6. It turns out that the energy associated with the source term for the shape mode, $E_5 = -a_5(X)A$, may absorb much energy when X becomes negative

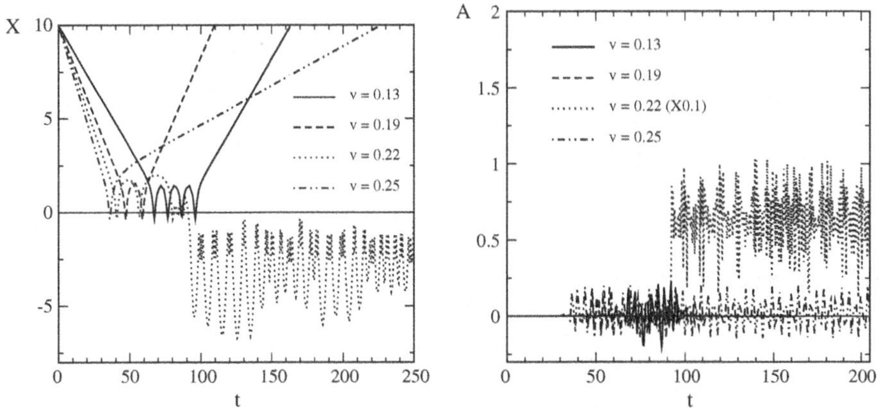

Fig. 3.6 Solution to the collective coordinate equations of motion in the ϕ^4 model. Left panel: kink-antikink separation, right panel: amplitude of shape mode (note the change of scale in the $v = 0.22$ entry)

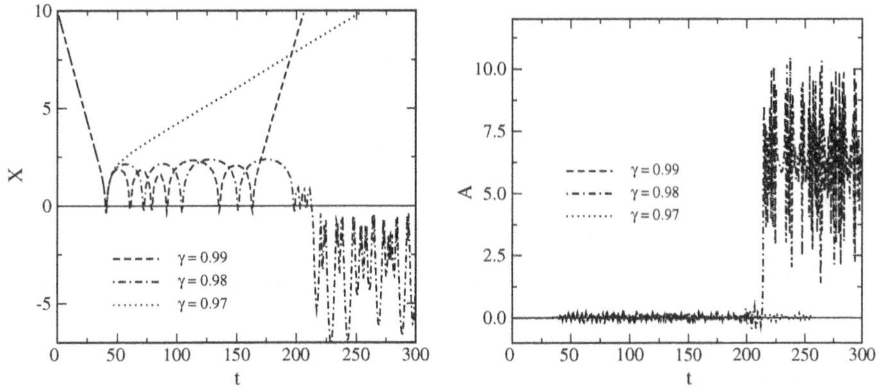

Fig. 3.7 Collective coordinates from scaled source, (3.16), for $v = 0.22$

and trapping type solutions with large amplitudes of the shape mode emerge. Hence the inclusion of the non-harmonic terms does not produce an a posteriori justification for implementing the harmonic approximation. The entry with $v = 0.22$ in Fig. 3.6 is a typical example thereof.

To further reflect on the quality of the collective coordinate formulation we again apply the scaling of (3.16) to one of the worst results to this case. The resulting time dependence of the separation variable is shown in Fig. 3.7. Obviously there is quite a sensitivity with respect to the strength of the source term. It is unlikely that the collective coordinate approach is preceptive enough to properly account for such fine details. This also concerns the predictions for the outgoing velocities.

Most notably, however, we find that without the many approximations (and $\gamma = 1$), the collective coordinate approach predicts a critical velocity of $v_c = 0.247$ above which trapping or bounce type solution cease to exist. This still compares favorably to the result from the full field equations of 0.260 [11, 12]. Yet, in view of the various discrepancies between the two approaches below v_c, this may be a mere artifact.

3.5 Modifications

Obviously negative $X(t)$, in particular with a large modulus, should not appear.[3] We have seen that when $X(t)$ is trapped at (large) negative values, even with the non-harmonic terms included, the amplitude of the shape mode contradicts the assumption for the small amplitude approximation which has been the point of departure for the shape mode and the subsequent use of its amplitude as collective coordinate.

Originally, taking X as a collective coordinate was motivated from the configuration containing widely separated kink and antikink structures. Of course, such a starting point does not determine a unique collective coordinate parameterization as we can always add any contribution that vanishes as $X \to \infty$. Here we will explore one such modification.

3.5.1 Kink-Antikink Penetration

The collective coordinate parameterization of (3.8) turns into a vacuum configuration of zero energy when $X \sim 0$. This causes attraction for small and moderate separation and eventually produces solutions that are trapped with negative X and correspond to (un-desired) configurations in which kink and antikink have penetrated each other. We thus modify the collective coordinate description and introduce a variational parameter q

$$\varphi_c(x,t) = \varphi_K(\xi_+) + \varphi_{\overline{K}}(\xi_-) - \tanh(qX) + \sqrt{\frac{3}{2}}A(t)\left[\chi(\xi_-) - \chi(\xi_+)\right] . \quad (3.18)$$

For $q \gg 1$ this agrees with the original formulation, (3.8) when $X \geq 0$ which embraces the desired initial condition. In contrast to (3.8), however, it is also a solution to the field equations when $X \to -\infty$ (with $A = 0$). Actually the configuration, (3.18) is anti-symmetric under $X \to -X$.

With the $\tanh(qX)$ term, the $X \sim 0$ configuration ceases to be a vacuum configuration. In turn this produces a repulsive potential, $a_2(X)$ for $X \to 0$. When q is large,

[3] The comparison with the solutions to the full field equations proceeds via $\langle x \rangle_t$ which is positive definite, cfr. (3.5). Hence any such comparison is bound to fail for negative $X(t)$. Therefore, we will also consider the actual field configurations in Sect. 3.5.4.

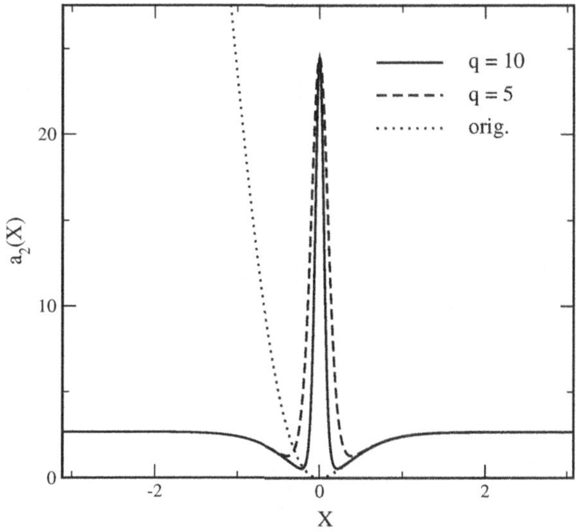

Fig. 3.8 Potential for the separation collective coordinate from the modified parametrization (3.18) with $v = 0.2$. The entry 'orig.' refers to the background of (3.8)

the repulsion resembles a peak on top of the intermediate attraction of the original parametrization. For moderate and small q it essentially removes that attraction. This scenario is pictured in Fig. 3.8.

We note that the parametrization (3.18) is different from the one used in [22] since here we immediately impose equal amplitudes for the two shape modes via (3.15). Though that equality also results from the equations of motion associated to (3.18), various approximations/modifications may yield different numerical results (most of the numerical simulations in [22] assumed the $X \to \infty$ values for the harmonic terms).

There is no fundamental principle in choosing a value for q other than that (3.18) resembles initially a well separated kink antikink system, i.e. $qX_0 \gg 1$. Later we will see that there is not much variation in the structure of the soliton with q once it is taken large enough. A guiding principle to select q could, for example, be to tune it for any given value of v such as to achieve a maximal agreement with the solution from the field equations. This was done in [30], but in the context of the ϕ^6 model, cfr. Sect. 3.6.

3.5.2 Improved Agreement with the Solution to Field Equations

We discuss the numerical solutions for the collective coordinate description of (3.18) and compare typical results with those from the full calculation in Fig. 3.9. The two graphs on top originate from taking the coefficients of the quadratic terms, both for the kinetic (a_3) and the potential (a_4), constant at their $X \to \infty$ values. This

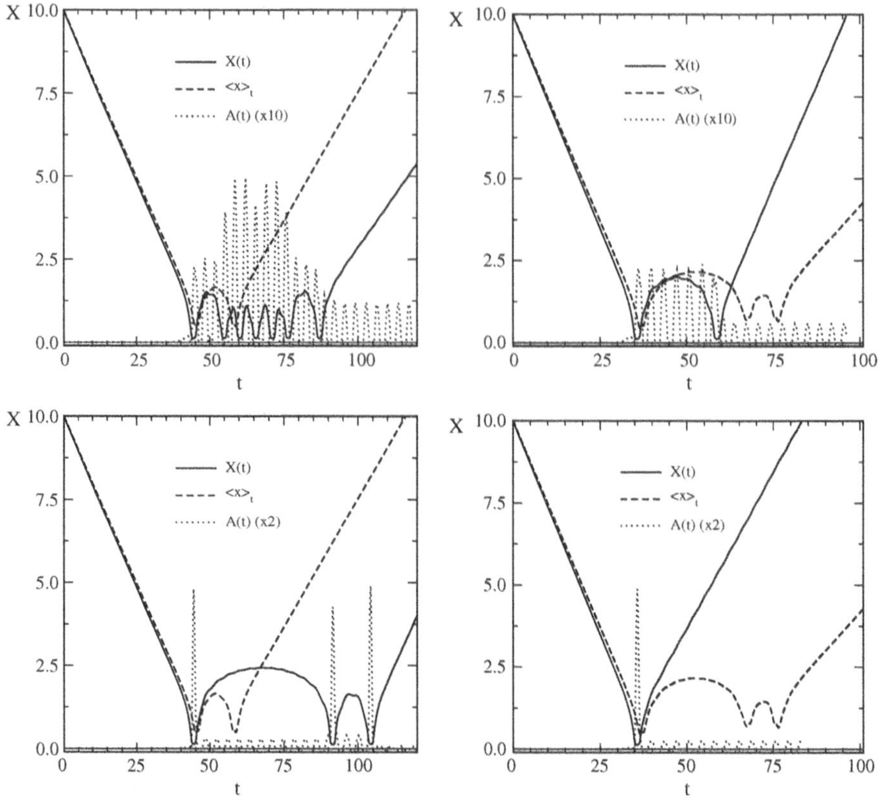

Fig. 3.9 Collective coordinate calculations with $q = 10$ in (3.18) with all terms from the collective coordinate Lagrangian included. Top graphs: Coefficients of harmonic terms kept constant at their $X \to \infty$ values. Bottom graphs: X dependence of harmonic terms included. Left panel: $v = 0.201$, right panel: $v = 0.251$. Note the different scales for $A(t)$

approximation is adapted from earlier attempts to avoid the null-vector problem. We see that this collective coordinate approach indeed describes a finite number of bounces and that these bounces come together with measurable excitations of the shape mode. However, neither the number of bounces nor the time scale at which these bounces occur are properly reproduced by the collective coordinate calculation. Also there is no systematics in the deviation such as, for example, that the collective coordinate approach would always underestimate the number of bounces or that it would always predict too large a time interval during which bounces occur.

The results shown in the two bottom pictures of Fig. 3.9 abandon all approximations. In that case the null-vector problem may occur. However, numerically this does not happen. For the cases displayed in Fig. 3.9 we always find that $X \geq 0.1$. Of course, this value is correlated to $1/q$ that characterizes the regime of the short range repulsion. Though the substitution $1 \longrightarrow \tanh(qX)$ affects the coefficients

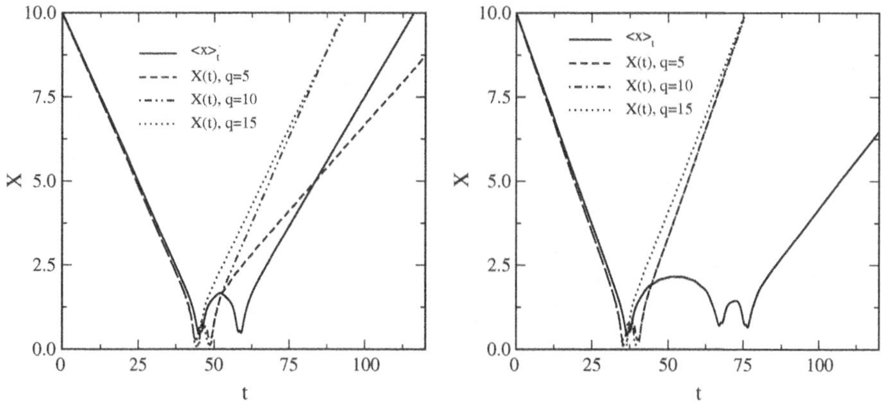

Fig. 3.10 Effect of tuning q in (3.18) for two velocities, $v = 0.201$ (left panel) and $v = 0.251$ (right panel)

$a_i(X)$ only in the moderate regime $|X| \lesssim 3$, significant changes for $X(t)$ and $A(t)$ are measured. Yet, the number of changes as well as the interaction times vary with v and q. Thus much of the predictive power of these calculations is lost.

It has become apparent that the quality of the collective coordinate approach strongly depends on the applied approximations and no rigorous conclusion can be drawn. Similarly to [30] one might consider the new parameter q a tunable variable. Results for such simulations are compared to the mean value of the full solution in Fig. 3.10. While there are some variations with q, the main structure of several bounces over a long time interval cannot be adjusted.

3.5.3 *Extraction of Critical Velocities*

Once kink and antikink start to separate, the major share of the energy is stored in the translational motion. This raises expectations that the predictions for the critical velocities above which no bounces occur agree for the two approaches. However, this is only partially the case. The exact value for the critical velocity from the field equations of 0.260 [11, 12] is underestimated within the collective coordinate approach of (3.18) to be 0.204 when $q = 10$. However, that particular value again changes with q. For example, bounces are observed for $v = 0.4$ when $q = 5$.

We compare the initial and final velocities above critical velocities as extracted from $\frac{\partial \langle x \rangle_t}{\partial t}$ and $\dot{X}_\infty = \frac{\partial X(t)}{\partial t}$ as $t \to \infty$. Subsequently the collective coordinate velocity is written in relativistic kinematics $v_f = \frac{\dot{X}_\infty}{\sqrt{1+\dot{X}_\infty^2}}$. Since there are still oscillations of the shape mode on top of the translation those velocities are not constant and the results listed in Table 3.1 are obtained from averaging over numerous such oscillations.

Table 3.1 Comparison of the predicted final velocities, v_f. The entry PDE refers to $\frac{\partial \langle x \rangle_t}{\partial t}\Big|_\infty$ and the q columns contain $\dot{X}_\infty$ originating from (3.18)

v	PDE	Collective coordinates		
		$q = 5$	$q = 10$	$q = 15$
0.4	0.279	0.216	0.399	0.340
0.5	0.390	0.447	0.496	0.404
0.6	0.494	0.569	0.577	0.388
0.7	0.595	0.700	0.612	0.699
0.8	0.697	0.818	0.806	0.808
0.9	0.797	0.897	0.879	0.906

Again we see that the collective coordinate approach reproduces the exact results only qualitatively and that particularities (here represented by a strong q dependence) matter. We also observe from the PDE entry of Table 3.1 that for large initial velocities the full field equations predict reduced final velocities. That is, even without bounces, some energy is stored in modes other than the translation.

3.5.4 *Mapping Collective Coordinates and Solution to Full Field Equation*

By pure definition, cfr. (3.5), $\langle x \rangle_t$ is non-negative. Thus a direct comparison with $X(t)$ may be misleading. We therefore attempt to rebuild the time dependent field from (3.18) with $q = 10$ by substituting $X(t)$ and $A(t)$ and then compare that configuration to the solution from the field equations (3.3) with the same initial velocity.

We analyze the (dis)agreement in two ways. First, in Fig. 3.11 we consider the field at the center ($x = 0$) as a function of time and, second, we contrast the configurations along the coordinate axis at different times in Fig. 3.12. As in Fig. 3.9 (with $v = 0.201$) we observe that the time between the first two collisions of kink and antikink is overestimated by the collective coordinate description. We also see that the collective coordinate approach yields two nearby bounces at late times when the field equations predict well separated kink-antikink structures. Similarly the number and positions of bounces are not correct for $v = 0.251$ either. Interestingly enough, the collective coordinate result for $v = 0.201$ agrees in shape (though not with respect to the time scale) with the exact result of $v = 0.251$. This again suggests that details of the parameterization matter and that the velocities should not be literally compared.

A further striking difference between the two approaches seen in Fig. 3.12 is that, at particular times, the peak amplitudes of the field configurations differ significantly. Of course, this just reflects that at those particular times which exhibit significant differences one approach produces a bounce but the other does not.

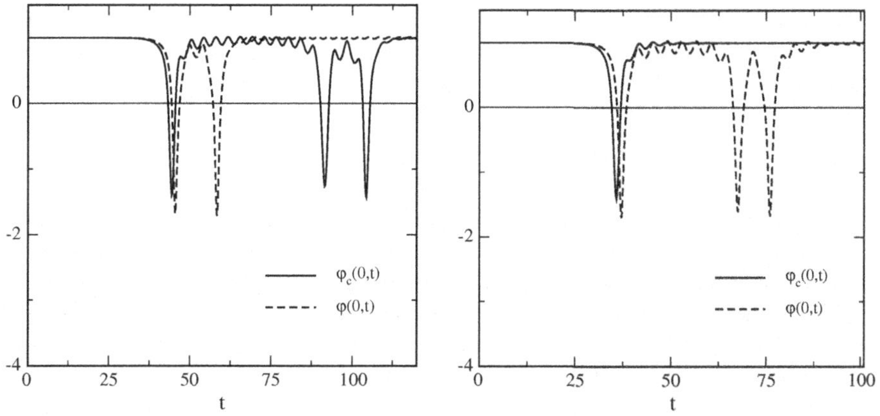

Fig. 3.11 Time dependence of the field at the origin. Left panel: $v = 0.201$, right panel: $v = 0.251$. Collective coordinate results are without any approximation, i.e. similarly to the bottom graphs of Fig. 3.9

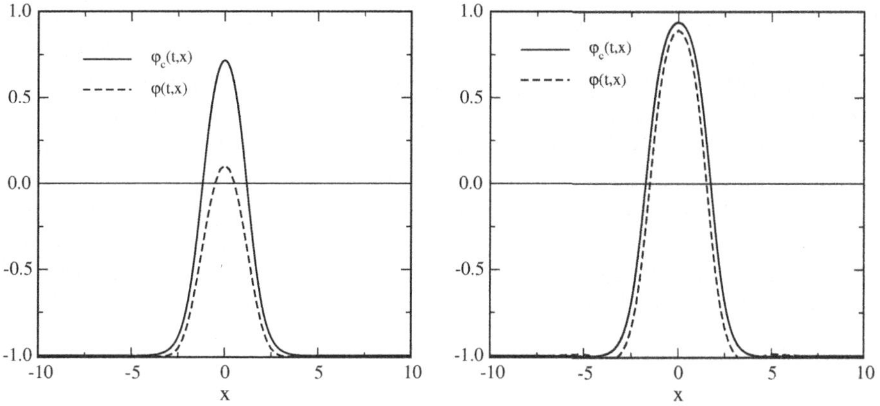

Fig. 3.12 Fields at different times for $v = 0.201$. Left panel: $t = 47$, right panel: $t = 50$

3.6 Comparison: ϕ^6 Model

Without going into much detail, we briefly reflect on related studies within the ϕ^6 model. It is interesting because bounce structures have been observed for the solutions to the field equations (analog to (3.3)) [31] even though no internal shape mode exists in the fluctuation spectrum of the soliton(s). This is a further indication that the shape mode is not the (only) explanation for bounces in kink-antikink scattering.

This model is defined by the (scaled) Lagrangian

$$\mathcal{L}_6 = \frac{1}{2}\left[\dot{\phi}^2 - \phi'^2\right] - \frac{1}{2}\left(\phi^2 + a^2\right)\left(\phi^2 - 1\right)^2 , \tag{3.19}$$

with the real parameter a. For $a \neq 0$ there are two vacua $\phi_{\text{vac}} = \pm 1$ and the soliton solution mediates between them. We will only consider the $a = 0$ case when an additional vacuum, $\phi_{\text{vac}} = 0$, emerges. Then the soliton solutions

$$\phi_{K,\overline{K}}(x) = \left[1 + \exp(\pm 2x)\right]^{-\frac{1}{2}} \tag{3.20}$$

mediate between 0 and 1 (or -1 when changing the overall sign). The small amplitude bound state spectrum for both solitons only contains the translational zero mode.

With two distinct soliton solutions available there are two independent initial conditions relating interactions between kink and antikink. They are usually referred to as kink-antikink ($K\overline{K}$) and antikink-kink ($\overline{K}K$) systems and the full field equations produce bounce structures for both systems [31].

Though there is no shape mode in the spectrum of the small amplitude fluctuations, its inclusion as in (3.15) may serve as working hypothesis for the collective coordinate approach to analyze the temporal storage of energy during the bounces. Furthermore a Fourier analysis of $\phi(0, t)$, the solution of the field equations at the origin, in the $\overline{K}K$ system shows large amplitudes at frequencies below threshold [31], suggesting that the energy is indeed stored in localized modes. Studies based on this hypothesis have been reported in [10, 22] and we reproduce typical results for the $K\overline{K}$ and $\overline{K}K$ systems in Figs. 3.13 and 3.14, respectively. Again, the null-vector problem has been circumvented by approximating the coefficients of the terms quadratic in A by their asymptotic values. As is by now standard, $\langle x \rangle_t$ has been extracted via (3.5) from the time dependent energy density deducted from (3.19) with $a = 0$.

For $v \leq 0.289$ the $K\overline{K}$ is always trapped [31]. Above that velocity the two structures always reflect without any bounce. With the separation as the only collective coordinate the former feature can, by construction, not be reproduced (see [30] for more details on this modification). Calculations with the shape mode added are shown

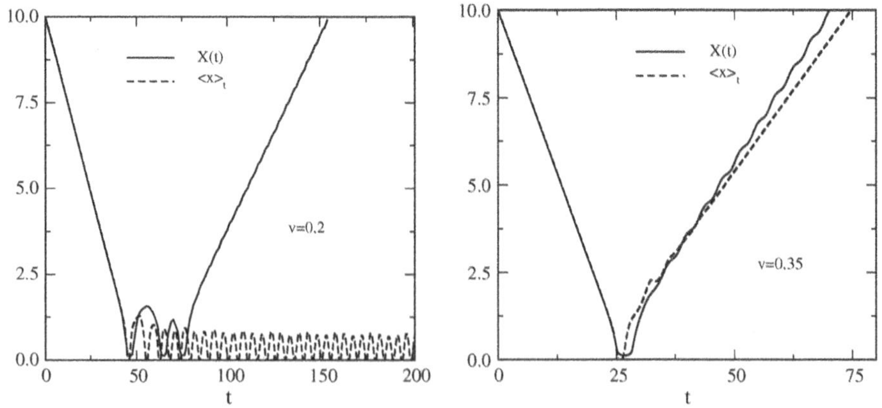

Fig. 3.13 Kink-antikink scattering in the ϕ^6 model: comparison of full and collective coordinate solutions for particular initial velocities

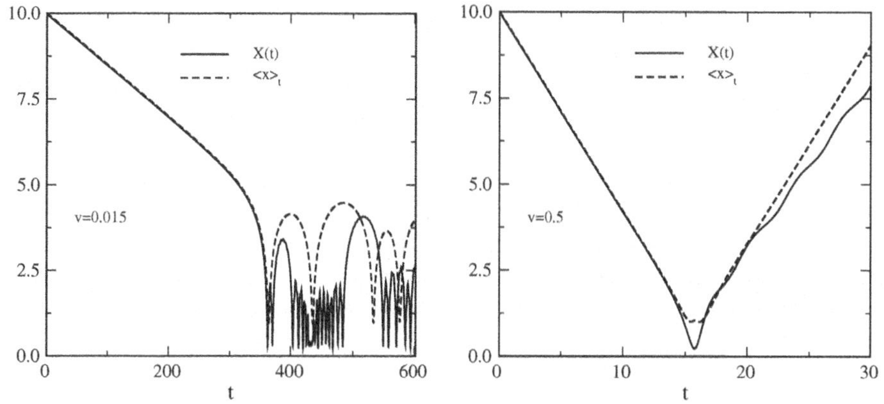

Fig. 3.14 Same as Fig. 3.13 but for antikink-kink scattering in the ϕ^6 model

in Fig. 3.13. Solutions with bounces are now produced for small enough velocities. In contrast to the full solution trapping, typically does not occur in the collective coordinate method. This is neither accomplished by modifications similar to (3.18) [22, 23], although the bounces are maintained by those modifications. On the other hand, for larger velocities the two calculations yield similar results.

In Fig. 3.14 the two methods are compared for the $\overline{K}K$ system, for which the critical velocity is much smaller: $v_c = 0.046$ [31]. There are obvious discrepancies below that velocity, while agreement is again observed for larger values.

Interestingly enough despite all the obstacles observed, the ordering relations for the critical velocities

$$v_c(\phi^6, \overline{\mathrm{K}}\mathrm{K}) \ll v_c(\phi^4, \mathrm{K}\overline{\mathrm{K}}) \lesssim v_c(\phi^6, \mathrm{K}\overline{\mathrm{K}})$$

in the different models are nevertheless observed in both the exact and the collective coordinate approaches [22]. The first relation is expected because those velocities differ by an order of magnitude and even a coarse approximation should reproduce it. The second relation may be a mere artifact.

3.7 Conclusion and Critical Analysis

The ϕ^4 model in one space and one time dimensions has localized static solutions, so-called (anti)kinks. Appropriate initial conditions for integrating the full field equations allow to simulate kink-antikink scattering as a prototype of particle interactions in the soliton picture. A number of interesting features emerge from these simulations as the relative initial velocity is varied. In particular, below a critical velocity (multiple) bounces occur between kink and antikink.

The collective coordinate method has been introduced as (i) a manageable approximation to the full field equations to simplify PDEs to ODEs and (ii) a sensible means to identify important modes in the kink-antikink interaction. In particular the degrees of freedom that are bound states of the single soliton have been considered. In the case of the ϕ^4 model these are the translational zero mode and so-called shape mode.

Here we have confronted the collective coordinate method with the exact treatment of the field equations in the ϕ^4 model. The collective coordinate approach reproduces well the first bounce observed in kink antikink collisions. This remarkably includes the acceleration shortly before kink and antikink sit on top of each other. However, when the separation is tiny, the non-linear equations of motion are particularly sensitive to small changes and the detailed pattern cannot be accommodated by the collective coordinate approach.

In fact, straightforward implementations of collective coordinates do not produce acceptable agreement with the solutions of the field equations. Intricate adjustments are needed to improve on the solutions. Even then significant discrepancies emerge. It is thus difficult to draw conclusions from the collective coordinate approach on the relevance of particular excitations, at least quantitatively.

Qualitatively the collective coordinate approach supports, to some extent, the conjecture that energy storage in the shape mode excitation leads to bouncing kink-antikink configurations. However, in that approach the amplitude of the shape mode is inflated compared to the exact solution. This suggests that other modes play a decisive role as well. It is also important to mention that within the interaction regime, i.e. when the distance between kink and antikink is not large, the shape mode ceases to be a solution to the fluctuation spectrum of $\varphi_{\rm cl}$. Though considering the linearized field equations is fully consistent only when the background is an extremum of the action — the kink-antikink system is not — it may clarify whether the shape mode maintains its unique role from the single kink system. Such studies [32, 33] show that the fluctuation spectrum of two (even widely) separated, stationary (anti)kinks differs significantly from that of a single kink. For fixed $X \gtrsim X_c \approx 0.37$ the zero mode acquires negative squared frequency, indicating instability since two localized (anti)kinks at finite separation do not solve the field equations. On the other hand, for $X \lesssim X_c$ the shape mode ceases to be bound which questions its incorporation as collective mode from the beginning. Also, the shape mode is a solution within the small amplitude approximation for deviations from a single kink. Smallness of this amplitude is not supported by the collective coordinate approach (when computing scattering phase shifts about a single kink from the full non-linear field equations, amplitudes as small as $A \sim 0.1$ yield results that deviate from the small amplitude approximation [34]). Hence most of the arguments leading to the incorporation of the shape mode are not rigorous in the regime of kink-antikink interactions. Another major issue in comparing the solutions of the two approaches is the fact that the relative velocities between the outgoing kink and antikink do not exactly match. This is most apparent from the different slopes of $X(t)$ and $\langle x \rangle_t$ in the above figures.

In conclusion, we see that though the shape mode has its share in producing bouncing configurations in the kink-antikink scattering, the collective coordinate approach based on this mode is not sensible enough to reproduce the scattering process quantitatively. It is thus very likely that other modes are equally relevant for the temporary storage of energy. To clarify whether these are just a few modes one could perform a thorough Fourier analysis in space and time of the difference

$$\varphi(x,t) - \left[\varphi_{\overline{K}}(x - \langle x\rangle_t) + \varphi_K(x + \langle x\rangle_t) - 1\right] ,$$

with $\langle x\rangle_t$ computed via (3.5) from $\varphi(x,t)$, the solution from the field equations. The result from that analysis may motivate more suitable collective coordinate parameterizations.

Acknowledgements The author gratefully acknowledges helpful contributions from A.M.H.H. Abdelhady and I. Takyi. This work is supported in part by the National Research Foundation of South Africa (NRF) by grant 109497.

References

1. R. Rajaraman, *Solitons and Instantons* (North Holland, Amsterdam, 1982)
2. Yu.S. Kivshar, B.A. Malomed, Rev. Mod. Phys. **61**, 763 (1989). [Addendum: Rev. Mod. Phys. **63**, 211 (1991)]
3. T. Vachaspati, *Kinks and Domain Walls: An Introduction to Classical and Quantum Solitons* (Cambridge University Press, Cambridge, 2010)
4. A. Vilenkin, E.P.S. Shellard, *Cosmic Strings and Other Topological Defects* (Cambridge University Press, Cambridge, 2010)
5. B. Ivanov, A. Kichiziev, Y.N. Mitsai, Sov. Phys. JETP **75**, 319 (1992)
6. A.R. Bishop, J.A. Krumhansl, S.E. Trullinger, Phys. D **1**, 1 (1980)
7. H. Weigel, Lect. Notes Phys. **743**, 1 (2008)
8. N.S. Manton, P. Sutcliffe, *Topological Solitons* (Cambridge University Press, Cambridge, 2004)
9. R. Vinh Mau, M. Lacombe, B. Loiseau, W.N. Cottingham, P. Lisboa, Phys. Lett. B **150**, 259 (1985)
10. H. Weigel, J. Phys. Conf. Ser. **482**, 012045 (2014)
11. M. Moshir, Nucl. Phys. B **185**, 318 (1981)
12. D.K. Campbell, J.F. Schonfeld, C.A. Wingate, Physica D **9**, 1 (1983)
13. T.I. Belova, A.E. Kudryavtsev, Physica D **32**, 18 (1988)
14. P. Anninos, S. Oliveira, R.A. Matzner, Phys. Rev. D **44**, 1147 (1991)
15. R.H. Goodman, R. Haberman, SIAM J. App. Dyn. Syst. **4**, 1195 (2005)
16. T.I. Belova, A.E. Kudryavtsev, Phys. Usp. **40**, 359 (1997)
17. N. Graham, M. Quandt, H. Weigel, Lect. Notes Phys. **777**, 1 (2009)
18. T.H.R. Skyrme, Proc. R. Soc. Lond. A **260**, 127 (1961)
19. G.S. Adkins, C.R. Nappi, E. Witten, Nucl. Phys. B **228**, 552 (1983)
20. A.E. Kudryavtsev, JETP Lett. **22**, 82 (1975)
21. T. Sugiyama, Prog. Theor. Phys. **61**, 1550 (1979)
22. I. Takyi, H. Weigel, Phys. Rev. D **94**, 085008 (2016)
23. I. Takyi, Collective coordinate description of kink-antikink interaction, MSc thesis, Stellenbosch University (2016), http://scholar.sun.ac.za/handle/10019.1/97877
24. J.G. Caputo, N. Flytzanis, Phys. Rev. A **44**, 6219 (1991)
25. R.H. Goodman, R. Haberman, Phys. Rev. Lett. **98**, 104103 (2007)
26. M. Peyrard, D.K. Campbell, Physica D **9**, 33 (1983)
27. Z. Fei, Yu.S. Kivshar, L. Vázquez, Phys. Rev. A **45**, 6019 (1992)
28. R.H. Goodman, R. Haberman, Physica D **195**, 303 (2004)

29. R.H. Goodman, A. Rahman, M.J. Bellanich, C.N. Morrison, Chaos **25**, 043109 (2015)
30. A. Demirkaya, R. Decker, P.G. Kevrekidis, I.C. Christov, A. Saxena, JHEP **2017**, 071 (2017)
31. P. Dorey, K. Mersh, T. Romanczukiewicz, Y. Shnir, Phys. Rev. Lett. **107**, 091602 (2011)
32. Z. Lee, Quantum corrections to the kink-antikink potential, MSc thesis, Stellenbosch University (2017), http://scholar.sun.ac.za/handle/10019.1/100796
33. N. Graham, R.L. Jaffe, Phys. Lett. B **435**, 145 (1998)
34. A.M.H.H. Abdelhady, H. Weigel, Int. J. Mod. Phys. A **26**, 3625 (2011)

Chapter 4
Mathematical Analysis of Fractal Kink-Antikink Collisions in the ϕ^4 Model

R. H. Goodman

To the memory of Richard Haberman

Abstract We analyze the fractal structure seen in kink-antikink collisions of the ϕ^4 equation. The analysis is based on qualitative ODE models related to those obtained using the variational approximation. We derive a discrete-time iterated map approximation to the dynamics, from which many features of the model ODE system may be derived. We discuss the problems with Sugiyama's variational model that have been pointed out recently by Takyi and Weigel.

4.1 Introduction

Intriguing behavior has long been observed in the dynamics of collisions between kinks and antikinks of the ϕ^4 model

$$\phi_{tt} - \phi_{xx} - \phi + \phi^3 = 0. \tag{4.1}$$

This equation possesses a family of traveling-wave solutions called kinks

$$\phi(x,t) = \phi_{\mathrm{K}}(x - vt) = \tanh(\xi/\sqrt{2})$$

where $\xi = (x - x_0 - vt)/\sqrt{1-v^2}$ for any velocity v, $-1 < v < 1$, and another family called antikinks $\phi_{\bar{\mathrm{K}}} = -\phi_{\mathrm{K}}$.

Consider a solution described by a kink and an antikink propagating toward each other from $\pm\infty$ with equal and opposite velocities. Researchers began simulations of this situation in the 1970s. Kudryavstev and Sugiyama separately found, through

R. H. Goodman (✉)
Department of Mathematical Sciences, New Jersey Institute of Technology, University Heights, Newark, NJ 07102, USA
e-mail: goodman@njit.edu

P. G. Kevrekidis and J. Cuevas-Maraver (eds.), *A Dynamical Perspective on the ϕ^4 Model*, Nonlinear Systems and Complexity 26,
https://doi.org/10.1007/978-3-030-11839-6_4

simulations with a small number of incident velocities, that sufficiently slow kinks and antikinks could be captured into a localized bound state [1, 2]. Ablowitz, Ladik, and Kruskal simulated a slightly larger number of initial conditions and made the surprising discovery that at one particular velocity, the kink and antikink collided, began to move apart, turned around, collided a second time, and then escaped [3]. More systematic numerical experiments by Campbell et al. showed a rich structure in the dynamics following such collisions [4, 5]. Solitary wave pairs with speed above some critical value v_c reflected off each other and escaped after colliding once. Most initial speeds below v_c led to capture: the kink and antikink form a bound state and never escape, although the localized solution subsequently decays due to the escape of radiation. However there exists a sequence of intervals such that, for initial speeds lying in such an interval, the kink and antikink collide, begin to separate, and then return to collide a second time before escaping. They called these intervals *two-bounce resonance* windows. More recent numerical simulations, e.g. [6], have shown that in between the two-bounce resonance windows are successively narrower windows where the solitary waves collide three, four, or even more times before escaping. This sensitive dependence of the output state on the input state is known as chaotic scattering [7]. These window intervals are interwoven in a manner that has been described as fractal. One such computation is shown in Fig. 4.1. Plots of various solutions are shown in Fig. 4.2. Careful inspection of panels (a) and (c) shows that in each successive two-bounce window, the dynamics contains one more internal oscillation than in the previous window. The initial condition in (c) was chosen to miss the peak of the resonance window, and the escaping solution is clearly more oscillatory than in panel (a).

Campbell and his collaborators hypothesized that the two-bounce resonance phenomenon is due to a resonant exchange of energy between the propagating solitary waves and an internal mode, i.e. an eigenfunction corresponding to a discrete eigenvalue in the linearized operator of the ϕ^4 equation (4.1) about ϕ_K. The authors ran a series of numerical experiments in similar nonlinear KG systems, which allowed them to investigate this hypothesis. One such system, the double sG equation possesses a tunable parameter η. The existence/non-existence of the internal mode depends on the value of this parameter, and the authors of the paper concluded that the existence of the two-bounce resonance phenomenon depends on the internal mode.

This conclusion has, in recent years, been reversed by a group that includes Dorey, Romańczukiewicz, Shnir and collaborators [8–10], who have shown that the phenomenon exists in systems with no internal mode. This is covered in more detail in Chap. 2 of this volume.

The collective coordinates method, usually derived using the variational approximation (VA), has been the most common tool in analyzing this phenomenon. These methods go back at least to Bondeson et al. [11], and are well-described in the review article of Malomed [12]. Such methods are applied to PDEs such as (4.1) describing the time evolution of a spatially dependent field $\phi(x, t)$ that minimizes a Lagrangian action

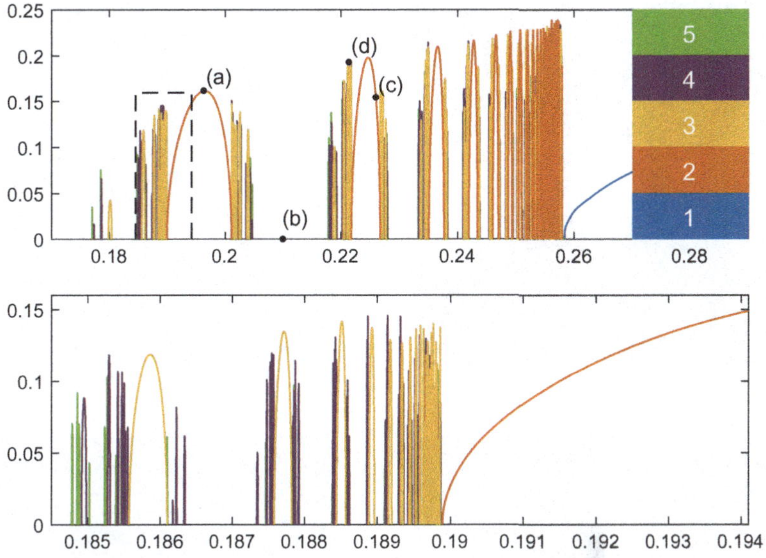

Fig. 4.1 (Top) The input velocity versus the output velocity, showing solutions that escaped after five or fewer collisions. Other initial conditions either led to capture or to escape after more than five collisions. (Bottom) Magnification of the interval indicated by the dashed line in the top image

$$\mathcal{A} = \iint \mathcal{L}(\phi) \mathrm{d}x \, \mathrm{d}t.$$

Instead of minimizing the action over all functions ϕ in the appropriate space, one chooses a solution ansatz dependent on a finite number of time-dependent parameters and minimizes the Lagrangian over all functions in that finite-dimensional family of functions. After integrating out the spatial dependence, one arrives at a finite-dimensional Lagrangian, whose Euler-Lagrange equations describe the evolution of the parameters. It is common to neglect higher-order terms that appear in this approximation.

A variational model was derived for the kink-antikink collision by Sugiyama in 1979 [2], and was later studied by Anninos et al. [13]. We presented a thorough analysis of this model in 2005, which we thought, together with some followups, would close the books on this problem [6, 14, 15].

Recent work by Weigel and his students has cast doubt on the validity of the VA models [16, 17]; this is covered in more detail in Chap. 3 of this volume, so the account here will be brief. There are two main issues. The first, initially noted by Caputo and Flytzanis [18] is that the mass matrix for the VA equations is singular at the instant the kink and antikink collide. An attempt to remedy this is discussed in [16]. The second is a mistake in the form of the term coupling the traveling wave to the internal mode in the VA approximation derived by Sugiyama and used, without correction in

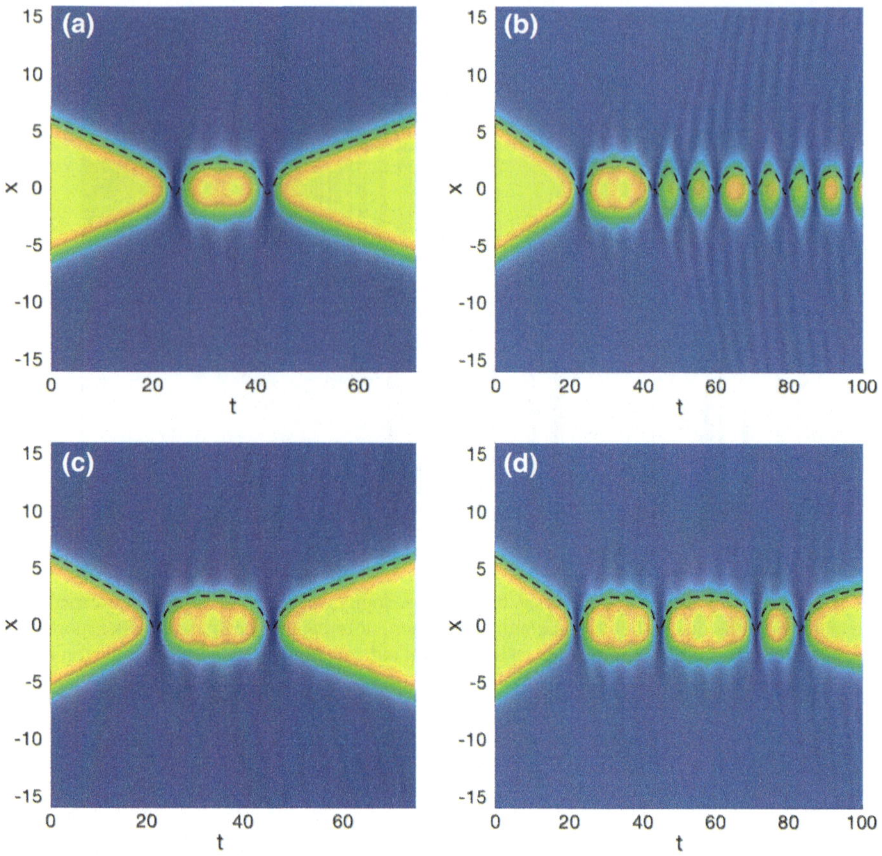

Fig. 4.2 Kink-antikink solutions to the ϕ^4 equation, with initial velocities demonstrating **a**, **c** two bounces, **b** capture, and **d** three bounces, corresponding to marked points in Fig. 4.1

many subsequent studies, including [6]. Most damningly, they show that replacing the incorrect term with its correct value further degrades the agreement between the full system and its VA model. They then show that the inclusion of higher-order terms ignored by Sugiyama, far from improving the model's agreement with PDE simulations, leads to additional disagreement, both quantitative and qualitative.

In addition, the inadequacy of the model equations has long been apparent due to the energy conservation. While the ϕ^4 equation conserves energy, the total energy available to the solitary waves after they collide is reduced by the propagation of radiation away from the location of the collision, as is clearly visible in Fig. 4.2 as ripples in the blue region emanating from the location of the collision. As a result of this energy loss, the outgoing velocity of the escaping solitons shown in Fig. 4.1 is never more than about ninety percent of that of the incoming soliton. By contrast, there is no radiative mode to carry energy away in the collective-coordinate model derived via the VA method. The condition used to define the resonant initial velocities is that the incoming and outgoing speeds are precisely equal. Clearly if energy loss due to radiation plays such a large role, then correct the form of the conservative collective coordinates model will be insufficient to account for its role.

4.2 Analysis of Collective-Coordinate Models

Work described in the previous section and elsewhere in this volume casts serious doubt on the correctness of the collective coordinates model derived by Sugiyama and analyzed by Goodman and Haberman. Nonetheless, in some cases the variational models have been quite effective in reproducing dynamics seen in the PDE from which they were derived. For example, Fei et al. derived a collective coordinate model that describes the behavior of solitons colliding with localized defects in the medium through which they propagate [19]. As in the case for the ϕ^4 equation, the approximations used in deriving the reduced model are unjustified and probably unjustifiable.

However the models derived by VA are important as phenomenological models. They distill an essential feature of such systems: the interaction between the kink and antikink is well described by a localized potential that decays at infinity, and the interaction between a kink and its internal mode is mediated by the presence of a nearby antikink. Therefore analyzing such models can provide insight into the dynamics of collisions even if the model equations do not describe the full dynamics in the sense of a convergent approximation.

4.2.1 The ODE Model

Sugiyama's VA model is derived using the ansatz

$$\phi_{\text{ansatz}}(x,t) = \phi_{\text{K}}(x + X(t)) - \phi_{\text{K}}(x - X(t)) + 1 + A(t)\big(\chi_1(x + X(t)) - \chi_1(x - X(t))\big),$$

where $X(t)$ is the undetermined position of the kink and $-X(t)$ that of the antikink, with the internal modes at the same location and with amplitude $A(t)$. The internal mode has the form

$$\chi_1(\xi) = \left(\frac{3}{\sqrt{2}}\right)^{1/2} \tanh\xi \operatorname{sech}\xi.$$

The resulting system has a mass matrix that depends on the position $X(t)$, and which is singular at $X = 0$, when the kink and antikink profiles cancel each other exactly. We present a qualitative model that retains the essential features of such a model while avoiding this difficulty [14, 15, 20]. It consists of a system of second order ODE

$$m\ddot{X} + U'(X) + \epsilon F'(X)A = 0; \tag{4.2a}$$

$$\ddot{A} + \omega^2 A + \epsilon F(X) = 0. \tag{4.2b}$$

This conserves an energy

$$H = \frac{m}{2}\dot{X}^2 + U(X) + \frac{1}{2}(\dot{A}^2 + \omega^2 A^2) + \epsilon F(X)A. \tag{4.3}$$

It is assumed that the potential $U(X)$ and the coupling function $F(X)$ decay rapidly as $X \to \infty$ and that $U(X)$ grows rapidly with negative X to prevent the waves from passing through each other. The other essential ingredient is that $U(X)$ has a minimum, so that, when uncoupled from the secondary oscillator A, the kink and antikink may oscillate around this minimum and be trapped, or, if they have sufficient energy, escape to infinity. Between these two regimes lies the *separatrix* curve, which is an orbit homoclinic to the point at infinity. We will let $U(X)$ be the Morse potential and $F(X)$ be a simple decaying exponential:

$$U(X) = e^{-2X} - e^{-X}; \;\; F(X) = e^{-X}. \tag{4.4}$$

Setting $\epsilon = 0$ decouples the dynamics, with X conserving an energy $E = \frac{m}{2}\dot{X}^2 + U(X)$. The phase plane for this system is plotted in Fig. 4.3. The separatrix corresponds to the level set $E = 0$ and to trapped orbits to negative energy. Also plotted in this figure is the projection into the $(X, \dot{X})$ plane of one solution to the equations with $\epsilon = \frac{1}{4}$.

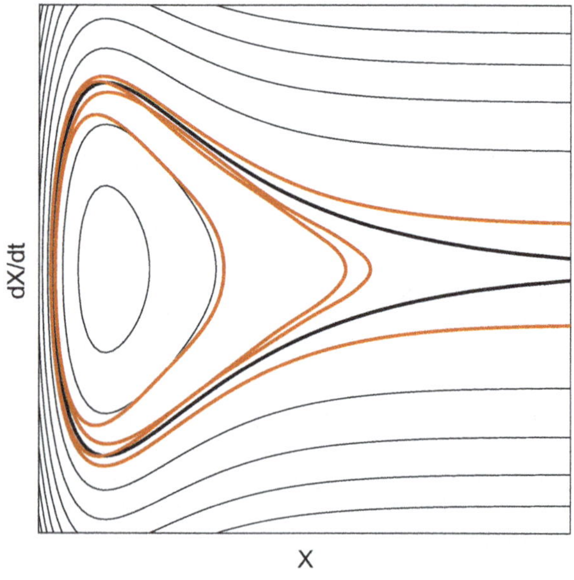

Fig. 4.3 The $(X, \dot{X})$ phase plane, and the projection of a solution to system (4.2) that begins in the lower-right-hand corner, crosses to the inside of the separatrix, makes four bounces, and then crosses out again at the upper-right

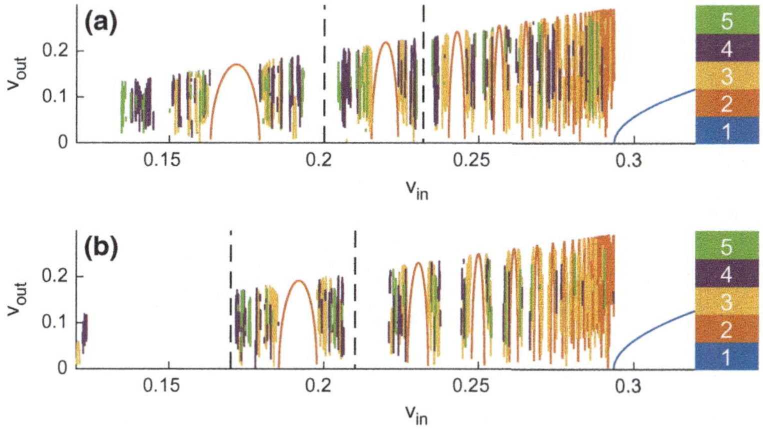

Fig. 4.4 **a** Numerical simulation of chaotic scattering in the model ODE system (4.2) with parameters $m = 1$, $\omega = 1$, and $\epsilon = \frac{1}{4}$. **b** Numerical simulation of the iterated map model (4.10)

Figure 4.4a shows the result of solutions to the model ODE system with initial conditions

$$X(0) = X_{\max} \gg 1; \;\; \dot{X}(0) = -v_{\text{in}}; \;\; a(0) = 0; \;\; \dot{a}(0) = 0$$

with 50,001 evenly-spaced spaced values of v_{in} and parameters $m = 1$, $\omega = 1$, and $\epsilon = \frac{1}{4}$. The color indicates how many collisions, i.e. minima of $X(t)$ occurred before escape, determined by $X(t) > X_{\max}$. True capture is possible for the ϕ^4 system, as evidenced by Fig. 4.1d. In these cases energy that escapes to infinity in the form of radiation prevents the solitary waves from escaping. By contrast, in the model ODE system (4.2), eventually, enough energy must be returned to the propagating mode $X(t)$ for it to escape, except for a measure-zero set of initial conditions that lie on the stable manifold of a bounded orbit. Therefore, capture is not observable in this figure. Note that this figure captures much of the structure of the PDE computation, although not the quantitative details.

4.2.2 *Derivation of the Discrete-Map Approximation*

The chaotic scattering is analyzed by deriving a discrete map approximation to the dynamics. Referencing Fig. 4.5, we see that $X(t)$ approaches $-\infty$ as $t \to \infty$, achieves local minima at a discrete sequence of "bounce" times t_j, and following the last minimum, at $t = t_4$, escapes. Also defined are the times $\tilde{t}_j$ at which the solitary waves achieve their largest separations. By definition $\tilde{t}_0 = -\infty$ and the final separation time $\tilde{t}_4 = \infty$. The second panel shows the energy $E(t)$ in the $(X, \dot{X})$

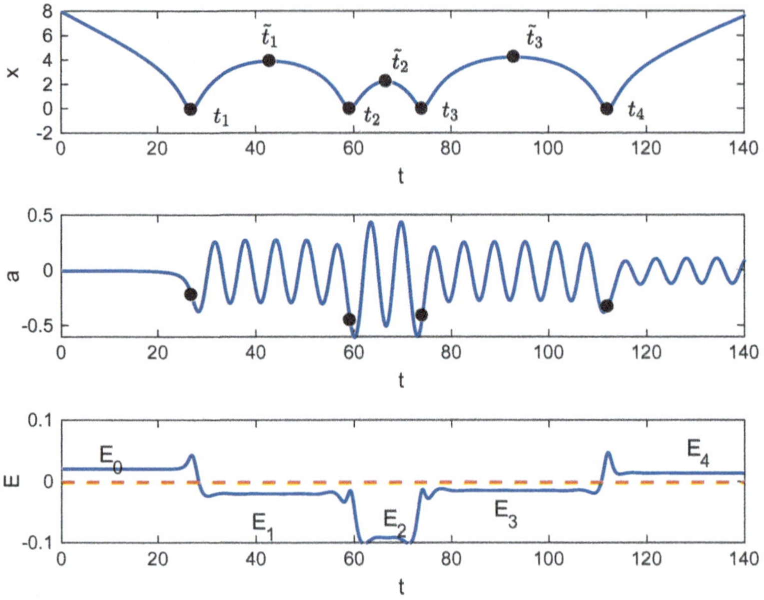

Fig. 4.5 The solution to (4.2) plotted in Fig. 4.3. Top: $X(t)$, showing the four bounce times and three near-infinity approach times. Middle: $a(t)$ showing that the amplitude and phase of the oscillator a change at each bounce. Bottom: The energy, which is positive when the solution is outside the separatrix, and negative inside

component, which is positive when the solution is outside the separatrix, and which jumps at each bounce. The third panel shows that $a(t)$ oscillates sinusoidally, with a jump in amplitude and phase at each bounce. The map we derive measures how the energy E, and the amplitude and phase of a change between the bounces.

Rather than describing $a(t)$ by the amplitude and phase, we instead assume that before bounce j,

$$A(t) \sim C\left(c_j \cos\omega(t-t_j) + s_j \sin\omega(t-t_j)\right) \text{ as } t-t_j \to -\infty$$

where the constant C must be chosen to normalize the variables (see below). This represents the value of $a(t)$ when X is large and the coupling is smallest. The map we derive takes the form

$$(c_{j+1}, s_{j+1}, E_{j+1}) = \mathcal{F}(c_j, s_j, E_j)$$

where E_j is the plateau energy level preceding the jth bounce time t_j.

Near time t_j, $X(t)$ can be represented by the solution along the separatrix $X_S(t-t_j)$, and we can solve (4.2b) for $A(t)$ by variation of parameters:

$$A(t) \sim C\left(c_j \cos\omega(t-t_j) + s_j \sin\omega(t-t_j)\right) \\ -\frac{\epsilon}{\omega}\sin\omega(t-t_j)\int_{-\infty}^{t} F(X_S(t-t_j))\cos\omega(t-t_j)\mathrm{d}t \\ +\frac{\epsilon}{\omega}\cos\omega(t-t_j)\int_{-\infty}^{t} F(X_S(t-t_j))\sin\omega(t-t_j)\mathrm{d}t \tag{4.5}$$

Since X_S is an even function of t, this gives, as $t-t_j \to \infty$,

$$A(t) \sim C\left(c_j \cos\omega(t-t_j) + s_j \sin\omega(t-t_j)\right) \\ -\frac{\epsilon}{\omega}\sin\omega(t-t_j)\int_{-\infty}^{\infty} F(X_S(t-t_j))\cos\omega(t-t_j)\mathrm{d}t$$

Defining

$$C = \frac{\epsilon}{\omega}\int_{-\infty}^{\infty} F(X_S(t))\cos\omega t\mathrm{d}t = \frac{\epsilon}{\omega}\int_{-\infty}^{\infty} F(X_S(t))e^{i\omega t}\mathrm{d}t, \tag{4.6}$$

we may rewrite this as

$$A(t) \sim C\left(c_j \cos\omega(t-t_j) + (s_j - 1)\sin\omega(t-t_j)\right) \text{ as } t - t_j \to \infty.$$

However, this should be written instead in terms of $t - t_{j+1}$. Note that the value of C is the first instance in this calculation where the particular form of the coupling function $F(X)$ is important.

Letting $\theta_{j+1} = -\omega(t_{j+1} - t_j)$, then $\omega(t-t_j) = \omega(t-t_{j+1}) - \theta_{j+1}$ and using angle addition formulas, we arrive at

$$A(t) \sim C\left(c_{j+1}\cos\omega(t-t_{j+1}) + s_{j+1}\sin\omega(t-t_{j+1})\right),$$

where

$$\begin{pmatrix} c_{j+1} \\ s_{j+1} \end{pmatrix} = \begin{pmatrix} \cos\theta_{j+1} & -\sin\theta_{j+1} \\ \sin\theta_{j+1} & \cos\theta_{j+1} \end{pmatrix} \begin{pmatrix} c_j \\ s_j - 1 \end{pmatrix}.$$

We introduce a complex variable $z_j = c_j + is_j$, which renders this as

$$z_{j+1} = e^{i\theta_{j+1}}\left(z_j - i\right). \tag{4.7}$$

Now it remains to approximate $(t_{j+1} - t_j)$. This requires matching a solution in a neighborhood of the bounce, where $X(t)$ is approximated by the separatrix orbit, with the solution near $X = \infty$. Here the leading-order behavior of $U(X)$ for large X is sufficient to describe the dynamics. The most important quantity for describing the large-X behavior is the energy in the X-component of the solutions

$$E = \frac{m}{2}\dot{X}^2 + U(X).$$

The time $t_{j+1} - t_j$ is related to the energy on the jth approach to infinity, so we must calculate how the energy changes. To do so, we integrate $\frac{dE}{dt}$ between the two near-infinity approach times. The trajectory is assumed to be in a neighborhood of the separatrix, so that the collision times are far apart, and the domain of integration can be replaced by $(-\infty, \infty)$. An integral of this type is known as a *Melnikov integral* [21].

$$\begin{aligned} E_{j+1} - E_j &= \int_{\tilde{t}_{j-1}}^{\tilde{t}_j} \frac{dE}{dt} dt \approx \int_{-\infty}^{\infty} \frac{dE}{dt} dt \\ &= \int_{-\infty}^{\infty} \left(\ddot{X} + U'(X)\right) \dot{X} dt = \int_{-\infty}^{\infty} -\epsilon A F'(X) \dot{X} \, dt = \epsilon \int_{-\infty}^{\infty} F(X) \dot{A} \, dt, \end{aligned}$$

and the integration by parts is justified since $F(X(t))$ vanishes sufficiently rapidly as $|t| \to \infty$.

In the above calculation, the only assumption so far is that the domain of integration can be approximated by the whole line, so the integrand must vanish sufficiently rapidly as $|t| \to \infty$. We further approximate $X(t)$ in the last expression by $X_S(t - t_j)$. The rest of the calculation depends on the specific form of functions $F(X)$ and $U(X)$ in the Hamiltonian (4.3). For the particular form of $U(X)$ specified in (4.4), the separatrix orbit is $X_S(t) = \log\left(1 + \frac{t^2}{2m}\right)$ and $F(X_S) = \frac{2m}{2m+t^2}$.

Using this form for $F(X)$, calculating $\dot{A}$ using (4.5), and integrating, taking advantage of symmetries, we arrive at

$$E_{j+1} - E_j = \frac{C^2\omega^2}{2} \left(2s_j - 1\right). \tag{4.8}$$

We define a rescaled energy by $E_j = \frac{C^2\omega^2}{2} \mathcal{E}_j$. Equations (4.7) and (4.8) then imply the conservation of the quantity

$$\mathcal{H} = \mathcal{E}_j + \left|z_j\right|^2 \tag{4.9}$$

A delicate matching procedure that depends on $U(X)$ is needed to determine the functional form of $\theta_{j+1} = \Theta(\mathcal{E}_{j+1})$. Using (4.7) and the conservation law (4.9), the map becomes

$$z_{j+1} = e^{i\Theta(\mathcal{H} - \left|z_j - i\right|^2)}(z_j - i).$$

The integral (4.6) may be evaluated by residues and yields

$$C = \frac{\epsilon\pi\sqrt{2m}}{\omega} e^{-\sqrt{2m}\omega}.$$

When $E_{j+1} < 0$, the time between bounces $t_{j+1} - t_j$ can be calculated via a matched asymptotic procedure between the near-saddle approaches, described by the separatrix orbit and the near-infinity approaches defined using the large X asymptotic expansion of $U(X)$, following [15]. Since the phase of θ_{j+1} enters the computation, it is necessary to approximate this quantity to $o(1)$. For this potential $U(X)$ given by (4.4), the time $(t_{j+1} - t_j)$ is simply the period of an orbit with energy $E_{j+1} < 0$, yielding

$$\Theta(\mathcal{E}_j) = -\frac{2\pi}{C}\sqrt{\frac{m}{-\mathcal{E}_j}}.$$

Finally, following Stolovitsky et al. [22], we define a new variable

$$Z_j = z_j - \frac{i}{2}$$

and a parameter $\alpha = 2\pi\sqrt{M}/C = \sqrt{2}\omega e^{2m\omega}/\epsilon$. With respect to this coordinate, the final form of the *separatrix map* is

$$\mathcal{F}(Z_j) = e^{-i\alpha/\sqrt{|Z_j - \frac{i}{2}|^2 - \mathcal{H}}}\left(Z_j - \frac{i}{2}\right) - \frac{i}{2}, \tag{4.10}$$

with inverse

$$\mathcal{F}^{-1}(Z_j) = e^{i\alpha/\sqrt{|Z_j + \frac{i}{2}|^2 - \mathcal{H}}}\left(Z_j + \frac{i}{2}\right) + \frac{i}{2}.$$

Despite being written in terms of a complex variable, these maps are non analytic as they involve absolute values. Defining ρ to be the complex conjugation operator $\rho(z) = z^*$, the two maps are related by conjugation,

$$\mathcal{F}^{-1} = \rho^{-1}\mathcal{F}\rho.$$

4.2.3 *Interpretation and Analysis of the Map*

The traditional set of tools may be applied to the map (4.10) such as the enumeration of fixed points, period-n points, and stable and unstable manifolds. Much of this is done in [15]. Here we concentrate on using the map to explain the structure found in Fig. 4.4. Each point in this figure corresponds to a different initial velocity, and thus a different value of $\mathcal{H}$, and a different map, so computing such quantities does not seem to be the most useful calculation.

For $\mathcal{H} > 0$, the domain of $\mathcal{F}$ excludes the closed disk

$$\mathcal{D}_{\text{out}} = \left\{Z : \left|Z - \frac{i}{2}\right| \leq \mathcal{H}\right\}.$$

Its range, and the domain of $\mathcal{F}^{-1}$ is the complex plane excluding the closed disk

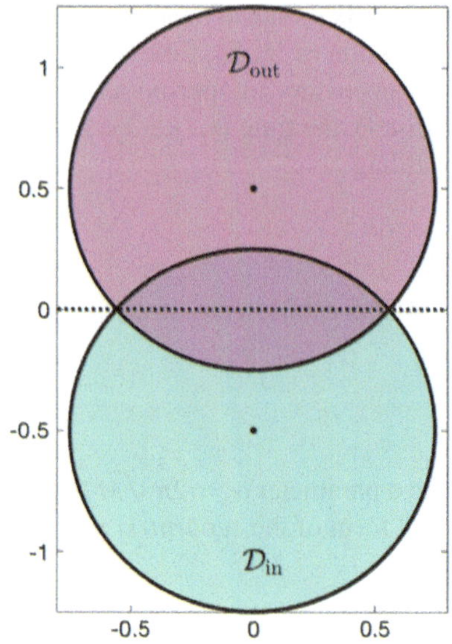

Fig. 4.6 The two disks on which the map $\mathcal{F}$ and its inverse are undefined. On the intersection, neither is defined

$$\mathcal{D}_{\text{in}} = \left\{ Z : \left| Z + \frac{i}{2} \right| \leq \mathcal{H} \right\}.$$

The two disks are shown in Fig. 4.6. These sets have an important and simple interpretation for the dynamics. For $Z_j \in \mathcal{D}_{\text{out}}$, $\mathcal{E}_{j+1} > 0$ is well-defined, but t_{j+1} is not, because solutions with positive energy escape to infinity and there is no $(j+1)$th bounce. Similarly points in $\mathcal{D}_{\text{in}}$ have no pre-image under the map and correspond to kink-antikink pairs arriving from infinity.

We assume that the internal mode is initially unexcited, i.e. $c_0 = s_0 = 0$ or to $Z_0 = -\frac{i}{2}$, the point at the center of the disk $\mathcal{D}_{\text{in}}$. If $\mathcal{H}$ is sufficiently large, then the point $Z_0 = -\frac{i}{2} \in \mathcal{D}_{\text{out}}$ and the solitary waves escape each other's embrace after just on interaction. Therefore the critical velocity is that such that $\mathcal{H}|_{v=v_c} = 1$, i.e. that

$$\mathcal{H}(v_c) = \mathcal{E} = \frac{2}{\omega^2 C^2} E = \frac{2}{\omega^2 C^2} \cdot \frac{m v_c^2}{2},$$

or

$$v_c = \frac{\omega C}{\sqrt{m}}. \tag{4.11}$$

This gives a useful relation

$$\mathcal{H} = \frac{v^2}{v_c^2}. \tag{4.12}$$

We will illustrate two features of this dynamics: the n-bounce resonant solutions and the hierarchy of the resonance windows. The n-bounce resonant solutions satisfy

$Z_0 = -\frac{i}{2}$ and $Z_{n-1} = \frac{i}{2}$,

$$\mathcal{F}^{n-1}\left(-\frac{i}{2}\right) = \frac{i}{2}.$$

If $n = 2m$ is even then

$$\mathcal{F}^{m}\left(-\frac{i}{2}\right) = \mathcal{F}^{-(m-1)}\left(\frac{i}{2}\right) = \rho\mathcal{F}^{(m-1)}\left(-\frac{i}{2}\right),$$

i.e.

$$\mathcal{F}(Z_m) = Z_m^*. \tag{4.13}$$

If $n = 2m + 1$ is odd then

$$\mathcal{F}^{m}\left(-\frac{i}{2}\right) = \mathcal{F}^{-m}\left(\frac{i}{2}\right) = \rho\mathcal{F}^{m}\left(-\frac{i}{2}\right),$$

i.e.

$$Z_m \in \mathbb{R}. \tag{4.14}$$

The edges of the n-bounce resonance windows correspond to the points that land on the boundary of the disk $\mathcal{D}_{\text{out}}$ after $n - 1$ iterates, i.e. those for which

$$\left|Z_{n-1} - \frac{i}{2}\right|^2 = \mathcal{H}. \tag{4.15}$$

Given $Z_0 = -\frac{i}{2}$, then

$$Z_1 = -ie^{-i\phi_1} - \frac{i}{2} \text{ and } Z_2 = -ie^{-i\phi_2}\left(e^{-i\phi_1} + 1\right) - \frac{i}{2},$$

where

$$\phi_1 = \frac{\alpha}{\sqrt{1-\mathcal{H}}} = \frac{2\pi\omega}{\sqrt{v_c^2 - v^2}} \text{ and } \phi_2 = \frac{\alpha}{\sqrt{4\cos^2\frac{\phi_1}{2} - \mathcal{H}}} = \frac{2\pi\omega}{\sqrt{4v_c^2\cos^2\frac{\pi\omega}{\sqrt{v_c^2-v^2}} - v^2}}.$$

These have all been put into a form where the dependence of the solution on the initial velocity v has been made explicit.

The condition for a two-bounce resonant initial velocity (4.13) is then just $Z_1 = i$, which can be simplified to $e^{i\phi_1} = -1$, or

$$v_{2,n} = \sqrt{v_c^2 - \frac{4\omega^2}{(2n-1)^2}}. \tag{4.16a}$$

Since ϕ_1 diverges as $v \nearrow v_c$, there is an infinite number of such solutions, indexed by n, which here specifies the number of oscillations of the secondary oscillator $A(t)$ between the two collisions. This formula indicates that such a solution only exists for

$$n > \frac{1}{2}\left(\frac{2\omega}{v_c} + 1\right).$$

The edges of the two-bounce windows satisfy (4.15) with $n = 2$ which can be simplified to

$$v^2 = 4v_c^2 \cos^2 \frac{\pi\omega}{\sqrt{v_c^2 - v^2}}.$$

The three-bounce resonant solutions solve (4.14) with $m = 1$, i.e. that $Z_1 \in \mathbb{R}$. This can be simplified to $\cos\phi_1 = \frac{-1}{2}$, or

$$v_{3,n\pm} = \sqrt{v_c^2 - \frac{4\omega^2}{\left(2n - 1 \pm \frac{1}{3}\right)^2}}. \tag{4.16b}$$

Again, there is an infinite number of such solutions, immediately to the right and left of each two-bounce window. The edges of the three-bounce windows satisfy (4.15) with $n = 3$, which can be simplified to

$$v^2 = \left(1 + 8\cos\frac{\phi_1}{2}\cos\frac{\phi_2}{2}\cos\frac{\phi_1+\phi_2}{2}\right)v_c^2.$$

Since $\phi_2 \to \infty$ as v approaches the edge of each two-bounce window, there must be an infinite number of three bounce windows accumulating outside of each two-bounce window.

We find the four-bounce resonant solutions using (4.13) with $m = 2$. This is more simply written as $e^{i(\phi_1+\phi_2)} = -1$ or

$$\phi_1 + \phi_2 = (2n-1)\pi, \tag{4.16c}$$

but of course we can expand the LHS as a function of v. The five-bounce resonant solutions solve (4.14) with $n = 2$, which requires

$$\cos\phi_2 + \cos(\phi_1 + \phi_2) = -\frac{1}{2}. \tag{4.16d}$$

Here again, the divergence of ϕ_2 guarantees that there is an infinite number of four- and five-bounce resonant solutions. With a little bit more work we can show that an infinite sequence of $n+1$-bounce windows accumulates at the left and right edges of each n-bounce resonance. However, beyond this point, it becomes difficult to develop such explicit formulas.

4.3 Computational Exploration of Map (4.10)

Figure 4.4b shows the chaotic scattering diagram for solutions to the map (4.10) with initial condition $Z_0 = -\frac{i}{2}$. It is impossible to resolve this figure, since the angle $\Theta(Z_j)$ diverges as Z_j approaches the boundary of its domain, i.e. the circle $\partial\mathcal{D}_{\text{out}}$, which implies that the full diagram consists of an infinite number of windows of infinitesimal width.

In this figure, (4.12) is used to replace the asymptotic value of the critical velocity (4.11) by the numerical value obtained from numerical simulation of the ODE model ($v_c \approx 0.294$ from simulation of (4.2), while formula (4.11) gives $v_c \approx 0.279$). Numerical comparisons in [14] show that as $\epsilon \to 0$ (or as $\omega \to \infty$) the accuracy of the formula (4.11), and the positions of the map's predictions for resonance windows improves. However, we elect to use a relatively large value of ϵ in this figure, as the window widths are smaller for small ϵ and it is harder for a reader to visualize the intricate structure.

In Fig. 4.7, we show three zoomed-in views of the ODE simulation shown in Fig. 4.4a, with each figure showing the input velocities between the dashed black lines in the previous figure. Surrounding each n-bounce window, we can see a sequence of $(n+1)$-bounce windows, accumulating at its edges. These windows have a clearly visible envelope, including a "tallest" window which may have a resonant solution at its center. It is impossible to tell from the ODE simulations whether these are truly resonances. The best we can do is to find the input velocities which locally minimize $v_{\text{in}} - v_{\text{out}}$.

An analogous set of zooms is shown for the map (4.10) in Fig. 4.8, zooming in beginning between the dashed vertical lines in Fig. 4.4b. For the map, the conditions (4.13) and (4.14), or more specifically (4.16) demonstrate which of the apparent n-bounce resonant velocities are and which are not actual resonances. In subplot (a), a two-bounce solution satisfying (4.16a) is flanked by two three-bounce solutions satisfying (4.16b). In subplot (b) a non-resonant three-bounce window has four-bounce

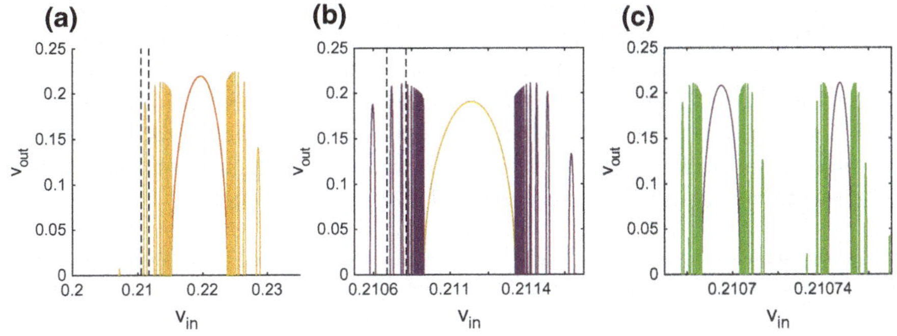

Fig. 4.7 Three zooms of the ODE simulation shown in Fig. 4.4a. From the figure it is not possible to determine which of the windows contains a resonance

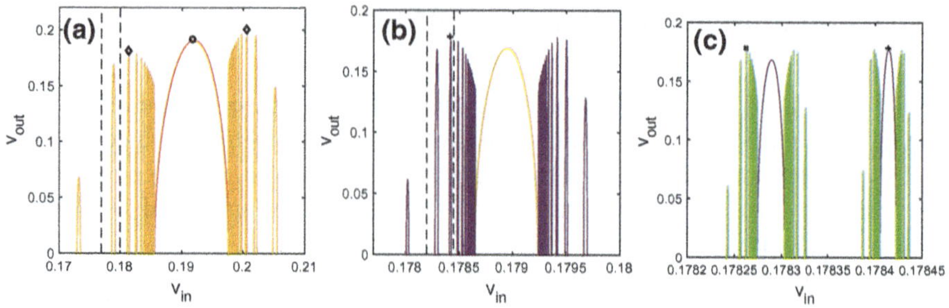

Fig. 4.8 Three zooms of the map simulation shown in Fig. 4.4b. The marked points indicate resonant solutions as determined by (4.16). Several of the points that appear to be resonances in this figure can be rejected as resonances by calculating that the quantities appearing on the LHS of the (4.16) are nonzero, indeed not close to zero

windows on both sides, but the only four-bounce window satisfying condition (4.16c) is on its right. Subplot (c) shows two four bounce windows, each surrounded by five bounce windows, but the only five bounce window satisfying (4.16d) is to the left of a non-resonant four-bounce window.

4.4 Conclusions and Outlook

To conclude, the assumptions underlying the derivation of model ODE systems like (4.2) have been shown, and their numerical infidelity to the underlying PDE system has likewise been highlighted. Nonetheless, as qualitative models, they can be used to guide thinking about the types of behaviors seen in solitary wave collisions. These models are amenable to a deep analysis that allows for the explanation of many of the features seen in numerical simulations.

Clearly a collective coordinates model that more completely captures the dynamics of the full PDE will be needed. Such a model should capture at least some of the energy lost due to radiation and will, ideally, be amenable to analysis like that presented here. An interesting twist is that the map will no longer obey a conservation law like (4.9), and thus not be reducible from three dimensions to two. An ODE model was derived to include radiative dissipation in the interaction between sG solitons and localized defects in [23]. However the additional terms did not allow us to extend the analytic methods described in the present chapter.

Acknowledgements Many of the results in this chapter were obtained in collaboration with Dr. Richard Haberman and funded by the NSF under grants DMS-0204881, DMS-0506495, and DMS-0807284.

References

1. A.E. Kudryavtsev, JETP Lett. **22**, 82 (1975)
2. T. Sugiyama, Prog. Theor. Phys. **61**, 1550 (1979)
3. M.J. Ablowitz, M. Kruskal, J. Ladik, SIAM J. Appl. Math. **36**, 428 (1979)
4. D.K. Campbell, J.F. Schonfeld, C.A. Wingate, Phys. D **9**, 1 (1983)
5. D.K. Campbell, M. Peyrard, Phys. D **18**, 47 (1986)
6. R.H. Goodman, R. Haberman, SIAM. J. Appl. Dyn. Sys. **4**, 1195 (2005)
7. E. Ott, T. Tél, Chaos **3**, 417 (1993)
8. P. Dorey, K. Mersh, T. Romanczukiewicz, Y. Shnir, Phys. Rev. Lett. **107**, 091602 (2011)
9. P. Dorey, T. Romanczukiewicz, Phys. Lett. B **779**, 117 (2018)
10. A. Halavanau, T. Romanczukiewicz, Y. Shnir, Phys. Rev. D **86**, 085027 (2012)
11. A. Bondeson, M. Lisak, D. Anderson, Phys. Scr. **20**, 479 (1979)
12. B.A. Malomed, Prog. Opt. **43**, 71 (2002)
13. P. Anninos, S. Oliveira, R. Matzner, Phys. Rev. D **44**, 1147 (1991)
14. R.H. Goodman, R. Haberman, Phys. Rev. Lett. **98**, 104103 (2007)
15. R.H. Goodman, Chaos **18**, 023113 (2008)
16. I. Takyi, H. Weigel, Phys. Rev. D **94**, 085008 (2016)
17. H. Weigel, J. Phys. Conf. Ser. **482**, 012045 (2014)
18. J.G. Caputo, N. Flytzanis, Phys. Rev. A **44**, 6219 (1991)
19. Z. Fei, Y.S. Kivshar, L. Vázquez, Phys. Rev. A **45**, 6019 (1992)
20. R.H. Goodman, A. Rahman, M.J. Bellanich, C.N. Morrison, Chaos **25**, 043109 (2015)
21. G. Zaslavsky, *The Physics of Chaos in Hamiltonian Systems* (Imperial College Press, London, 2007)
22. G. Stolovitzky, T.J. Kaper, L. Sirovich, Chaos **5**, 671 (1995)
23. R.H. Goodman, P.J. Holmes, M.I. Weinstein, Phys. D **161**, 21 (2002)

Chapter 5
Stochastic Dynamics of ϕ^4 Kinks: Numerics and Analysis

Grant Lythe

Abstract The stationary density of the overdamped ϕ^4 SPDE corresponds to a mean number of kinks and antikinks that is maintained by a balance between nucleation of new kink-antikink pairs and annihilation whenever a kink and antikink collide. We consider numerical methods for solution of the SPDE, and a definition of the location of the centre of a kink using a smoothing function that is the derivative of the function describing the shape and energy of an isolated kink. This definition allows calculation of the diffusivity of a kink as a function of noise intensity or "temperature" and defines the parameter characterising the "small-noise" régime. In the reaction-diffusion description of the dynamics (where kink-antikink pairs are nucleated with rate Γ, diffuse with diffusivity, D and annihilate on collision) the number of kinks per unit length in the steady-state has a simple exact expression.

5.1 Introduction

Extended nonlinear systems often exhibit localized structures, such as moving domain walls, that move about under the influence of perturbations [1–6]. The ϕ^4 equation can be used in multiple dimensions [7, 8], but we restrict ourselves to one space dimension, where the localised structures are known as kinks and antikinks [2]. Consider the SPDE for a field whose value at position $x \in [0, L]$ and time t is denoted $\mathbf{\Phi}_t(x)$ [9–18]:

$$\frac{\partial}{\partial t}\mathbf{\Phi}_t(x) - \frac{\partial^2}{\partial x^2}\mathbf{\Phi}_t(x) = \frac{1}{\alpha}(\mathbf{\Phi}_t(x) - \mathbf{\Phi}_t^3(x)) + (2KT)^{1/2}\,\eta_t(x) \tag{5.1}$$

where the last term in (5.1) is space-time white noise [19–28]:

$$\mathbb{E}(\eta_t(x)\eta_{t'}(x')) = \delta(x - x')\delta(t - t'). \tag{5.2}$$

G. Lythe (✉)
School of Mathematics, University of Leeds, Leeds, UK
e-mail: grant@maths.leeds.ac.uk

P. G. Kevrekidis and J. Cuevas-Maraver (eds.), *A Dynamical Perspective on the ϕ^4 Model*, Nonlinear Systems and Complexity 26,
https://doi.org/10.1007/978-3-030-11839-6_5

The dynamics can be thought of as that of a string subject to a double-well on-site potential (imagine two parallel trenches) and additive white noise all along its length. The amplitude of the noise is $(2KT)^{1/2}$ where T has the interpretation of temperature and K is Boltzmann's constant. We will use the notation $\beta = (KT)^{-1}$. The field $\mathbf{\Phi}_t$ is nondimensionalized in such a way that, on the RHS of (5.1), $\mathbf{\Phi}_t^3$ has the same coefficient as $\mathbf{\Phi}_t$. The distance x has been scaled so that the Laplacian term (second term on the LHS of (5.1)) has coefficient 1. The parameter α can be removed by rescaling time, but we retain it because several different conventions exist in the literature and because it provides a useful parameter (with dimensions of time) in analytical calculations. In numerical work, we typically adopt $\alpha = 1$, use periodic boundary conditions on $[0, L]$ and take very large values of L.

An alternative notation to (5.1) is the following:

$$\mathrm{d}\mathbf{\Phi}_t(x) = \frac{1}{\alpha}\left(\mathbf{\Phi}_t(x) - \mathbf{\Phi}_t^3(x) + \alpha\frac{\partial^2}{\partial x^2}\mathbf{\Phi}_t(x)\right)\mathrm{d}t + (2KT)^{1/2}\,\mathrm{d}\mathbf{B}_t(x), \tag{5.3}$$

where

$$\mathrm{d}\mathbf{B}_t(x)\mathrm{d}\mathbf{B}_t(x') = \delta(x - x')\mathrm{d}t. \tag{5.4}$$

5.1.1 Some Definitions

We may define the potential function

$$V(\Phi) = \frac{1}{\alpha}\left(-\frac{1}{2}\Phi^2 + \frac{1}{4}\Phi^4\right), \tag{5.5}$$

and the energy functional [29]

$$\mathcal{E}[f] = \int \left(V(f(x)) + \frac{1}{2}\left(\frac{\partial}{\partial x}f(x)\right)^2\right)\mathrm{d}x. \tag{5.6}$$

The kink shape function, $\phi(y)$, is defined as the solution of the ODE

$$\phi''(y) = V'(\phi(y)), \qquad \lim_{x\to-\infty}\phi(x) = -1 \text{ and } \lim_{x\to\infty}\phi(x) = 1. \tag{5.7}$$

With $V(\Phi)$ as in (5.5), the solution of (5.7) is [30]

$$\phi(y) = \tanh(y/\sqrt{2\alpha}). \tag{5.8}$$

(The function describing the shape of an antikink is $1 - \phi(y)$.) The energy or "mass" of a kink is

$$
\begin{aligned}
E_{\rm k} = \mathcal{E}[\phi] &= \int_{-\infty}^{\infty} \left(V(\phi(y)) + \frac{1}{2}\phi'(y)^2 \right) \mathrm{d}y \\
&= \int_{-\infty}^{\infty} \phi'(y)^2 \mathrm{d}y \\
&= -\int_{-\infty}^{\infty} \phi''(y)\phi(y)\mathrm{d}y \\
&= \left(\frac{8}{9\alpha}\right)^{1/2} .
\end{aligned} \tag{5.9}
$$

The ϕ^4 SPDE is primarily of interest in the "low noise" régime, where the typical distance between kinks is much greater than the width of a kink. The square root of the ratio of the thermal energy KT and $E_{\rm k}$ is the corresponding small parameter:

$$
\epsilon = \left(\frac{KT}{E_k}\right)^{\frac{1}{2}} . \tag{5.10}
$$

The notation (5.3) expresses, in a natural way, the simplest algorithm that may be used to solve the SPDE numerically [12, 31, 32]. Under finite differences, a numerical solution is generated on a grid of points separated by Δx. That is, $x = i\Delta x$, $i = 1, \ldots, N$. The discretized version is a system of N SDEs:

$$
\mathrm{d}\tilde{\mathbf{\Phi}}_t(i) = \frac{1}{\alpha}\left(-V'(\tilde{\mathbf{\Phi}}_t(i)) + \alpha(\Delta x)^{-2}\tilde{\Delta}\tilde{\mathbf{\Phi}}_t(i)\right)\mathrm{d}t + (2KT/\Delta x)^{1/2}\,\mathrm{d}\mathbf{W}_t(i), \tag{5.11}
$$

where

$$
\tilde{\Delta}\tilde{\mathbf{\Phi}}_t(i) = \tilde{\mathbf{\Phi}}_t(i+1) + \tilde{\mathbf{\Phi}}_t(i-1) - 2\tilde{\mathbf{\Phi}}_t(i) \quad \text{and} \quad \mathbb{E}(\mathrm{d}\mathbf{W}_t(i)\mathrm{d}\mathbf{W}_t(i')) = \delta_{i-i'}\,\mathrm{d}t.
$$

5.1.2 *Structure of the Chapter*

The ϕ^4 SPDE (5.1) can be solved as an initial value problem. On inspecting $\mathbf{\Phi}_t(x)$ as a function of x at a fixed t (that is, inspecting "configurations"), with $L \gg 1$, we observe wide regions where $\mathbf{\Phi}_t(x)$ is close to either 1 or -1, separated by narrow regions called kinks ($+1$ to the right) or antikinks ($+1$ to the left). These kinks and antikinks have some of the properties of particles: they are created in pairs, follow Brownian paths, and annihilate on collision. In the same way as the the series of configurations evolves towards a stationary density on the space of continuous functions, the number of kinks and antikinks per unit length evolves towards a well-defined mean value. As well as numerical methods for solving the SPDE itself, this chapter will be concerned with extracting the density of kinks from the stationary density, with the practical question of how to locate the positions of kinks and antikinks in a configuration, with the question of the diffusivity of a kink, and with the rate of nucleation of kink-antikink pairs.

5.2 Stationary Densities

The stationary density of the SPDE [33–39] is the limit as $\Delta x \to 0$ of the stationary density of the set of SDEs (5.11):

$$\rho(\Phi(1), \ldots, \Phi(N)) = Z_N \exp(-\beta \Delta x E(\Phi(1), \ldots \Phi(N))) \tag{5.12}$$

where the energy function of the discretized system is, with $\alpha = 1$,

$$E(\Phi(1), \ldots, \Phi(N)) = \sum_{i=1}^{N} \left(V(\Phi(i)) + \frac{1}{2} \left(\frac{\Phi(i) - \Phi(i-1)}{\Delta x} \right)^2 \right) \tag{5.13}$$

and the partition function [40–42] is

$$Z_N = \int_{-\infty}^{\infty} \cdots \int_{-\infty}^{\infty} \exp(-\beta \Delta x E(\Phi(1), \ldots \Phi(N))) \mathrm{d}\Phi(1) \ldots \mathrm{d}\Phi(N). \tag{5.14}$$

Using the "transfer integral" [10, 12, 41, 43, 44], it is possible to calculate Z_N in the continuum limit, $N \to \infty$ with $L = N\Delta x$ fixed:

$$Z = \lim_{N\to\infty} Z_N = \lim_{N\to\infty} \int_{-\infty}^{\infty} \cdots \int_{-\infty}^{\infty} \prod_{i=1}^{N} T(\Phi(i), \Phi(i+1)) \mathrm{d}\Phi(1) \ldots \mathrm{d}\Phi(N), \tag{5.15}$$

where

$$T(\Phi(i), \Phi(i+1)) = \exp\left(-\frac{1}{2} \beta \Delta x \left(\left(\frac{\Phi(i+1) - \Phi(i)}{\Delta x} \right)^2 + V(\Phi(i)) + V(\Phi(i+1)) \right) \right)$$

and $\Phi(N+1) = \Phi(1)$. To proceed, suppose we can find a suitable set of eigenfunctions ψ_n and corresponding eigenvalues t_n satisfying

$$\int_{-\infty}^{\infty} T(\Phi(i), \Phi(i+1)) \psi_n(\Phi(i)) \mathrm{d}\Phi(i) = t_n \psi_n(\Phi(i+1)). \tag{5.16}$$

Then the integral in (5.15) is written as

$$Z_N = \int_{-\infty}^{\infty} \cdots \int_{-\infty}^{\infty} \prod_{i=1}^{N} \left(\sum_n t_n \psi_n(\Phi(i)) \psi_n(\Phi(i+1)) \right) \mathrm{d}\Phi(1) \ldots \mathrm{d}\Phi(N) = \sum_n t_n^N.$$

Now $Z = \lim_{N\to\infty} \sum_n t_n^N = t_0^N$, where t_0 is the maximum eigenvalue. The functions ψ_n are solutions of the well-known equation [12, 45, 46]

$$\left(-\frac{1}{2\beta^2}\frac{\partial^2}{\partial\Phi^2} + V(\Phi)\right)\psi_n = \epsilon_n\psi_n, \qquad \text{and } t_n = \mathrm{e}^{-\beta\Delta x\epsilon_n}. \tag{5.17}$$

Now we are in a position to use this information to calculate quantities of interest, such as long-time mean values of the field. The mean value of $f(\mathbf{\Phi}_t(x))$ as $t \to \infty$ is independent of x and given by

$$\lim_{t\to\infty} \mathbb{E}(f(\mathbf{\Phi}_t(x))) = \frac{1}{Z}\lim_{N\to\infty}\int_{-\infty}^{\infty}\cdots\int_{-\infty}^{\infty} f(\Phi(j))\prod_{i=1}^{N} T(\Phi(i),\Phi(i+1))\mathrm{d}\Phi(1)\ldots\mathrm{d}\Phi(N),$$

where $x = j\Delta x$. Thus

$$\lim_{t\to\infty} \mathbb{E}(f(\mathbf{\Phi}_t(x))) = \int_{-\infty}^{\infty} f(\Phi)\psi_0(\Phi)^2\mathrm{d}\Phi. \tag{5.18}$$

That is, the square of the zeroth eigenfunction of the transfer operator is the "one-point density function", which has peaks near ± 1 but is not simply related to $V(\Phi)$ [10]. Knowing the one-point density function permits fully nonlinear calculation of numerical values, for example of the mean value of $\mathbf{\Phi}_t(x)$, to arbitrary precision, for comparison with the results of direct simulation of the SPDEs [9, 11]. The results from the transfer integral depend only on β in the limit $\Delta x \to 0$. However, because the transfer integral is explicitly written with the form of the numerical discretization, quantities such as the one-point density and correlation length can be calculated for finite Δx [12].

When kinks are the most interesting aspect of the system, the most illuminating consequence of the stationary density is the correlation function, which is the mean of the product of values at two different spatial points, as a function of their separation:

$$\begin{aligned} c(|x-y|) &= \lim_{t\to\infty} \mathbb{E}(\mathbf{\Phi}_t(x)\mathbf{\Phi}_t(y)) \\ &= \frac{1}{Z}\lim_{N\to\infty}\int_{-\infty}^{\infty}\cdots\int_{-\infty}^{\infty} \Phi(k)\Phi(j)\prod_{i=1}^{N} T(\Phi(i),\Phi(i+1))\mathrm{d}\Phi(1)\ldots\mathrm{d}\Phi(N), \end{aligned}$$

where $x = j\Delta x$ and $y = k\Delta x$. Thus the spatial correlation function is a sum of exponentials:

$$c(|x-y|) = t_0^{-|k-j|}\sum_n s_n^2 t_n^{|k-j|} = \sum_n s_n^2 \exp(-\beta x(\epsilon_n - \epsilon_0)). \tag{5.19}$$

The "correlation length" is given by

$$\lambda = -\lim_{x\to\infty} \frac{c(x)}{c'(x)} = \frac{1}{\beta(\epsilon_1 - \epsilon_0)}. \tag{5.20}$$

As $x \to \infty$, $c(x) \to s_1^2 \exp(-x/\lambda)$. If kinks and antikinks are randomly distributed in space, then the density of kinks is simply $1/(4\lambda)$ [9].

The existence of *exact* quantities that can be used to benchmark numerical results and analytical approximations is of immense value. For example, the correct values of λ can be evaluated to arbitrary precision by numerical solution of (5.17). For some choices of $V(\phi)$ and some parameter values, analytical results are known [47–49].

There are interesting relations between the stationary density of an SPDE and the Markov chain Monte Carlo (MCMC) method in computational Bayesian statistics, which is a procedure for sampling by constructing a Markov process whose invariant density is equal to a target density [50]. The configurations (functions of space) making up the stationary density of the SPDE sample from the same density as the sample paths (functions of time) of a suitably-constructed SDE. In diffusion-bridge sampling, the required processes satisfy SPDEs of the form (5.1), having unique invariant measures that are ergodic [34, 51, 52].

5.3 Numerical Solution

As the notation (5.11) suggests, in the Euler algorithm, the change between time t and time $t + \Delta t$ at spatial location i is

$$\tilde{\mathbf{\Phi}}_{t+\Delta t}(i) - \tilde{\mathbf{\Phi}}_t(i) = \left(-V'(\tilde{\mathbf{\Phi}}_t(i)) + \alpha(\Delta x)^{-2}\tilde{\Delta}\tilde{\mathbf{\Phi}}_t(i)\right)\frac{\Delta t}{\alpha} + \left(\frac{2KT}{\Delta x}\right)^{1/2} \Delta\mathbf{B}_t(i), \tag{5.21}$$

where $\Delta\mathbf{B}_t(i)$ is Gaussian random variable at each i and t with mean zero and

$$\mathbb{E}(\Delta\mathbf{B}_t(i)\Delta\mathbf{B}_{t'}(i')) = \begin{cases} \Delta t & i = i' \text{ and } t = t', \\ 0 & \text{otherwise}. \end{cases}$$

Perhaps the only counter-intuitive aspect of the algorithm is the scaling of the noise with Δx in (5.21).

It is straightforward to use the Heun algorithm instead of (5.21), so that errors in quantities such as the mean value at a spatial point are proportional to Δt^2 [10, 12]. In practice, an important part of designing a numerical experiment is choosing L, initial conditions, and run time [53]. The lower the temperature, the easier it is to identify and track kinks. However, genuine thermodynamic simulations with properly equilibrated kink populations require larger values of L, and longer run times, as larger values of β are chosen [9, 10].

An easier numerical task is to measure the diffusivity of an isolated kink: small spacetime domains lengths L can be used, with initial conditions chosen to have a single kink-antikink pair [54]. A consistent way to define and measure the position of (the centre of) a kink is then crucial. It is also possible to establish the rate of nucleation of new kink-antikink pairs by performing numerical experiments, with a kinkless initial configuration, at small and intermediate values of L [13].

It is a remarkable fact that the stationary density of the second-order-in-time SPDE

$$\frac{\partial^2}{\partial t^2}\mathbf{\Phi}_t(x) + \eta\frac{\partial}{\partial t}\mathbf{\Phi}_t(x) = \frac{\partial^2}{\partial x^2}\mathbf{\Phi}_t(x) + \frac{1}{\alpha}(\mathbf{\Phi}_t(x) - \mathbf{\Phi}_t^3(x)) + (2\eta KT)^{1/2}\ \eta_t(x), \tag{5.22}$$

is independent of the damping η and identical to that of (5.1), although the dynamics depends strongly on η. Kink-antikink pairs do not necessarily annihilate on collision; they bounce inelastically, producing trapping and resonance phenomena [55] that are the analogues of breathers [56, 57] found in noiseless systems. The code used in this chapter, to solve the first-order equation (5.1), is reproduced in the Appendix. A starting point for numerical solution of the second-order equation (5.22) could be to treat the system as $2N$ SDEs that can be solved using Euler or Heun timestepping. The dependence of the stationary density on the spatial discretization is unaffected of course. Evaluation of the accuracy of timestepping methods, on the other hand, is dramatically different. Rather than treating position and velocity variables on an equal footing, it is more efficient to use algorithms that explicitly respect the "partitioned" nature of the system. The advantanges are magnified in stochastic systems because only the velocity variables contain noise terms [58–65]. Mannella's modification of the leapfrog method is easily implemented and has the virtues of giving an error in the velocity variables that is independent of η. Of all known explicit two-step methods that use one random variable per timestep per gridpoint, the "reverse leapfrog" method [66] performs best in the position variable and the velocity correlation function at a separation of two grid points. The characteristic feature of the reverse leapfrog method is that the velocity variables are the last ones produced at each step. Both the reverse leapfrog and the Mannella method are symplectic in a deterministic setting.

5.4 Brownian Motion of Kinks

We cannot count kinks simply by counting zero crossings, because configurations are jagged on small scales. However, we can do so after convolving the configuration with a smoothing function [67]. Many symmetric functions with weight concentrated near zero could be used; here, we use the derivative of the function ϕ. That is, kinks and antikinks are located at values of x such that $h(x, \mathbf{\Phi}_t) = 0$ where

$$h(x, \mathbf{\Phi}_t) = \int \mathbf{\Phi}_t(u)\phi'(u-x)\mathrm{d}u. \tag{5.23}$$

If the position of a kink at time t is $\mathbf{X}_t$ then the configuration in the vicinity of $\mathbf{X}_t$ region can be written as

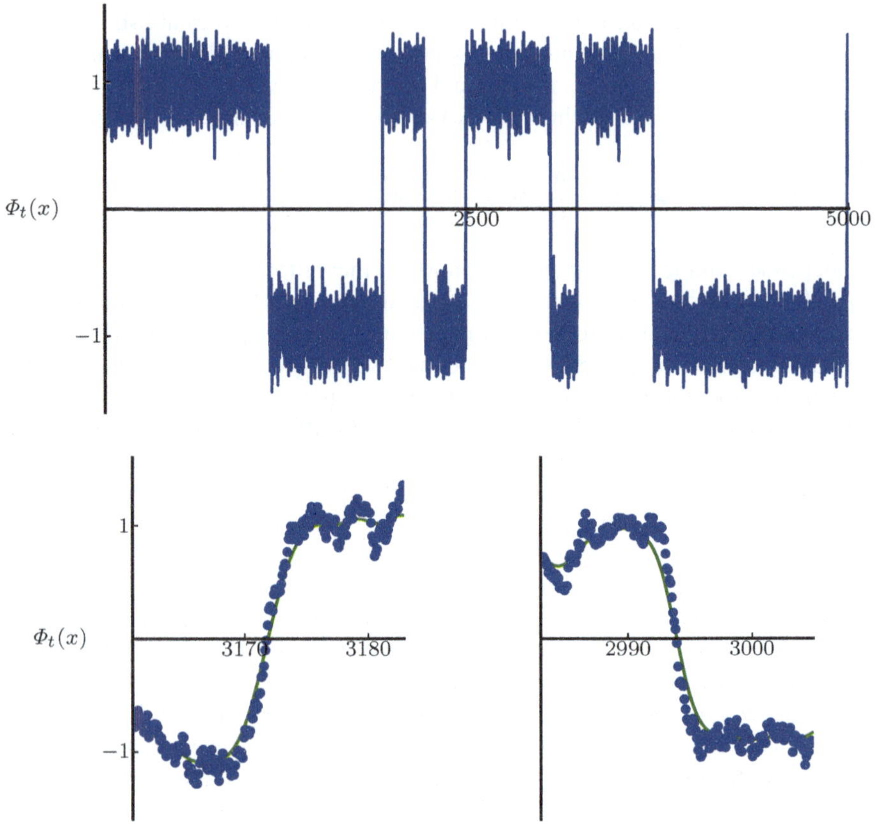

Fig. 5.1 Top: a configuration from one numerical realisation. Bottom: zoomed-in views of two small regions, containing a kink and an antikink. The green line is the function $h(x, \mathbf{\Phi}_t)$, obtained using the smoothing function $\phi'(y)$. The parameters are $\alpha = 1$ and $\beta = 20$, the numerical simulation was carried out with $\Delta x = 0.1$ and $\Delta t = 0.002$

$$\mathbf{\Phi}_t(x) = \phi(x - \mathbf{X}_t) + \chi_t(x - \mathbf{X}_t). \tag{5.24}$$

By introducing the coordinate

$$y = x - \mathbf{X}_t, \tag{5.25}$$

the field χ_t has stationary statistical properties [54, 67]. The separation (5.24) has been introduced by several authors [68, 69] and the assumption made that $\chi_t(y)$ is "small" in a sense that we can now make precise, using the parameter ϵ defined in (5.10). The choice of the smoothing function in (5.23) is illustrated in Fig. 5.1.

We proceed to derive an SDE for the evolution of the kink position. The general form for such an SDE is as follows [30]:

$$\mathrm{d}\mathbf{X}_t = a(\mathbf{\Phi}_t, t)\mathrm{d}t + b(\mathbf{\Phi}_t, t)\mathrm{d}\mathbf{W}_t^{(X)}. \tag{5.26}$$

We have chosen to define the kink's position $\mathbf{X}_t$ by the condition

$$\int_{-\infty}^{\infty} \mathbf{\Phi}_t(x)\phi'(x - \mathbf{X}_t)\mathrm{d}x = 0. \tag{5.27}$$

Similarly, $\mathbf{X}_{t+\Delta t}$ is the value of x such that $h(x, \mathbf{\Phi}_{t+\Delta t}) = 0$. That is

$$\int_{-\infty}^{\infty} (\mathbf{\Phi}_t(x) + \Delta\mathbf{\Phi}_t(x))\phi'(x - \mathbf{X}_{t+\Delta t})\mathrm{d}x = 0, \tag{5.28}$$

where $\Delta\mathbf{\Phi}_t(x) = \mathbf{\Phi}_{t+\Delta t}(x) - \mathbf{\Phi}_t(x)$. Letting $\Delta t \to 0$, we can restate (5.28) in the form of a stochastic differential:

$$\mathrm{d}h(\mathbf{X}_t, \mathbf{\Phi}_t) = 0. \tag{5.29}$$

The Ito formula [70, 71] is used to expand (5.29), using (5.3) and (5.26):

$$\begin{aligned} 0 = &\int_{-\infty}^{\infty} \left(\mathrm{d}\mathbf{\Phi}_t(x)\phi'(x - \mathbf{X}_t)\right) \mathrm{d}x \\ &- \left(\int_{-\infty}^{\infty} \phi''(x - \mathbf{X}_t)\mathbf{\Phi}_t(x)\mathrm{d}x\right) \mathrm{d}\mathbf{X}_t \\ &+ \left(\int_{-\infty}^{\infty} \phi'''(x - \mathbf{X}_t)\mathbf{\Phi}_t(x)\mathrm{d}x\right) \frac{1}{2}b^2(\mathbf{\Phi}_t)\mathrm{d}t \\ &- \int_{-\infty}^{\infty} ((2KT)^{1/2}\, b(t, \mathbf{\Phi}_t)\mathrm{d}\mathbf{W}_t^{(X)}\mathrm{d}\mathbf{B}_t(x)\phi''(x - \mathbf{X}_t))\mathrm{d}x. \end{aligned} \tag{5.30}$$

The second term on the RHS of (5.30) is $\mathbf{M}_t\mathrm{d}\mathbf{X}_t$ where, using (5.24) and (5.9),

$$\mathbf{M}_t = -\int_{-\infty}^{\infty} \phi''(x - \mathbf{X}_t)\mathbf{\Phi}_t(x)\mathrm{d}x = E_\mathrm{k} + \int_{-\infty}^{\infty} \phi''(x - \mathbf{X}_t)\chi_t(x)\mathrm{d}x. \tag{5.31}$$

Thus we find, performing the integral in the first term on the RHS of (5.30),

$$b(\mathbf{\Phi}_t, t) = \mathbf{M}_t^{-1}\,(2KTE_\mathrm{k})^{1/2}. \tag{5.32}$$

Thus, at lowest order in ϵ (5.10), $\mathbf{M}_t = E_\mathrm{k}$ and an isolated kink's motion is Brownian with diffusivity ϵ^2, that is, with mean-squared displacement at time t given by $2\epsilon^2 t$.

In order to calculate higher-order terms, we use the operator $\mathcal{L}$ that acts on a function $f(y)$ as follows

$$\mathcal{L}f(y) = \alpha f''(y) + f(y) - 3\phi^2(y)f(y). \tag{5.33}$$

The operator $\mathcal{L}$ is obtained from the equation of motion (5.1) by linearizing about the single-kink solution $\phi(y)$ [2]. The eigenvalue equation,

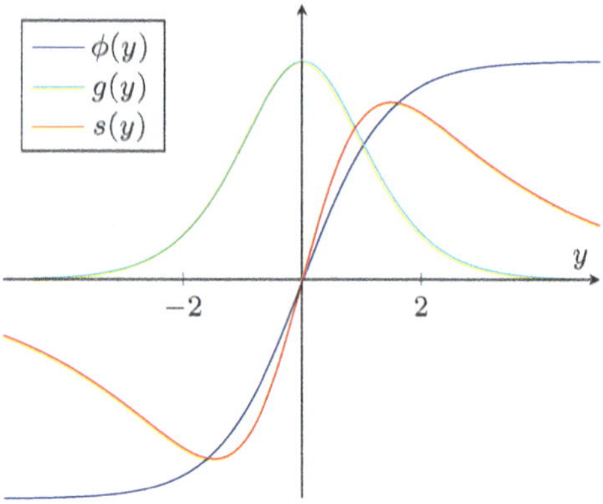

Fig. 5.2 The kink shape function, $\phi(y)$, is shown in blue. Also shown are the functions $g(y)$ and $s(y)$, eigenfunctions of the operator $\mathcal{L}$ with eigenvalues 0 and $-\frac{3}{2}$. $\alpha = 1$

$$\mathcal{L}f(y) = -\lambda f(y), \tag{5.34}$$

has two discrete solutions and also has an infinite series of solutions with $\lambda > 2$. The discrete solutions [2]

$$g(y) = \sqrt{2\alpha}\phi'(y) = \frac{1}{\cosh^2(y/\sqrt{2\alpha})} \tag{5.35}$$

and

$$s(y) = \left(-\alpha\phi''(y)\phi(y)\right)^{1/2} = \frac{\sinh(y/\sqrt{2\alpha})}{\cosh^2(y/\sqrt{2\alpha})}, \tag{5.36}$$

known as the "translation mode" and "shape mode" [72], are plotted in Fig. 5.2. $\mathcal{L}g(y) = 0$ and $\mathcal{L}s(y) = -\frac{3}{2}s(y)$. The eigenfunctions with $\lambda > 2$ are extended or "phonon" modes.

The inhomogeneous stochastic field $\chi_t(y)$ may now be decomposed using the eigenfunctions of $\mathcal{L}$. As $t \to \infty$, at first order in ϵ, $\chi_t(y)$ is Gaussian with mean zero; the amplitudes of the extended modes are independent Ornstein–Uhlenbeck processes with variance proportional to ϵ^2. At second order, the mean amplitude of the shape mode is nonzero. So too is the antisymmetric part of every second extended mode. Exchange of energy between these modes adds terms proportional to ϵ^2 to the diffusivity of a kink.

5.5 Nucleation of Kink-Antikink Pairs

New kinks and antikinks never appear alone, but are created in pairs in "nucleation" events, when a fluctuation causes part of a configuration to surmount the energy barrier separating the two wells at -1 and $+1$. Similarly, kinks and antikinks disappear in pairs, when their Brownian wanderings intersect. How frequent are nucleation events? We seek the rate Γ of nucleation events, per unit length and time.

It is possible, in principle, to calculate the rate of occurrence, per unit length in a large system, of fluctuations large enough to provoke nucleation [73, 74]. Intuitively, the rate should be proportional to $\exp(-2\beta E_k)$ because the nucleation event results in the creation of two localised structures, each with energy E_k [9, 73, 75–82]. Alternatively, it is illuminating to consider the balance between nucleation and annihilation events:

$$\Gamma\tau = \rho_k, \tag{5.37}$$

where τ is the mean lifetime of a kink and ρ_k is the number of kinks per unit length. The RHS of (5.37) is known from the stationary density of the SPDE and the transfer integral that allows the correlation length to be evaluated (Sect. 5.2): $\rho_k \propto \exp(-\beta E_k)$. The mean lifetime τ can be calculated by considering the life histories of kinks and antikinks, created in pairs and then diffusing until colliding and annihilating [83–87]. The diffusivity D is also well understood (Sect. 5.4). If the kink-antikink separation at the time of nucleation, b, is independent of β and much smaller than the typical distance between nearest-neighbour kinks and antikinks, then most annihilation events are, in fact, recombination events between kink-antikink pairs that were nucleated together [9, 83]. Exact expressions exist for this diffusion-limited reaction scenario. The steady-state density in terms of the parameters Γ, D and b is given by [86]

$$\rho_k \to \left(\frac{b\Gamma}{2D}\right)^{1/2} \quad \text{as} \quad \left(\frac{2\Gamma}{D}\right)^{1/3} b \to 0.$$

Thus, because the steady-state density of kinks is proportional to $\exp(-\beta E_k)$, we conclude that the nucleation rate of kink-antikink pairs is proportional to $\exp(-2\beta E_k)$, consistent with intuition. Independently, the appropriate value for the parameter b has been estimated, from first-passage type numerical experiments on small-to-intermediate-sized domains, as $8E_k$ [13]. It is also possible to perform numerical experiments to evaluate the nucleation rate directly, as a function of β, to measure kink lifetimes and to distinguish recombinant and non-recombinant annihilation events [9].

Numerical experiments on large domains and over long times, while maintaining accurate spatial and temporal resolution, were only possible on computer clusters 10–20 years ago [9, 88] but are more accessible to desktop experiments nowadays. An example is given in Fig. 5.3, where numerical counts of numbers of kinks per unit length are compared with the predictions obtained using (5.20). The larger the value of β, the smaller the steady-state number of kinks per unit length, and the longer the equilibration time.

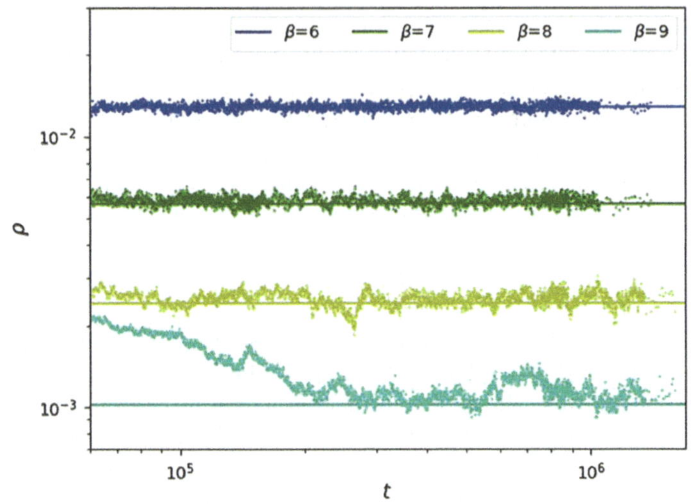

Fig. 5.3 Density of kinks versus time at different values of β. The solid lines are the predictions using (5.20). Dotted lines are numerical solutions with $\alpha = 1$, $\Delta x = 0.4$ and $\Delta t = 0.032$

5.6 Discussion

On the one hand, it is indispensable to perform numerical simulations of a detailed model of a physical system, on the largest domain and with the highest spatial resolution feasible, for as long a time as possible. Such numerical work is complemented by developing theoretical methods that efficiently identify the structures of interest and predict their number, form, dynamics and interactions. Some questions one can ask are: How many coherent structures are there, on average, in the long term? How do they move and interact with their environment and with each other? How strongly do these properties answers depend on parameters such as the temperature? In such circumstances, exact theoretical results are extremely valuable as benchmarks for numerical algorithms.

On the other hand, many theoretical results are approximations whose accuracy increases as the temperature decreases. Unfortunately, low temperatures require large amounts of computer time because the timescales of the system typically increase as temperature decreases (so one must perform runs of increasing duration to sample adequately) and because the density of coherent structures also decreases with decreasing temperature (so larger systems are needed). Often, even a small decrease in temperature results in an order-of-magnitude increase in computing time required. The continuing increase in computing power available year by year, impressive though it is, will not be sufficient to break new ground without the assistance of new theoretical insights.

Understanding of the noisy ϕ^4 equation in multiple space dimensions [7, 89, 90] is advancing. There are surely many interesting dynamical features of coherent structures in two and three space dimensions waiting to be discovered. Systems with multiplicative noise [91] tend to exhibit intermittent behaviours that are challenging for the theorist, numericist and physicist.

Acknowledgements I am grateful for many years of friendship and scientific discussions with Salman Habib and Franz Mertens. I am grateful to the Isaac Newton Institute programme *Stochastic dynamical systems in biology: numerical methods and applications* (2016).

Appendix: Python Code

Python code for counting kinks

```
#  GDL 2018. python 3
# Phi4 SPDE with (additive) spacetime white noise
# Measure <phi> and correlation function.
# Count kinks using smoothing function h(x).
#  example command lines
#   i) to start from t=0:
#   python Count06.py 50000 100000 0.2 0.01 10.0
#   ii) to start from written configuration:
#   python Count06.py Config10_20_10_100.dat 10000
import numpy as np
import os, sys

if len(sys.argv) ==6:
    # start from t=0
    N,tmax = int(sys.argv[1]),float(sys.argv[2])
    dx,dt = float(sys.argv[3]),float(sys.argv[4])
    beta = float(sys.argv[5])
    t = 0.0
    tstart = min(tmax/2,1000)
    phi = np.zeros(N)
else:
    # start from existing file
    startfile = open(sys.argv[1],'r')
    print(len(sys.argv))
    items = sys.argv[1].split('_')
    print(items)
    dx,dt = int(items[2])*0.01,int(items[3])*0.0001
    beta = float(items[0].replace('Config',''))
    t = int(items[4].split('.')[0])*1000.0
    tmax = t + float(sys.argv[2])
    print(dx,dt,beta,t,tmax)
    for line in startfile:
        oldconfig = line.split()
    phi = np.array([float(p) for p in oldconfig])
    N = len(phi)
    tstart = t
```

```
tint = max((tmax/1000),100)
k,fac,nc = 1.0/(dx*dx),np.sqrt(2*dt/(beta*dx)),int(100/dx)

def phicub(ph):
    ''' cube of phi '''
    return ph*ph*ph

def lap2(phi):
    ''' laplacian of phi '''
    return np.append(phi[1:], phi[0]) + \
           np.append(phi[-1], phi[:-1]) - 2*phi

def myra2(N, mysd):
    ''' N draws from gaussian distribution '''
    return np.random.normal(0,mysd,N)

def euler(phi,g,dt,k):
    ''' one Euler step '''
    return phi + dt*(phi-phicub(phi)+k*lap(phi)) + g

def heun2(phi,g,dt,k):
    ''' one Heun step, keep intermediate to avoid recalculation'''
    phit = phi.copy()
    pinc = phi - phicub(phi) + k*lap2(phi)
    phit = phi + dt*(pinc) + g
    phi = phi + dt*(pinc +  phit - phicub(phit) + k*lap2(phit))/2
    return phi + g



def phip(x):
    ''' derivative of kink shape function '''
    return 1.0/(np.cosh(x/np.sqrt(2)))**2

def separation(i,ii):
    ''' modulus N '''
    j = abs(i-ii)
    if j > N/2:
        j = N-j
    return j

def makeh():
    ''' h(x) is smoothed version of phi(x) '''
    phipl = [phip(i*dx) for i in range(N)]
    h = []
    w = int(30/dx) # width for integration
    for i in range(N):
        h.append(sum([phi[ii%N]*phipl[separation(i,ii)] \
                 for ii in range(i-w,i+w)]))
    return [hh*dx*np.sqrt(2)/4 for hh in h]

def cx(phi,i):
    ''' sum of product of phi(x) and phi(x+i) '''
    return np.sum(phi*np.roll(phi,i))

pid = os.getpid()
sdt,sdx = str(int(dt*10000)),str(int(dx*100))
stm,sL = str(int(tmax/1000)),str(int(N*dx/1000))
sfname = 'Phis'+str(int(beta))+'_'+str(pid)+'_'+ \
        sdx+'_'+sdt+'_'+stm+'.dat'
sfile = open(sfname,'w')
```

```
lfname = 'Length'+str(int(beta))+'_'+str(pid)+'_'+ \
         sdx+'_'+sdt+'_'+stm+'_'+sL+'.dat'
lfile = open(lfname,'w')
cfname = 'Config'+str(int(beta))+'_'+str(pid)+'_'+ \
         sdx+'_'+sdt+'_'+stm+'.dat'
cfile = open(cfname,'w')



def writestuff(sum2,num2,sumc,sumhist):
    '''write to files'''
    h = makeh()
    nk = len([i for i in range(N-1) if h[i]*h[i+1]<0])
    print(int(t+0.4),nk,len([i for i in range(N-1) \
                              if phi[i]*phi[i+1]<0]))
    sfile.write('%f\n'%(sum2/num2))
    lfile.write('%.0f %.2f %.2f %i\n' \
                %(t,np.mean(-dxc/np.diff(np.log(sumc))),
             np.std(-dxc/np.diff(np.log(sumc))),nk))
    sum2,num2 = 0.0,0
    sumc,sumhist = np.zeros(len(ilist)),np.zeros(len(thishist))
    ttt = dt/100
    for thisfile in (sfile,lfile,cfile):
        thisfile.flush()

host = os.getenv('HOSTNAME').split('.')[0]
print('N =', N,'tmax =', tmax)
print('dx =', dx,'dt =', dt,'beta =',beta)
sfile.write("# %i %.0f %.4f %.5f %s\n"%(N,tmax,dx,dt,host))
lfile.write("# %i %.0f %.2f %s\n"%(N,tmax,tint,host))
xmin,xmax = 10.0,40.0
di,dxc = int(1.0/dx),int(1.0/dx)*dx
ilist = []
i = int(xmin/dx)+1
while i*dx < xmax:
    ilist.append(i)
    i += di



mybins=np.linspace(-2.0,2.0,401,endpoint=True)
tt,ttt = dt/100,dt/100
sum2,num2 = 0.0,0
sumc,sumhist = np.zeros(len(ilist)),np.zeros(len(mybins)-1)

############# main loop ################################
while t < tmax-dt:
    g = myra2(N,fac)
    phi = heun2(phi,g,dt,k)
    t += dt
    tt += dt
    if tt > tint:
        tt = dt/100
        if t > tstart:
            thissums = np.sum(phi*phi)
            sum2 += thissums/N
            num2 += 1
            for i in range(len(ilist)):
                sumc[i]+=cx(phi,ilist[i])/thissums
            ttt += tint
            thishist=np.histogram(phi,bins=mybins,normed=True)[0]
```

```
            sumhist += thishist
            writestuff(sum2,num2,sumc,sumhist)
        else:
            print(int(t+0.4),len([i for i in range(N-1) if phi[i]*phi[i+1]<0]))
######################################################

for i in range(N):
    cfile.write(' %.3f'%phi[i])
for thisfile in (sfile,lfile,cfile):
    thisfile.close
```

References

1. A.C. Scott, F. Chu, D. McLaughlin, Proc. IEEE **61**, 1443 (1973)
2. A. Bishop, J. Krumhansl, J. Schrieffer, Phys. D **1**, 1 (1980)
3. A. Sánchez, A. Bishop, SIAM Rev. **40**, 579 (1998)
4. M.J. Ward, Stud. Appl. Math. **91**, 51 (1994)
5. T. Dauxois, M. Peyrard, *Physics of Solitons* (Cambridge University Press, Cambridge, 2006)
6. J. García-Ojalvo, J. Sancho, *Noise in Spatially Extended Systems* (Springer Science + Business Media, New York, 1999)
7. M.D. Ryser, N. Nigam, P.F. Tupper, J. Comput. Phys. **231**(6), 2537 (2012)
8. J.C. Mourrat, H. Weber, Commun. Pure Appl. Math. **70**(4), 717 (2017)
9. S. Habib, G. Lythe, Phys. Rev. Lett. **84**, 1070 (2000)
10. G. Lythe, S. Habib, Comp. Phys. Commun. **142**, 29 (2001)
11. G. Lythe, S. Habib, Comput. Sci. Eng. **8**, 10 (2006)
12. L. Bettencourt, S. Habib, G. Lythe, Phys. Rev. D **60**, 105039 (1999)
13. M. Castro, G. Lythe, SIAM, J. Appl. Dyn. Syst. **7**, 207 (2008)
14. T. Shardlow, Electr. J. Differ. Equ. **47**, 1 (2000)
15. R.V. Kohn, F. Otto, M.G. Reznikoff, E. Vanden-Eijnden, Commun. Pure Appl. Math. **LIX**, 0001 (2006)
16. M.A. Katsoulakis, G.T. Kossioris, O. Lakkis, Interfaces Free. Bound **9**, 1 (2007)
17. F. Barret, A. Bovier, S. Méléard et al., Electron. J. Probab. **15**, 323 (2009)
18. G.J. Lord, C.E. Powell, T. Shardlow, *An Introduction to Computational Stochastic PDEs* (Cambridge University Press, Cambridge, 2014)
19. J.B. Walsh, in *Ecole d'été de probabilités de St-Flour XIV, 1984*, vol. 1180, Lecture Notes in Mathematics, ed. by P.L. Hennequin (Springer, Berlin, 1986), p. 265
20. C.R. Doering, Commun. Math. Phys. **109**, 537 (1987)
21. T. Funaki, Nagoya Math. J. **89**, 129 (1983)
22. H. Kunita, *Stochastic Flows and Stochastic Differential Equations* (Cambridge University Press, Cambridge, 1997)
23. G.D. Prato, J. Zabczyk, *Stochastic Equations in Infinite Dimensions* (Cambridge University Press, Cambridge, 1992)
24. C. Prévôt, M. Röckner, *A Concise Course on Stochastic Partial Differential Equations*, vol. 1905, Lecture Notes in Mathematics (Springer, Berlin, 2007)
25. A. Jentzen, P.E. Kloeden, Proc. R. Soc. A **465**, 649 (2009)
26. A. Jentzen, P.E. Kloeden, *Taylor Approximations for Stochastic Partial Differential Equations* (SIAM, Philadelphia, 2011)
27. P.E. Kloeden, G. Lord, A. Neuenkirch, T. Shardlow, J. Comput. Appl. Math. **235**, 1245 (2011)
28. D. Khoshnevisan, *Analysis of Stochastic Partial Differential Equations* (American Mathematical Society, Providence, 2014)
29. W. Faris, G. Jona-Lasinio, J. Phys. A **15**, 3025 (1982)

30. T. Funaki, Probab. Theory Relat. Fields **102**, 221 (1995)
31. I. Gyongy, Potential Anal. **9**, 1 (1998)
32. I. Gyongy, Potential Anal. **11**, 1 (1999)
33. M. Büttiker, R. Landauer, J. Phys. C **13**, L325 (1980)
34. A. Stuart, J. Voss, P. Wiberg, Commun. Math. Sci. **2**, 585 (2004)
35. M.G. Reznikoff, E. Vanden-Eijnden, Comptes Rendus Acad. Sci. **1340**, 305 (2005)
36. H. Weber, Commun. Pure Appl. Math. **63**, 1071 (2010)
37. A. Abdulle, G. Vilmart, K. Zygalakis, SIAM, J. Numer. Anal. **52**, 1600 (2014)
38. M. Tretyakov, Z. Zhang, SIAM J. Numer. Anal. **51**, 3135 (2013)
39. F. Otto, H. Weber, M.G. Westdickenberg et al., Electron. J. Probab. **19**, 1 (2014)
40. A. Seeger, P. Schiller, in *Physical Acoustics: Principles and Methods*, ed. by W.P. Mason (Academic Press, New York, 1966), p. 361
41. D. Scalapino, M. Sears, R. Ferrell, Phys. Rev. B **6**, 3409 (1972)
42. F.J. Alexander, S. Habib, Phys. Rev. Lett. **71**, 955 (1993)
43. G. Parisi, *Statistical Field Theory* (Addison-Wesley, Reading, 1988)
44. M. Creutz, *Quarks, Gluons, and Lattices* (Cambridge University Press, Cambridge, 1983)
45. J. Currie, J. Krumhansl, A. Bishop, S. Trullinger, Phys. Rev. B **22**, 477 (1980)
46. G. Lythe, S. Habib, in *Proceedings of the IUTAM Symposium on Nonlinear Stochastic Dynamics, Illinois, U.S.A., 26–30 August 2002*, vol. 110, Solid Mechanics and its Applications, ed. by N. Namachchivaya, Y.K. Lin (Kluwer, Dordrecht, 2003), p. 435
47. A. Khare, S. Habib, A. Saxena, Phys. Rev. Lett. **79**, 3797 (1997)
48. A. Saxena, S. Habib, Phys. D **107**, 338 (1997)
49. S. Habib, A. Khare, A. Saxena, Phys. D **123**, 341 (1998)
50. W. Gilks, S. Richardson, D. Spiegelhalter (eds.), *Markov Chain Monte Carlo in Practice* (Chapman & Hall, London, 1996)
51. M. Hairer, A. Stuart, J. Voss, P. Wiberg, Commun. Math. Sci. **3**, 587 (2005)
52. A. Apte, M. Hairer, A. Stuart, J. Voss, Phys. D **230**, 50 (2007)
53. R.D. Skeel, J. Numer. Anal. Ind. Appl. Math. **5**, 103 (2010)
54. G. Lythe, F. Mertens, Phys. Rev. E **67**, 027601 (2003)
55. D.K. Campbell, J.F. Schonfeld, C.A. Wingate, Phys. D **9**, 1 (1983)
56. H. Segur, M.D. Kruskal, Phys. Rev. Lett. **58**, 747 (1987)
57. P. Kevrekidis, A. Saxena, A. Bishop, Phys. Rev. E **64**, 026613 (2001)
58. W. Wang, R.D. Skeel, Mol. Phys. **101**, 2149 (2003)
59. H. Schurz, Stoch. Anal. Appl. **17**, 463 (1999)
60. H. Schurz, in *Handbook of Stochastic Analysis and Applications*, ed. by D. Kannan, V. Lakshmikantham (Marcel Dekker, New York, 2002), p. 237
61. B. Leimkuhler, S. Reich, *Simulating Hamiltonian dynamics* (Cambridge University Press, Cambridge, 2004)
62. K. Burrage, G. Lythe, SIAM J. Numer. Anal. **47**, 1601 (2009)
63. J. Voss, Commun. Math. Sci. **10**, 1143 (2012)
64. R. Mannella, Phys. Rev. E **69**, 041107 (2004)
65. R. Mannella, SIAM J. Sci. Comput. **27**, 2121 (2006)
66. K. Burrage, I. Lenane, G. Lythe, SIAM J. Sci. Comput. **29**, 245 (2007)
67. D. Kaup, Phys. Rev. B **27**, 6787 (1983)
68. E. Tomboulis, Phys. Rev. D **12**, 1678 (1975)
69. H. Segur, J. Math. Phys. **24**, 1439 (1983)
70. C.W. Gardiner, *Handbook of Stochastic Methods for Physics, Chemistry and the Natural Sciences*, 3rd edn. (Springer, Berlin, 2004)
71. S. Karlin, H.M. Taylor, *A Second Course in Stochastic Processes* (Academic Press, New York, 1981)
72. Y.S. Kivshar, D.E. Pelinovsky, T. Cretegny, M. Peyrard, Phys. Rev. Lett. **80**, 5032 (1998)
73. M. Büttiker, R. Landauer, Phys. Rev. Lett. **43**, 1453 (1979)
74. F. Barret, Ann. Inst. H. PoincarÃl' Probab. Statist. **51**, 129 (2015)
75. R. Landauer, J. Swanson, Phys. Rev. **121**, 1668 (1961)

76. J. Lothe, J.P. Hirth, Phys. Rev. **115**, 543 (1959)
77. M. Büttiker, R. Landauer, Phys. Rev. A **23**, 1397 (1981)
78. M. Büttiker, T. Christen, Phys. Rev. Lett. **75**, 1895 (1995)
79. R.S. Maier, D.L. Stein, Phys. Rev. Lett. **87**, 270601 (2001)
80. W.E.W. Ren, E. Vanden-Eijnden, Commun. Pure Appl. Math. **LVII**, 0637 (2004)
81. R.V. Kohn, M.G. Reznikoff, Y. Tonegawa, Calc. Var. **25**, 503 (2006)
82. N. Berglund, B. Gentz, J. Phys. A: Math. Theor. **42**, 052001 (2009)
83. T. Christen, M. Büttiker, Phys. Rev. E **58**, 1533 (1998)
84. S. Habib, K. Lindenberg, G. Lythe, C. Molina-París, J. Chem. Phys. **115**, 73 (2001)
85. T.O. Masser, D. ben Avraham, Phys. Rev. E **63**, 006108 (2001)
86. G. Lythe, Phys. D **222**, 159 (2006)
87. H.C. Fogedby, J. Hertz, A. Svane, Europhys. Lett. **62**, 795 (2003)
88. F.J. Alexander, S. Habib, A. Kovner, Phys. Rev. E **48**, 4284 (1993)
89. M. Hairer, M.D. Ryser, H. Weber, Electron. J. Probab. **17**, 1 (2012)
90. J.C. Mourrat, H. Weber, Ann. Probab. **45**, 2398 (2017)
91. D. Khoshnevisan, K. Kim, Y. Xiao, Commun. Math. Phys. **360**, 307 (2018)

Chapter 6
Discrete Variants of the ϕ^4 Model: Exceptional Discretizations, Conservation Laws and Related Topics

Sergey V. Dmitriev and Panayotis G. Kevrekidis

Abstract Exceptional dicretizations of the ϕ^4 model are reviewed, corresponding conservation laws are reported, and the properties of static and moving discrete kinks are discussed. Different approaches to producing such discretizations are given and unifying perspectives thereof are brought forth. It is also demonstrated that the high kink mobility in the exceptional dicretizations makes it possible to analyze kink-antikink collisions in the regime of high discreteness.

6.1 Introduction

Localized nonlinear excitations such as topological solitons and breathers play an important role in many areas of physics and very often they are considered in discrete media. For example, in solid state physics they are used to describe domain walls, dislocations, and crowdions in crystals [1–5]; they also arise in macroscopic models of coupled pendulums [6], in granular crystals [7], in arrays of electric circuits [8] and micromechanical cantilevers [9], among many others. In all these applications the mobility of solitary waves is an important issue, especially given that a typical discretization breaks the translational invariance of the continuum model and thus renders the discrete case far less amenable to genuine traveling dynamics.

From the physical point of view, the analysis of soliton mobility can be done in two different directions. The first is the estimation of the minimal external force needed

S. V. Dmitriev (✉)
Institute for Metals Superplasticity Problems, Russian Academy of Sciences, Ufa 450001, Russia
e-mail: dmitriev.sergey.v@gmail.com

S. V. Dmitriev
National Research Tomsk State University, 36 Lenin Prospekt, 634050 Tomsk, Russia

P. G. Kevrekidis
Department of Mathematics and Statistics, University of Massachusetts, Amherst, MA 01003, USA
e-mail: kevrekid@math.umass.edu

P. G. Kevrekidis and J. Cuevas-Maraver (eds.), *A Dynamical Perspective on the ϕ^4 Model*, Nonlinear Systems and Complexity 26,
https://doi.org/10.1007/978-3-030-11839-6_6

to set a standing soliton in motion, and the second is the estimation of deceleration rate for a moving soliton in the absence of external forces.

Continuum KG equations such as ϕ^4 or sG support kinks that can be accelerated by any small external force, and moving kinks radiate no energy propagating at constant speed given the Lorentz invariance of such KG models (and of their D'Alembertian operators). This is typically not the case for the discrete KG equations, unless an integrable lattice is considered [10]. For standard (generic) discretizations, discrete kinks are found to possess two equilibrium configurations; the one having maximal potential energy is unstable, while the one with minimal energy is stable. The difference between kink energies in these two configurations defines the height of the so-called static Peierls–Nabarro potential (PNp) [11]. The maximal gradient of this potential defines the minimal force needed to accelerate a standing kink. In order to reduce the force required to accelerate a kink one has to reduce the maximal gradient of the static PNp, ideally making it zero. In a series of works by Speight and Ward, about 2 decades ago, one possible way to construct a discrete version of KG field with precisely zero static PNp was proposed [12–14]. Note that the absence of the static PNp does not ensure that a moving kink will propagate at constant velocity [15], which means that solution of the first mobility problem does not necessarily mean that the second problem will be solved automatically.

A moving discrete kink typically (i.e., in the case of generic discretizations) radiates small-amplitude wave packets losing its energy and being initially decelerated and eventually trapped by the lattice in a well of the PNp [16]. There exist numerous attempts to analyze and reduce the radiation from a moving kink [15–31]. The radiation is attributed to the resonance of moving kinks with small-amplitude phonons. It turns out that in the discrete KG systems supporting kinks, for any kink velocity there exists a phonon with the same phase velocity. Excitation of such phonons by a moving kink results in its deceleration. Nevertheless it has been shown that the radiation can vanish for a set of selected kink velocities [15, 16, 20, 22, 24, 25, 31]. There exists an example of a kink in a non-integrable lattice, radiating no energy while moving with an arbitrary speed [18]. Kink solutions with oscillating background (also known as nanopterons) have been found as permanent profile traveling waves moving with constant velocity [24].

In this work we focus on the first mobility problem by describing existing approaches to discretize the KG field in a way such that static kinks are not trapped by the lattice and they can be accelerated by any weak external force. Following [32], such discretizations will be called *exceptional*. The kink mobility will also be discussed, as well as the conservation laws valid for the different exceptional discretizations.

6.2 Klein-Gordon Field with ϕ^4 Potential and Its "Standard" Discretization

The KG model is described by the Hamiltonian

$$H = K + E = \frac{1}{2}\int_{-\infty}^{+\infty} \phi_t^2 \mathrm{d}x + \int_{-\infty}^{+\infty} \left[\frac{1}{2}\phi_x^2 + V(\phi)\right]\mathrm{d}x, \tag{6.1}$$

where the first term represents the kinetic energy, K, and the second one the potential energy, E, with $\phi(x,t)$ being the unknown scalar field of spatial and temporal coordinates x and t, respectively. Differentiation with respect to x and t is denoted by the corresponding subscripts. The function $V(\phi)$ defines the on-site potential. The following equation of motion can be derived from (6.1) using Hamilton's equations:

$$\phi_{tt} + \frac{\mathrm{d}V}{\mathrm{d}\phi} - \phi_{xx} = 0. \tag{6.2}$$

For static problems, the equation of motion (6.2) reduces to

$$\phi_{xx} = \frac{\mathrm{d}V}{\mathrm{d}\phi}. \tag{6.3}$$

The on-site potential of the ϕ^4 model has the form

$$V(\phi) = \frac{1}{2}(1-\phi^2)^2, \tag{6.4}$$

which has two energetically equivalent minima at $\phi = \pm 1$ separated by the potential barrier of height 1/2.

The ϕ^4 equation defined by (6.2) and (6.4) supports the (anti)kink solution

$$\phi(x,t) = \pm \tanh \frac{x - x_0 - vt}{\sqrt{(1-v^2)}}, \tag{6.5}$$

moving with any velocity $-1 < v < 1$ (given the Lorentz invariance) starting at $t = 0$ from $x = x_0$. The kink (antikink) interpolates between two minima of the on-site potential (6.4), from -1 to 1 (from 1 to -1). Kinks and antikinks propagate radiating no energy but their collisions are inelastic since the ϕ^4 equation is non-integrable, as is discussed in some detail in other chapters within this special volume. Due to translational invariance of the continuum nonlinear KG equation, a static kink ($v = 0$) can be placed at any $x = x_0$ and any small external force f in the equation of motion $\phi_{tt} + \mathrm{d}V/\mathrm{d}\phi - \phi_{xx} = f$ will accelerate the kink.

In order to discretize the KG equation, the lattice $x = nh$ is introduced, where $n = 0, \pm 1, \pm 2, \ldots$ and h is the lattice spacing. The most straightforward discretization of the ϕ^4 equation reads [33]

$$\begin{aligned}\ddot{\phi}_n &= \frac{1}{h^2}(\phi_{n-1} - 2\phi_n + \phi_{n+1}) - V'(\phi_n) \\ &= \frac{1}{h^2}(\phi_{n-1} - 2\phi_n + \phi_{n+1}) + 2\phi_n(1 - \phi_n^2),\end{aligned} \tag{6.6}$$

where $\phi_n(t) = \phi(nh, t)$ and overdot means differentiation with respect to time.

This discretization conserves the Hamiltonian (total energy)

$$\begin{aligned}H_1 &= \frac{h}{2}\sum_n \left[\dot{\phi}_n^2 + \left(\frac{\phi_{n+1} - \phi_n}{h}\right)^2 + V(\phi_n)\right] \\ &= \frac{h}{2}\sum_n \left[\dot{\phi}_n^2 + \left(\frac{\phi_{n+1} - \phi_n}{h}\right)^2 + (1 - \phi_n^2)^2\right].\end{aligned} \tag{6.7}$$

Model (6.6) does not support an exact moving kink solution. As for the static kinks, they exist only for the two symmetric configurations, centered either on a lattice site (unstable) or in the middle between two neighboring sites (stable). Such static solutions can be found numerically, e.g., by minimizing the potential energy of the solution with the use of the steepest gradient method, or by using fixed point iterations in the steady state form of (6.6). As a first approximation, one can use the following *approximate* static kink solution

$$\phi_n = \pm \tanh[h(n - x_0)]. \tag{6.8}$$

The minimization for $x_0 = 0$ ($x_0 = 1/2$) will produce the on-site (inter-site) kink. In Fig. 6.1a equilibrium kink profiles are shown for relatively high discreteness parameter of $h = 0.8$.

It is instructive to study small-amplitude vibrations of the lattice with a static kink. For this we introduce the ansatz $\phi_n(t) = \phi_n^0 + \varepsilon_n(t)$ (where ϕ_n^0 is an equilibrium kink solution and $\varepsilon_n(t)$ is a small perturbation), and linearize (6.6) with respect to ε_n. The result is

$$\ddot{\varepsilon}_n = \Delta_2 \varepsilon_n + 2\varepsilon_n - 6(\phi_n^0)^2 \varepsilon_n. \tag{6.9}$$

Here, $\Delta_2 \varepsilon_n = (1/h^2)(\varepsilon_{n+1} + \varepsilon_{n-1} - 2\varepsilon_n)$ stands for the discrete Laplacian. For the small-amplitude phonon waves, $\varepsilon_n = \exp(ikn - i\omega t)$, where ω is the frequency and k is the wave number, from (6.9) one finds the following dispersion relation:

$$\omega^2 = \frac{4}{h^2}\sin^2\left(\frac{k}{2}\right) - 2 + 6(\phi_n^0)^2. \tag{6.10}$$

From (6.10), the vacuum solution $\phi_n^0 = \pm 1$ has the spectrum

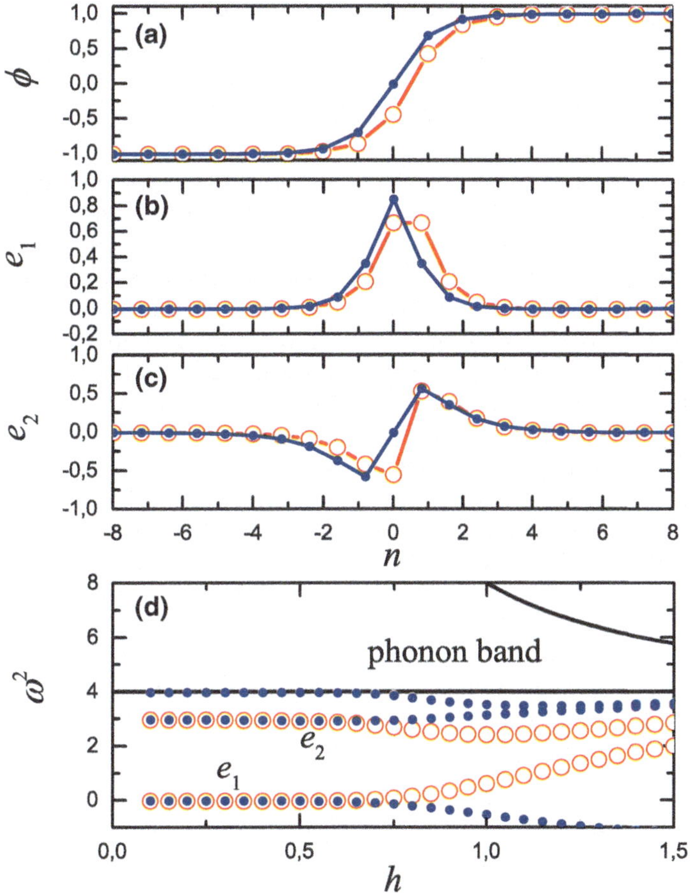

Fig. 6.1 Standard model defined by (6.6), (6.7): **a** Kink profile, **b** Goldstone (lowest frequency internal) mode associated with translation in the continuum limit, and **c** kink vibrational mode for $h = 0.8$. **d** Frequencies of the kink's internal modes for different magnitudes of the discreteness parameter h. Two solid lines show the borders of the spectrum of vacuum, (6.11). Results for the on-site and inter-site kinks are shown by dots and circles, respectively. The inter-site kink is stable, while the on-site kink is unstable because the vibrational spectrum contains one pair of imaginary frequencies

$$\omega^2 = 4 + \frac{4}{h^2} \sin^2\left(\frac{k}{2}\right). \tag{6.11}$$

We solve the eigenproblem (6.10) numerically for the chain of $N = 201$ particles with a static kink in the middle of the chain. The solution gives N eigenfrequencies ω_n^2 and N eigenvectors e_n. Most of the eigenmodes have frequencies coinciding with the phonon spectrum of vacuum, (6.11), as expected by Weyl's spectral theorem. There exist a few eigenmodes localized around the kink with frequencies below the phonon spectrum of vacuum. In Fig. 6.1b, c we show the two lowest frequency eigenmodes, e_1 and e_2, localized around the kink (dots for the on-site and circles for the inter-site kink). In Fig. 6.1d the upper and lower edges of the phonon spectrum of vacuum are shown by the solid lines as a function of the lattice spacing h. Dots and circles show the frequencies of the eigenmodes localized around the on-site and inter-site kink, respectively. The mode e_1 in (b) corresponds to the kink translational mode (associated with a symmetry and a so-called Goldstone mode corresponding to zero frequency in the continuum limit) and for small h it deviates exponentially weakly from the limit value, i.e., as $e^{-C/h}$ [34] for a suitable constant C. The lowest squared eigenfrequency of the on-site mode is negative, which becomes clear for increasing h in Fig. 6.1d. This indicates the instability of the on-site kink configuration. For the inter-site kink, the (positive) lowest squared frequency of this mode increases with h [34]. This frequency is the kink oscillation frequency in the well of the PNp. The mode e_2 in (c) is the well-known kink's internal mode which in the continuum ϕ^4 equation has frequency $\omega = \sqrt{3}$ [33, 35–38].

A direct calculation using the kink ansatz (cf. (6.8)) inside the energy of (6.7) shows that the energy for the discrete ϕ^4 kink in the PNp can be approximated by

$$H_0 = \frac{1}{2}M\dot{X}^2 + \frac{4}{3} + \alpha(h)\cos\left(\frac{2\pi X}{h}\right), \quad M = \frac{4}{3}, \quad \alpha(h) = \frac{8\pi^2}{3h^3}\frac{h^2+\pi^2}{\sinh(\pi^2/h)}, \tag{6.12}$$

where X and M are the coordinate of the center of mass and the mass of the kink, respectively. From this, one finds

$$E_0 = 2\alpha, \quad F_0 = \alpha\frac{2\pi}{h}, \quad \omega_0 = \frac{\pi}{h}\sqrt{3\alpha}, \tag{6.13}$$

which are the height of PNp, the maximal gradient of PNp, and the frequency of small-amplitude oscillations of the kink near the well of PNp, respectively.

For $h = 0.8$ from (6.12) and (6.13) we have $\alpha = 4.74 \times 10^{-3}$, $E_0 = 9.48 \times 10^{-3}$, $F_0 = 0.0372$, and $\omega_0 = 0.468$. Numerically we find that for $h = 0.8$ the on-site (inter-site) kink has potential energy 1.3058 (1.2973). The difference gives the height of PNp of 8.5×10^{-3}, which is only 10% smaller than the estimated value. With decreasing h the accuracy of the formula (6.12) rapidly increases. For example, for $h = 0.7$ the estimation (6.13) gives $E_0 = 2.39 \times 10^{-3}$, while the numerically found value is 2.33×10^{-3}, which is just 2.5% smaller. Numerical evaluation of the small-amplitude vibration frequency of the kink in the well of PNp for $h = 0.8$ gives the value of 0.435, which differs by 7% from the above estimation. For $h = 0.7$ the error of the frequency estimation reduces to 3.7%.

Let us estimate numerically the minimal external force needed for a kink to overcome the PNp. To the RHS of equation of motion (6.6) we add the external force f:

$$\ddot{\phi}_n = \frac{1}{h^2}(\phi_{n-1} - 2\phi_n + \phi_{n+1}) + 2\phi_n(1 - \phi_n^2) + f, \tag{6.14}$$

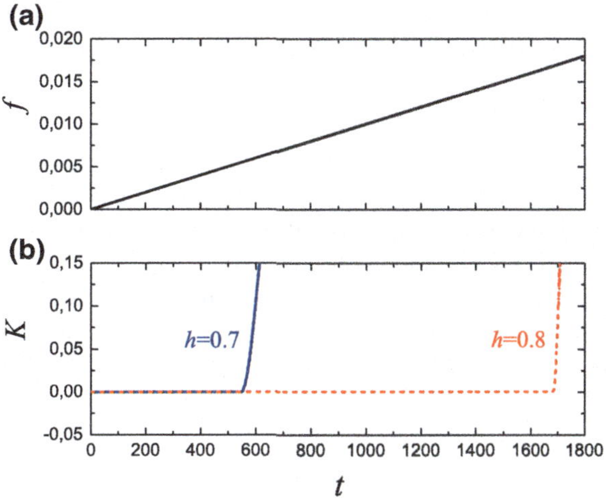

Fig. 6.2 Results for the classical model (6.6). **a** External force as a function of time. **b** Kink kinetic energy as the function of time. Results for $h = 0.7$ (solid line) and $h = 0.8$ (dashed line)

assuming that the external force increases linearly with time, $f = f_0 t$. To achieve a quasi-static loading we use a small value of $f_0 = 10^{-5}$. As initial condition, we use the equilibrium inter-site kink solution.

In Fig. 6.2a $f(t)$ is shown, while in (b) the kink kinetic energy is plotted as a function of time for the discreteness parameter $h = 0.7$ (solid line) and $h = 0.8$ (dashed line). While the kink is trapped in a well of the PNp, its kinetic energy is nearly zero. For $h = 0.8$, the kink starts to move at $t = 1690$ when the external force reaches the value of $f = 0.0169$. The value $2f = 0.0338$ gives the numerical estimation of the maximal gradient of PNp, which was found to be $F_0 = 0.0372$ from (6.13). The difference is 9% and for $h = 0.7$ the difference reduces to 3%. This agreement is exceptionally good, especially taking into consideration the non-stationary nature of the process.

From the above results, we have seen that if the ϕ^4 field is discretized as given by (6.6), a stable static kink solution exists only for the inter-site configuration and this kink is trapped by the lattice since the translational Goldstone (neutral) mode is destroyed. To set a standing kink in motion, an external force above a threshold value can be applied; see also [39, 40]. The effects of discreteness are rather small for $h < 0.5$ and they grow in the above mentioned exponential functional form with increasing h.

In the following section, the existing ways of obtaining exceptional discretizations with kinks not trapped by the lattice will be described.

6.3 Approaches to Derive Discrete ϕ^4 Models Free of the Static Peierls–Nabarro Potential

There are two main approaches to the derivation of exceptional discretizations of the KG field equation. The first concerns the discretization of the Hamiltonian (6.1) and derivation of the corresponding discrete equations of motion with the use of the well-known (Hamilton equation based) technique. In this case the energy-conserving models are obtained. The alternative way involves the discretization of the equation of motion (6.2) directly. The resulting discrete equations can be Hamiltonian or non-Hamiltonian (not conserving energy). On the other hand, such discrete models can be constructed so as to conserve momentum in the form [41]

$$P = \sum_n \dot{\phi}_n (\phi_{n+1} - \phi_{n-1}), \tag{6.15}$$

or in a different form.

The standard discretization of the ϕ^4 equation in the form of (6.6) employs a local discretization of the on-site potential (i.e., only the nth site is involved). In most of the exceptional models, $V(\phi)$ or $V'(\phi)$ will be discretized in a way involving three neighboring sites, so that the general form of the discretization will be

$$\ddot{\phi}_n = \frac{1}{h^2}(\phi_{n-1} - 2\phi_n + \phi_{n+1}) + F(\phi_{n-1}, \phi_n, \phi_{n+1}). \tag{6.16}$$

However, there will be an example of a Hamiltonian exceptional discretization with $F(\phi_{n-1}, \phi_n, \phi_{n+1}) = F(\phi_n)$.

6.3.1 *Speight-Ward Hamiltonian Model*

The potential energy of the ϕ^4 field (see (6.1) and (6.4)) can be written as

$$E = \int_{-\infty}^{+\infty} \left[\frac{1}{2}\phi_x^2 + \frac{1}{2}(1-\phi^2)^2\right] dx =$$

$$\int_{-\infty}^{+\infty} \frac{1}{2}[\phi_x - (1-\phi^2)]^2 dx + \int_{-\infty}^{+\infty} (1-\phi^2)\phi_x dx \equiv \tilde{E} + \frac{4}{3}, \tag{6.17}$$

where the second integral was calculated as follows $\int_{-1}^{1}(1-\phi^2)d\phi = 4/3$, taking into account the kink asymptotics $\phi \to -1$ as $x \to -\infty$ and $\phi \to 1$ as $x \to \infty$. Since the first integral has non-negative integrand, the minimal possible value of the potential energy is $E = 4/3$ and this happens when

$$\phi_x = 1 - \phi^2 \;\Rightarrow\; \phi = \tanh(x - x_0). \tag{6.18}$$

This fact that the minimal energy for the solution satisfying the kink asymptotics is actually the kink solution is often referred to as a Bogomol'nyi bound [42].

The idea of applying the Bogomol'nyi bound to the discrete case was proposed by Speight and Ward in a series of works [12–14]. They assumed that the on-site potential can be discretized using three neighboring nodes and considered a potential energy of the form

$$E = h \sum_n \left[\frac{1}{2} \left(\frac{\phi_{n+1} - \phi_n}{h} \right)^2 + \frac{1}{2} F^2 \right], \tag{6.19}$$

where the on-site potential (second term) is taken in the form that one can write

$$E = h \sum_n \frac{1}{2} \left(\frac{\phi_{n+1} - \phi_n}{h} - F \right)^2 + \sum_n (\phi_{n+1} - \phi_n) F. \tag{6.20}$$

If F is chosen in a way that the second sum in (6.20) is telescopic, then the Bogomol'nyi bound can be achieved via evaluating the sum at $n \to \pm\infty$. For the ϕ^4 model it was suggested to take $F = 1 - (\phi_{n+1}^2 + \phi_{n+1}\phi_n + \phi_n^2)/3$ because for this choice

$$\sum_n (\phi_{n+1} - \phi_n) F = \sum_n \left[(\phi_{n+1} - \phi_n) - \frac{1}{3} (\phi_{n+1}^3 - \phi_n^3) \right] = 2 - \frac{2}{3} = \frac{4}{3}. \tag{6.21}$$

In this case again the minimum energy configuration is the kink (antikink) which can be found from the two-point map

$$u \equiv \pm \frac{\phi_{n+1} - \phi_n}{h} - 1 + \frac{\phi_n^2 + \phi_n \phi_{n+1} + \phi_{n+1}^2}{3} = 0. \tag{6.22}$$

The Hamiltonian of this exceptional discretization has the form:

$$H_2 = \frac{h}{2} \sum_n \left(\dot{\phi}_n^2 + u^2 \right). \tag{6.23}$$

The corresponding equation of motion is

$$\begin{aligned} \ddot{\phi}_n &= \left(\frac{1}{h^2} + \frac{1}{3} \right) (\phi_{n-1} - 2\phi_n + \phi_{n+1}) + 2\phi_n \\ &- \frac{1}{9} \left[2\phi_n^3 + (\phi_n + \phi_{n-1})^3 + (\phi_n + \phi_{n+1})^3 \right]. \end{aligned} \tag{6.24}$$

The static kink solution can be found iteratively from the quadratic equation (6.22)

$$\phi_{n\pm1} = -\frac{\phi_n}{2} \mp \frac{3}{2h} \pm \frac{\sqrt{3}}{2} \sqrt{-\phi_n^2 \pm \frac{6}{h}\phi_n + \frac{3}{h^2} + 4}, \tag{6.25}$$

where one can take either upper or lower signs. The on-site kink can be obtained starting from the initial value $\phi_n = 0$, while for the inter-site kink the initial value is $\phi_n = 3/h - \sqrt{3+9/h^2}$ (given its placement around the origin, i.e., its anti-symmetry). A kink placed arbitrarily with respect to the lattice can be found starting from any initial value $|\phi_n| < 1$, and all such kinks will have exactly the same potential energy. This means that in the model (6.24) static kinks do not experience PNp.

Linearization around an equilibrium solution, ϕ_n^0, yields:

$$\ddot{\varepsilon}_n = \Delta_2\varepsilon_n + \frac{1}{3}\Big[1 - (\phi_n^0 + \phi_{n-1}^0)^2\Big]\varepsilon_{n-1} + \frac{1}{3}\Big[1 - (\phi_n^0 + \phi_{n+1}^0)^2\Big]\varepsilon_{n+1}$$
$$+ \frac{1}{3}\Big[4 - 2\phi_n^{02} - (\phi_n^0 + \phi_{n-1}^0)^2 - (\phi_n^0 + \phi_{n+1}^0)^2\Big]\varepsilon_n. \qquad (6.26)$$

Substituting here the ansatz $\varepsilon_n = \exp(ikn - i\omega t)$ one obtains the eigenvalue problem to find eigenfrequencies and eigenmodes of small-amplitude oscillations around the equilibrium solution ϕ_n^0. For the spectrum of the vacuum solution, $\phi_n^0 = \pm 1$, one finds

$$\omega^2 = 4 + 4\frac{1-h^2}{h^2}\sin^2\left(\frac{k}{2}\right). \qquad (6.27)$$

In Fig. 6.3 we plot the same as in Fig. 6.1, but for the model given by (6.24), taking the same, relatively large value of the discreteness parameter, $h = 0.8$. The results for the on-site and inter-site kinks are plotted by dots and circles, respectively. In (a) the kink profiles are shown. In (b) and (c) the two lowest frequency modes localized around the kinks are shown. The mode e_1 is the zero frequency Goldstone mode. The mode e_2, as before, represents the kink's internal vibration mode. In (d) solid lines give the lower and upper edges of the phonon band (6.27) as a function of the discreteness parameter. The symbols represent the frequencies of the modes localized around the kink. The presence of the zero-frequency translational mode for any h suggests that the kinks in model (6.24) are not trapped by the lattice and that the PNp is precisely zero.

PNp-free kinks are accelerated by any weak external force. This is illustrated in Fig. 6.4, which should be compared to Fig. 6.2b plotted for the standard discrete model (6.6). Now we apply external force f constant in time adding it to the RHS of (6.24). The kink kinetic energy as a function of time is presented in log-log scale. The results for $h = 0.7$ (solid line) and $h = 0.8$ (dashed line) practically coincide. The dotted line shows the slope equal to 2. It can be seen that $K \sim t^2$, which was expected for the uniformly accelerated kink motion with kink velocity $V_{\mathrm{K}} \sim t$ and kink kinetic energy $K \sim V_{\mathrm{K}}^2$.

The PNp-free kinks can be boosted with the help of the Goldstone mode. Suppose the equation of motion (6.24) is integrated numerically with the time step Δt. The initial conditions can be taken as follows. At $t = 0$ we set $\phi_n = \phi_n^0$, where ϕ_n^0 is the static kink solution, and at $t = \Delta t$ the Goldstone mode is added, $\phi_n = \phi_n^0 + s(e_1)_n$, with a small coefficient s which defines the speed of the boosted kink. It is, of course,

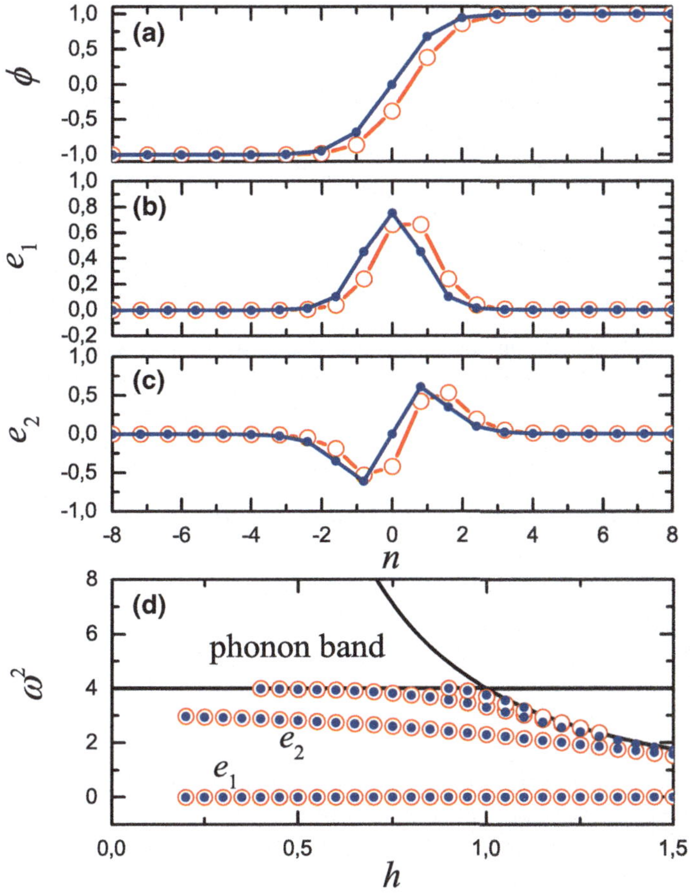

Fig. 6.3 Model (6.24): **a** Kink profile, **b** Goldstone mode, and **c** kink vibrational mode for $h = 0.8$. **d** Frequencies of the kink's internal modes for different magnitudes of the discreteness parameter h. Two solid lines show the borders of the spectrum of vacuum, (6.11). Results for the on-site and inter-site kinks are shown by dots and circles, respectively. For any h the kinks possess the zero frequency translational Goldstone mode in contrast to the kinks in standard model (6.6), see Fig. 6.1. This means that in model (6.24) the kinks are not trapped by the lattice and the PNp for them is precisely zero

possible to excite simultaneously the translational Goldstone and kink's internal vibrational mode by using at $t = \Delta t$: $\phi_n = \phi_n^0 + s(e_1)_n + v(e_2)_n$, with a small value of v, which defines the internal mode amplitude.

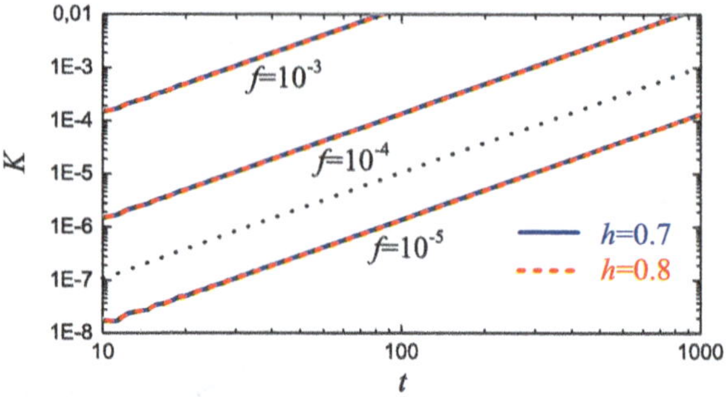

Fig. 6.4 Results for the Speight-Ward exceptional discretization (6.24). Kink kinetic energy as a function of time for constant in time external force f, as specified for each curve. The discreteness parameter is $h = 0.7$ (solid line) and $h = 0.8$ (dashed line). The dotted line shows the slope equal to 2

6.3.2 *Momentum Conserving Discretizations*

Two different approaches to the derivation of momentum-conserving models will now be described.

6.3.2.1 First Approach

In the previous section it was shown, following the works [12–14], that the telescopic summation was essential in the construction of the exceptional discretization. A different class of exceptional discretizations was offered in [41] and again the telescopic summation was successfully used. This discretization is constructed not for the Hamiltonian but for the equation of motion. This approach leads to the model conserving momentum (6.15), which is a discrete version of the continuum momentum $P = \int \phi_t \phi_x \mathrm{d}x$; note that this expression is often by convention used with a $(-)$ sign in front of the integral.

Conservation of momentum implies $\mathrm{d}P/\mathrm{d}t = 0$. Differentiation of (6.15) with respect to time gives

$$\frac{\mathrm{d}P}{\mathrm{d}t} = \sum_n \left[\ddot{\phi}_n(\phi_{n+1} - \phi_{n-1}) + \dot{\phi}_n(\dot{\phi}_{n+1} - \dot{\phi}_{n-1})\right] = 0. \tag{6.28}$$

Note that due to the telescopic summation the second sum is zero. Let us see what happens when the standard discretization of the ϕ^4 field (6.6) is substituted in (6.28). One obtains

$$\frac{dP}{dt} = \sum_n \left[\frac{1}{h^2}(\phi_{n+1}^2 - \phi_{n-1}^2) - \left(\frac{2}{h^2} - 2\right)(\phi_n\phi_{n+1} - \phi_n\phi_{n-1}) - 2(\phi_n^3\phi_{n+1} - \phi_n^3\phi_{n-1}) \right] \neq 0. \quad (6.29)$$

The first two sums vanish due to their telescopic nature. For the first sum this becomes clear after rewriting it in the form $[(\phi_{n+1}^2 + \phi_n^2) - (\phi_n^2 + \phi_{n-1}^2)]/h^2$. However the third term does not disappear and we conclude that the standard discretization destroys the conservation of momentum.

According to the work [41], the conservation of momentum can be achieved if the cubic term is discretized as $\phi^3 \to \phi_n^2(\phi_{n+1} + \phi_{n-1})/2$. The resulting exceptional discretization reads

$$\ddot{\phi}_n = \frac{1}{h^2}(\phi_{n-1} - 2\phi_n + \phi_{n+1}) + 2\phi_n - \phi_n^2(\phi_{n+1} + \phi_{n-1}). \quad (6.30)$$

The momentum is conserved because multiplying $\phi_n^2(\phi_{n+1} + \phi_{n-1})/2$ by $(\phi_{n+1} - \phi_{n-1})/2$ one gets $\phi_n^2(\phi_{n+1}^2 - \phi_{n-1}^2)/4$, which is telescopic under summation and disappears, hence $dP/dt = 0$.

The model (6.30) conserves momentum (6.15) but this is a non-Hamiltonian model and does not have a conserved energy. It was shown that standard nearest-neighbour discretizations of KG models cannot conserve energy and linear momentum simultaneously [43].

6.3.2.2 Second Approach

Another approach based on the methods of non-holonomic mechanics has been developed in [44]. Let us consider the Lagrangian of discretized KG equation in the form

$$L = \sum_n \left[\frac{1}{2}\dot{\phi}_n^2 - \left(\frac{1}{2h^2}(\phi_{n+1} - \phi_n)^2 + V(\phi_n) \right) \right]. \quad (6.31)$$

In order to enforce the conservation law (6.15) the following constraint in velocities is applied

$$\sum_n a_n(\boldsymbol{\phi})\dot{\phi}_n = b(\boldsymbol{\phi}) \quad \text{with} \quad a_n(\boldsymbol{\phi}) = \phi_{n+1} - \phi_{n-1}, \quad b(\boldsymbol{\phi}) = P, \quad (6.32)$$

where the notation $\boldsymbol{\phi} = (\phi_1, \phi_2, \ldots \phi_n)$ is introduced. This constraint, linear in velocity, is non-holonomic because it essentially depends on velocities and it cannot be integrated to contain only coordinates. The dynamics of the system under this type of constraints can be analyzed with the use of the Lagrange-d'Alembert principle [45, 46]. For the constraints of the type (6.32), the variations $\delta\phi_n$ are not arbitrary, but must satisfy the linear relations

$$\sum_n a_n(\boldsymbol{\phi})\delta\phi_n = \sum_n (\phi_{n+1} - \phi_{n-1})\delta\phi_n = 0. \tag{6.33}$$

If r is the Largange multiplier introduced to enforce the constraint (6.33), the critical action principle respecting the constraint gives

$$\int_{t_0}^{t_1} \left[\frac{\mathrm{d}}{\mathrm{d}t}\frac{\partial L}{\partial \dot{\phi}_n} - \frac{1}{h^2}(\phi_{n+1} + \phi_{n-1} - \phi_n) + V'(\phi_n) - r(\phi_{n+1} - \phi_{n-1}) \right] \delta\phi_n = 0, \tag{6.34}$$

provided $\delta\phi_n$ vanish at $t = t_{0,1}$ so that no boundary integration terms appear in the equations. Taking into account that $\delta\phi_n$ is arbitrary, the equations of motion for the model conserving momentum are

$$\ddot{\phi}_n = \frac{1}{h^2}(\phi_{n-1} - 2\phi_n + \phi_{n+1}) - V'(\phi_n) + r(\phi_{n+1} - \phi_{n-1}). \tag{6.35}$$

Note the appearance of the new term proportional to r on the RHS of (6.35), as compared to (6.6). The physical meaning of r is the constraint force that is chosen to ensure momentum conservation. One can find r by introducing (6.35) into (6.28). The linear terms vanish telescopically and the nonlinear term disappears when

$$r = \frac{\sum_n V'(\phi_n)(\phi_{n+1} - \phi_{n-1})}{\sum_n (\phi_{n+1} - \phi_{n-1})^2}. \tag{6.36}$$

We need to demonstrate that the last term in (6.35) vanishes in the continuum limit $h \to 0$. The continuum analog of $r(\phi_{n+1} - \phi_{n-1})$ is $\phi_x(\int V'(\phi)\phi_x \mathrm{d}x)/\int \phi_x^2 \mathrm{d}x$. The integral in the numerator can be integrated to give $V(\phi_2) - V(\phi_1)$ which is zero for the KG field supporting static kinks with the asymptotics ϕ_1 for $x \to -\infty$ and ϕ_2 for $x \to \infty$.

In the work [44] kink mobility was analyzed for model (6.35) and for the standard discretization (6.6). It was shown that even in the regime of high discreteness, where the standard discretization produces kinks that are trapped by PNp, kinks in model (6.35) are mobile.

Another remark is that model (6.35) does not conserve energy. Multiplying both sides of (6.35) with r defined in (6.36) by $\dot{\phi}_n$ and summing we find that instead of $\mathrm{d}H/\mathrm{d}t = 0$, which was true for the Hamiltonian (6.7), now the following holds [44]

$$\frac{\mathrm{d}H}{\mathrm{d}t} = rP. \tag{6.37}$$

6.3.3 Discretized First Integral Approach

We saw that the static ϕ^4 kink solutions for the exceptional discretizations derived in Sects. 6.3.1 and 6.3.2 can be found from a two-point map starting from any initial value $-1 < \phi_n < 1$. With the use of the discretized first integral (DFI) approach [47] we will obtain the map from which not only kinks but a one-parameter family of static solutions can be obtained.

We consider the static KG equation (6.3) and denote

$$D(\phi(x)) \equiv \phi_{xx} - V'(\phi) = 0\,. \tag{6.38}$$

The first integral of (6.38) is

$$U(x) \equiv \phi_x^2 - 2V(\phi) + C = 0\,, \tag{6.39}$$

where C is the integration constant. The first integral can also be taken in a modified form, e.g., as

$$u(x) \equiv \pm\phi_x - \sqrt{2V(\phi) - C} = 0\,. \tag{6.40}$$

We will consider discrete versions of (6.39) and (6.40) in the form

$$U(h, \phi_{n-1}, \phi_n) \equiv \frac{(\phi_n - \phi_{n-1})^2}{h^2} - 2V(\phi_{n-1}, \phi_n) + C = 0\,, \tag{6.41}$$

and

$$u(h, \phi_{n-1}, \phi_n) \equiv \pm\frac{\phi_n - \phi_{n-1}}{h} - \sqrt{2V(\phi_{n-1}, \phi_n) - C} = 0\,, \tag{6.42}$$

respectively, where we demand that $V(\phi_{n-1}, \phi_n) \to V(\phi)$ when $h \to 0$.

6.3.3.1 Momentum Conserving Model

Considering the first integral of the form (6.39) we calculate $\mathrm{d}U/\mathrm{d}x$ and multiply the result by $(\mathrm{d}x/\mathrm{d}\phi)/2$ to obtain

$$\frac{1}{2}\frac{\mathrm{d}U}{\mathrm{d}\phi} = D\,. \tag{6.43}$$

Substituting the LHS of (6.43) with its discrete version and restoring the dynamical term we obtain the discrete model after [41] which conserves momentum (6.15):

$$\ddot{\phi}_n = \frac{U(h, \phi_n, \phi_{n+1}) - U(h, \phi_{n-1}, \phi_n)}{\phi_{n+1} - \phi_{n-1}}\,. \tag{6.44}$$

Obviously, static solutions of this model can be found from the two-point map (6.41) which contains a free parameter C.

Particularly, considering the discrete version of the ϕ^4 potential (6.4) in the form

$$V(\phi_{n-1}, \phi_n) = \frac{1}{2}(1 - \phi_{n-1}\phi_n)^2, \tag{6.45}$$

from (6.41) and (6.44) we get the equation of motion of model (6.30) which does not include the integration constant C. All static solutions of model (6.30) can be found from the two-point map defined by (6.41) and (6.45). This map is a quadratic equation with the roots:

$$\phi_n = \frac{(1-h^2)\phi_{n-1} \pm \sqrt{\mathcal{D}}/2}{1 - h^2\phi_{n-1}^2}, \quad \mathcal{D} = 4h^2(1-\phi_{n-1}^2)^2 + 4h^2(h^2\phi_{n-1}^2 - 1)C, \tag{6.46}$$

where, due to the symmetry of the equation, one can interchange ϕ_n and ϕ_{n-1}.

For any chosen value of the integration constant C, starting from any initial value of ϕ_n for which $\mathcal{D} \geq 0$ and $h^2\phi_{n-1}^2 \neq 1$, one can obtain from (6.46) a static solution of model (6.30). The kink solution is found for $C = 0$ and any $-1 < \phi_n < 1$. Solutions for other values of C were analyzed in [47]. It was shown that those solutions can be expressed in terms of the Jacobi elliptic functions.

6.3.3.2 Energy Conserving Models

We now modify (6.44) as follows

$$\ddot{\phi}_n = e(h, \phi_n)\frac{U(h, \phi_n, \phi_{n+1}) - U(h, \phi_{n-1}, \phi_n)}{\phi_{n+1} - \phi_{n-1}}, \tag{6.47}$$

assuming that $e(h, \phi_n)$ is a continuous function that never vanishes and $e(h, \phi_n) \to 1$ in the continuum limit ($h \to 0$). If so, this multiplier will not affect the static solutions of this exceptional discretization (6.44) and will not change its continuum limit. For the particular choice $e(h, \phi_n) = 1/(1 - h^2\phi_n^2)$ one can obtain from (6.47), (6.41) and (6.45) the following model discovered in [48]

$$\ddot{\phi}_n = \frac{1}{h^2}(\phi_{n-1} - 2\phi_n + \phi_{n+1}) + \frac{2(\phi_n - \phi_n^3)}{1 - h^2\phi_n^2}, \tag{6.48}$$

which does not depend on C and possesses the Hamiltonian

$$H_4 = \frac{h}{2}\sum_n \left[\dot{\phi}_n^2 + \frac{(\phi_n - \phi_{n-1})^2}{h^2} + V(\phi_n)\right], \tag{6.49}$$

with the potential

$$V(\phi_n) = -\frac{1}{h^2}\left(\phi_n^2 + \frac{1-h^2}{h^2}\ln\left|1-h^2\phi_n^2\right|\right). \tag{6.50}$$

This energy-conserving exceptional discretization is of particular interest because here the on-site potential is discretized locally. Nevertheless, so far there does not appear to exist a general approach to the derivation of exceptional discretizations having this property.

Static solutions of model (6.48) coincide with those of momentum-conserving model (6.30) and can be found iteratively with the help of (6.46). On the other hand, dynamic properties of models (6.30) and (6.48) are different. Stability of static solutions of model (6.30) and model (6.48) was analyzed in [47]. The kink solution is stable in both models, yet other solutions can, in principle, be stable in one model and unstable in the other.

The ϕ^4 models (6.30) and (6.48) do not include the integration constant C even though their DFI does. Generally speaking, the DFI approach produces discrete models that include C. Let us demonstrate this for the Speight-Ward model [12–14]. We rewrite the potential energy of the Hamiltonian (6.1) with the use of (6.40) in the following form

$$E = \int_{-\infty}^{+\infty}\left[[u(x)]^2 + 2\phi_x\sqrt{2V(\phi)-C}\right]\mathrm{d}x, \tag{6.51}$$

where the constant term was omitted. Discretizing this potential energy and adding the kinetic energy term we obtain the Hamiltonian

$$H = \frac{h}{2}\sum_n\left[\dot{\phi}_n^2 + [u(h,\phi_{n-1},\phi_n)]^2 + 2\frac{\phi_n-\phi_{n-1}}{h}\sqrt{2V(\phi_{n-1},\phi_n)-C}\right]. \tag{6.52}$$

The on-site potential is discretized following the works [12–14] in the form

$$\sqrt{2V(\phi_{n-1},\phi_n)-C} = \frac{G(\phi_n)-G(\phi_{n-1})}{\phi_n-\phi_{n-1}},$$
$$\text{where}\quad G'(\phi) = \sqrt{2V(\phi)-C}. \tag{6.53}$$

With this choice the last term of (6.52) disappears in telescopic summation. The exceptional discretization will have the Hamiltonian

$$\tilde{H} = \frac{h}{2}\sum_n\left[\dot{\phi}_n^2 + [\tilde{u}(h,\phi_{n-1},\phi_n)]^2\right], \tag{6.54}$$

where according to (6.42) and (6.53)

$$\tilde{u}(h,\phi_{n-1},\phi_n) = \frac{\phi_n-\phi_{n-1}}{h} - \frac{G(\phi_n)-G(\phi_{n-1})}{\phi_n-\phi_{n-1}}. \tag{6.55}$$

This model and DFI $\tilde{u}(h, \phi_{n-1}, \phi_n) = 0$ include the integration constant C through the function G.

6.3.3.3 More Examples of Exceptional Discrete ϕ^4 Models

The following intriguing PNp-free model was derived by Barashenkov et al. [32]

$$\ddot{\phi}_n = \frac{1}{h^2}(\phi_{n-1} - 2\phi_n + \phi_{n+1}) + 2\phi_n - 2\phi_{n-1}\phi_n\phi_{n+1}\,, \tag{6.56}$$

and it was subsequently found [49] that it conserves momentum in the form

$$P = \sum_n \dot{\phi}_n(\phi_{n+2} - \phi_{n-2}), \tag{6.57}$$

which is different from (6.15).

All static solutions of (6.56) can be found from the following DFI [49]

$$\phi_{n\pm1} = (2 - \Lambda)\frac{Z\phi_n \pm \sqrt{\mathcal{D}}}{2 - \Lambda - \Lambda\phi_n^2}, \tag{6.58}$$

where

$$\mathcal{D} = \frac{\Lambda}{2 - \Lambda}(1 - X\phi_n^2 + \phi_n^4 - C), \quad \Lambda = 2h^2,$$
$$Z = \frac{1 - 2h^2 + Ch^4}{1 - h^2}, \quad X = \frac{(1 - C)\Lambda^2 + (2 - \Lambda)^2(1 - Z^2)}{\Lambda(2 - \Lambda)}, \tag{6.59}$$

and C is the integration constant. One can substitute $\phi_{n\pm1}$ with $\phi_{n\mp1}$ in the LHS of (6.58) due to the symmetry of the equation. For any pair of C and admissible initial value ϕ_0 the two-point map (6.58) produces a static solution to (6.56). The map has two different roots and for ϕ_{n+1} (or for ϕ_{n-1}) one must take the one different from ϕ_{n-1} (or from ϕ_{n+1}). Analysis of the static solutions to (6.56) and their expression in terms of the Jacobi elliptic functions can be found in [49].

6.3.4 *A Unifying View*

Several different approaches to derivation of exceptional discretizations have been described above, including ones conserving energy or momentum. Such models have different physical properties, but there should be something in common between them. As suggested in [32], all exceptional discretizations must have a static second order (three-point) difference problem reducible to a first-order (two-point) map.

Then a static solution $\phi_n = g(nh - x_0)$ which exists for an arbitrary central position x_0 can be derived iteratively. For the monotonically growing function g describing a kink solution this can be inverted to obtain $nh - x_0 = g^{-1}(\phi_n)$ so that

$$\phi_{n+1} = g(g^{-1}(\phi_n) + h). \tag{6.60}$$

This means that for a translationally invariant static solution a two-point map can be defined. Then the following strategy for the construction of exceptional discretizations has been offered [32], taking a two-point map in the form

$$\frac{\phi_{n+1} - \phi_n}{h} = F(\phi_{n+1}, \phi_n), \tag{6.61}$$

where for the ϕ^4 case (6.4) it is assumed that $F(\phi_{n+1}, \phi_n) \to 1 - \phi^2$ when $h \to 0$, to ensure the correct continuum limit

$$\phi_x = 1 - \phi^2. \tag{6.62}$$

The static kink solution (6.5) with $v = 0$ to (6.3) is also a static solution to its first integral (6.62). Now we square both sides of (6.61) and of the shift $(\phi_n - \phi_{n-1})/h = F(\phi_n, \phi_{n-1})$ and find their difference

$$\frac{\phi_{n-1} - 2\phi_n + \phi_{n+1}}{h^2} = \frac{F^2(\phi_{n+1}, \phi_n) - F^2(\phi_n, \phi_{n-1})}{\phi_{n+1} - \phi_{n-1}}. \tag{6.63}$$

This is an exceptional discretization and it is clear now that it has a common property with the energy conserving model (6.24) and momentum conserving model (6.30), namely, static translationally invariant solutions for all exceptional discretizations are derivable from two-point maps.

6.4 Conserved Quantities for Exceptional Discretizations of the ϕ^4 Field

In [32, 50] a generalized discretization of the ϕ^4 field was analyzed in the form

$$\begin{aligned}\ddot{\phi}_n = \frac{1}{h^2}(\phi_{n-1} - 2\phi_n + \phi_{n+1}) + 2\phi_n - A_1\phi_n^3 - \frac{A_2}{2}\phi_n^2(\phi_{n+1} + \phi_{n-1}) \\ -\frac{A_3}{2}\phi_n(\phi_{n+1}^2 + \phi_{n-1}^2) - A_4\phi_{n-1}\phi_n\phi_{n+1} \\ -\frac{A_5}{2}\phi_{n+1}\phi_{n-1}(\phi_{n+1} + \phi_{n-1}) - \frac{A_6}{2}(\phi_{n+1}^3 + \phi_{n-1}^3),\end{aligned} \tag{6.64}$$

where model parameters satisfy

$$\sum_{k=1}^{6} A_k = 2\,, \tag{6.65}$$

to ensure the correct continuum limit.

In this general form the model is not an exceptional discretization, but it includes the above mentioned exceptional discretizations as special cases, except for the Hamiltonian model (6.48), which is not a member of (6.64).

It can be straightforwardly verified that if

$$A_1 = \frac{A_3}{2} = A_4 = 2\delta, \quad A_2 = 2(1 - 4\gamma - 4\delta), \quad A_5 = A_6 = 4\gamma, \tag{6.66}$$

for arbitrary δ and γ, then the model (6.64) is a special discretization conserving momentum (6.15). All static solutions of this model can be found from the two-point map

$$U(\phi_{n-1}, \phi_n) = \frac{(\phi_n - \phi_{n-1})^2}{h^2} + 2\phi_n\phi_{n-1} - 2\gamma(\phi_n^4 + \phi_{n-1}^4)$$
$$-2\delta\phi_n\phi_{n-1}(\phi_n^2 + \phi_{n-1}^2) + 2(2\gamma + 2\delta - \frac{1}{2})\phi_n^2\phi_{n-1}^2 - C = 0, \tag{6.67}$$

where C is the integration constant. This is so because in this case (6.64) can be written in the form (6.44). The two-point map (6.67) is DFI, i.e., in the continuum limit it reduces to (6.39). Note that for $\delta = \gamma = 0$ (6.57) is the so-called Bender-Tovbis model, see [51].

As was already stated before, momentum (6.57) is conserved by (6.64) only if $A_4 = 2$ and all other A_k are zero [49].

Model (6.64) with parameters satisfying

$$A_1 = 8\alpha_1, \quad A_2 = 12\alpha_2, \quad A_3 = 8\alpha_3, \quad A_4 = A_5 = 0, \quad A_6 = 4\alpha_2,$$
$$\text{with} \quad \alpha_1 + 2\alpha_2 + \alpha_3 = 1/4, \tag{6.68}$$

conserves the total energy

$$H = h\sum_n \left[\frac{\dot{\phi}_n^2}{2} + \frac{(\phi_n - \phi_{n-1})^2}{2h^2} - \phi_n^2 + \alpha_1\phi_n^4 + \alpha_2\phi_n\phi_{n-1}(\phi_n^2 + \phi_{n-1}^2) + \alpha_3\phi_n^2\phi_{n-1}^2 \right]. \tag{6.69}$$

This model has two free parameters. The Hamiltonian Speight-Ward model (6.24) [12–14] corresponds to $\alpha_1 = \alpha_2 = 1/18$ and $\alpha_3 = 1/12$.

6.5 Exact Moving Kink Solutions for the Discrete ϕ^4 Models

In the work [52] exact moving kink solutions for the ϕ^4 discretization (6.64), (6.65) have been derived. The solution of the form

$$\phi_n(x,t) = \tanh[\beta(hn + hx_0 - vt)], \tag{6.70}$$

where v is the propagation velocity, β is the inverse width, and x_0 is the initial position of the kink, exists under the following constraints

$$\beta^2 = -\frac{A_1}{2v^2}, \tag{6.71}$$

$$2A_3 = -A_5(1+T), \tag{6.72}$$

$$A_6 = 0, \tag{6.73}$$

$$\frac{2T}{h^2} = A_2 + A_4(1+T) + \frac{A_5}{2}(1+2T-T^2), \tag{6.74}$$

$$\frac{2(1-h^2)T}{h^2} = -A_1 T + A_2(1-T) + A_4 + \frac{A_5}{2}(1+T), \tag{6.75}$$

where

$$T = \tanh^2(h\beta). \tag{6.76}$$

Interestingly, this exact kink solution can have an arbitrarily large velocity, though this is possible only for non-Hamiltonian models. To make this more transparent a particular case will be considered when only A_1, A_2, and A_4 are nonzero. Taking into account the continuum limit constraint (6.65), this model has a free parameter $A_1 < 0$. From (6.71) to (6.75) and (6.65) one has

$$\beta^2 = -\frac{A_1}{2v^2}, \quad T = \tanh^2(h\beta), \quad A_4 = \frac{2}{h^2} - \frac{2-A_1}{T}, \quad A_2 = 2 - A_1 - A_4. \tag{6.77}$$

It is clear that, for any $A_1 < 0$, one can take any value of v and then find from (6.77) parameters β, A_4, and A_2. From the first relation in (6.77) it is clear that the kink becomes wider for larger velocity, which is opposite to the continuum ϕ^4 kink (6.5).

Interestingly, the kink (6.70) with the parameters (6.77) can have arbitrarily large velocity for any small but finite value of h. Taking into account that for small h we have $T \approx h^2\beta^2$, the third expression in (6.77) assumes the form $A_4 = [2 - (2 - A_1)/\beta^2]/h^2$. From here it is clear that for $h \to 0$ the coefficients A_2 and A_4 diverge having opposite signs.

Recall that the model (6.64) is Hamiltonian when (6.68) is satisfied. From the analysis of (6.71)–(6.75) it is not difficult to come to the conclusion that among the models supporting moving kinks there are no Hamiltonian models.

6.6 Collision of Translationally Invariant ϕ^4 Kinks

Enhanced kink mobility in the models free of static PNp makes it possible to study kink collisions even in the regime of high discreteness, which is impossible in the standard discretization, given the deceleration and eventual pinning of the kinks in the latter due to the PNp. The translational Goldstone mode can be used for boosting the kinks, as was described in the last paragraph of Sect. 6.3.1.

Kink collisions in different discrete ϕ^4 models were analyzed in [53] for $h \leq 0.3$, i.e. close to the continuum limit. Collisions in the standard discrete model of (6.6) were compared to those in the PNp-free energy conserving models (6.24) and (6.48) and in two momentum conserving models. Overall it was concluded that kink collisions in the translationally invariant models are more elastic as compared to the standard discrete model, because the threshold escape velocity in the PNp-free models is always smaller.

To the best of our knowledge, kink-antikink collisions in the regime of high discreteness have not been analyzed because for $h \sim 1$ the PNp is typically so high that kinks can hardly propagate. Our aim here is to demonstrate the possibility of such studies for the exceptional discretizations, rather than to do a comprehensive analysis for any one particular model.

We choose the Hamiltonian model of (6.24) and compare kink-antikink collisions for $h = 0.5$ and $h = 1.5$. From Fig. 6.3d one can see that the lines showing the borders of the phonon spectrum cross each other at $h = 1$. That is why, for $h < 1$ kinks have smooth tails, while for $h > 1$ their tails oscillate near the asymptotic values ± 1, see Fig. 6.5a. Recall that the exact static kink solutions can be found from the two-point map (6.25). In Fig. 6.5b the Goldstone translational mode is shown for the kinks presented in (a). This mode is found by solving the eigenvalue problem for the equation of motion (6.26) linearized in the vicinity of static kink solution. The results for $h = 0.5$ are shown by squares and for $h = 1.5$ by circles.

The Goldstone mode is then used for kink boosting, with the moving kink colliding with its mirror image antikink. Collision results are presented in Fig. 6.6, where particles having total energy greater than $0.2E_{\max}$, where $E_{\max}$ is the maximal energy, are shown in the (x, t) plane. This way the trajectories of the colliding kinks and antikinks are visualized. Let us first discuss the results for relatively small discreteness, $h = 0.5$, shown in the top panels of Fig. 6.6. The collision velocity increases from the left to the right. In (a) kink and antikink collide with a small velocity of 0.03, which is below the escape threshold value. A bound state arises in the form of a bion, which gradually radiates its energy. In (b) the collision velocity is 0.073, which is close to the escape threshold. A three-bounce collision ends with kink separation with final velocity ≈ 0.018. This resonant effect in kink-antikink collisions is due

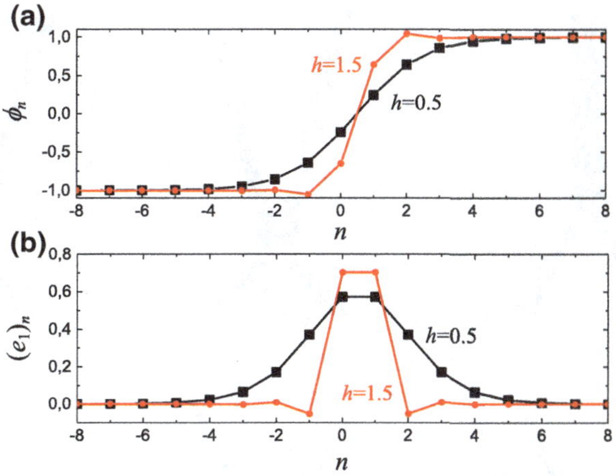

Fig. 6.5 Results for the model of (6.24). **a** Inter-site kink profile and **b** Goldstone mode for $h = 0.5$ (squares) and $h = 1.5$ (dots). For $h < 1$ kink tails are smooth but for $h > 1$ they oscillate near the asymptotic values ± 1

to the energy exchange between kink's translational and internal vibration modes, as described in a number of studies for the continuum ϕ^4 equation [33, 35–38] (and also in multiple other Chapters in this Special Volume). In (c) the collision velocity is 0.09, which is above the threshold escape velocity. This inelastic collision is accompanied by radiation of phonon wave packets and the final kink velocity reduces to 0.035.

The results for the case of high discreteness, $h = 1.5$, are shown in the bottom panels of Fig. 6.6. Here, as well, the collision velocity increases from the left to the right. It turns out that the kink and antikink having oscillatory tails are repulsive solitons in contrast to the mutually attractive kink and antikink with smooth tails. This is the reason of qualitatively different picture of collisions observed for large discreteness. In (d) the collision velocity is 0.01, which is not sufficient to overcome the apparent repulsion and the cores of the kink and antikink do not merge. As a result, the collision appears to be approximately elastic. In (e) the collision velocity is 0.15, which is close but still below the value needed to overcome mutual repulsion of the kink and antikink, and again the collision is practically elastic. Finally in (f), the collision velocity of 0.155 is sufficient to overcome the repulsion between colliding solitons and this collision is strongly inelastic resulting in a fast decay of the solitons into small-amplitude phonon wave packets.

It can be concluded that the PNp-free discrete models give the possibility to study collisions between kinks and antikinks in the regime of high discreteness and new effects can be observed in this regime, including the apparent kink repulsion and the potential kink destruction as a result of their collisions. Nevertheless, these preliminary observations clearly warrant further and more systematic study.

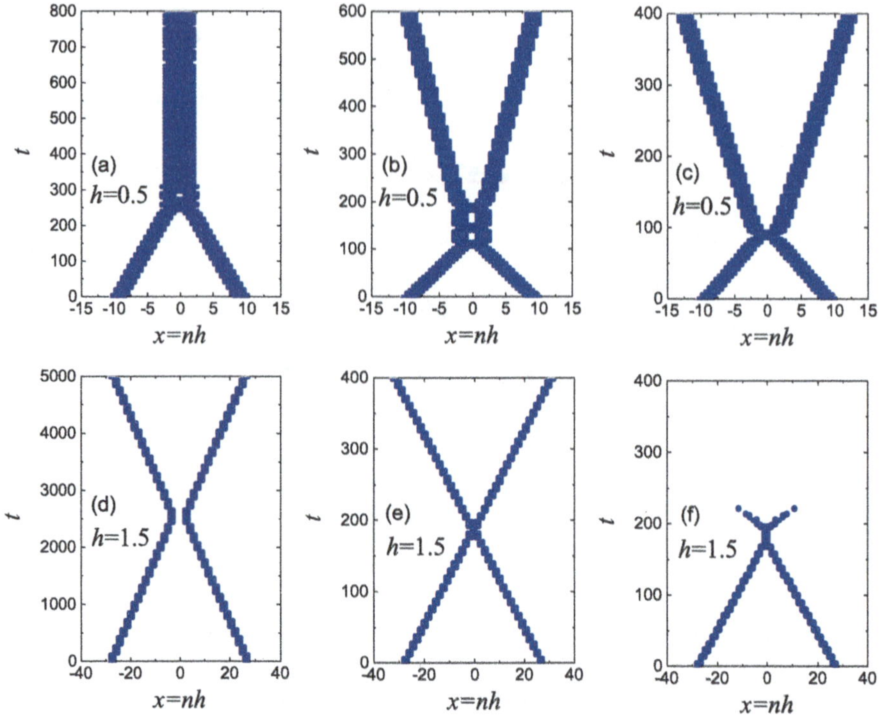

Fig. 6.6 Results for the model of (6.24). Kink-antikink collisions for **a–c** $h = 0.5$ and **d–f** $h = 1.5$. Shown are particles in the (x, t) plane having total energy greater than $0.2E_{\max}$, where $E_{\max}$ is the maximal energy. Kink velocity before the collision is: **a** 0.03, **b** 0.073, **c** 0.09, **d** 0.01, **e** 0.15, and **f** 0.155. In **a** kink-antikink annihilation with formation of a standing bion takes place. In **f** kink-antikink annihilation produces rapidly spreading radiation. Kink velocity after collision is **b** 0.018, **c** 0.035, **d** 0.01, **e** 0.15

6.7 Conclusions and Future Challenges

In the present review, we have revisited the different methods that can produce exceptional discretizations in ϕ^4 models, as a special case of more general KG type chains. We have illustrated the undesirable features of standard discretizations, such as the exponentially increasing (with the lattice spacing h) Peierls–Nabarro potential and the resulting deceleration, eventual trapping and pinning of kinks. A side product of this is the inability to consider kink-antikink collisions and their potential energy-exchange mechanisms. We saw that these features (including also the finite external dc force needed to de-pin the kinks in the above standard case) typically disappear in exceptional discretizations and provided a diverse array of methods that can produce such discretizations. These consisted of energy-conserving ones (including the ones produced in the work of Speight (and collaborators) [12–14] but also others such as [48]), as well as momentum conserving ones (including some of the work of the present authors); we also illustrated the connections between the two and the relevance of the discretized first integral approach. The resulting discretizations not only featured the presence of arbitrarily centered kinks and a neutral (Goldstone-like) eigendirection. They also enabled the consideration of collisions between kinks and antikinks even in the highly discrete realm.

At the same time, this discussion paved the way for a number of quite important questions. While the case of weak discreteness has seen some analysis at the level of kink-antikink interactions and how they compare for different models, the territory is far less charted in the case of strong discreteness and large values of h. We have only scratched the tip of the iceberg in that regard illustrating unusual features in the case of discretizations of the type of [13]. These included the non-exponential (spatial) decay of the kink for $h > 1$ in this model, as well as the detrimental effect of interactions for suitably large velocities. It would be particularly interesting to further explore these features and whether they may arise in other exceptional discretization models as well.

Most of the models considered herein have been motivated by the (in some ways, favorable) properties of the respective discretizations. However, there also exist discrete physical models supporting PNp-free kinks, e.g., a chain of electric dipoles rotating in a plane containing the chain [54] or kinks in topological mechanical chains [55]. The search for other applications of such discrete systems is of great importance and the more detailed consideration of the properties of such physically relevant models is of paramount importance in its own right.

Acknowledgements The work of S.V.D. was supported by the grant of the Russian Science Foundation (No. 16-12-10175). This material is based upon work supported 510 by the National Science Foundation under Grant No. DMS-1809074 (P.G.K).

References

1. J. Cuevas-Maraver, P.G. Kevrekidis, F. Williams (eds.), *The Sine-Gordon Model and Its Applications: From Pendula and Josephson Junctions to Gravity and High-Energy Physics* (Springer, Heidelberg, 2014)
2. O.M. Braun, Yu.S. Kivshar, *The Frenkel-Kontorova Model: Concepts, Methods, and Applications* (Springer, Berlin, 2004)
3. J.F.R. Archilla, Yu.A. Kosevich, N. Jimenez, V.J. Sanchez-Morcillo, L.M. Garcia-Raffi, Phys. Rev. E **91**, 022912 (2015)
4. S.V. Dmitriev, N.N. Medvedev, A.P. Chetverikov, K. Zhou, M.G. Velarde, Phys. Status Solidi RRL **11**, 1700298 (2017)
5. K.S. Sergeev, S.V. Dmitriev, E.A. Korznikova, A.P. Chetverikov, Nelineinaya Dinamika [Rus. J. Nonlinear Dyn.] **14**, 195 (2018)
6. A. Mehrem, L.J. Salmeron-Contreras, N. Jimenez, V.J. Sanchez-Morcillo, R. Pico, L.M. Garcia-Raffi, J.F.R. Archilla, Y.A. Kosevich, in *Nonlinear Systems*, vol. 2, ed. by J.F.R. Archilla, F. Palmero, M.C. Lemos, B. Sánchez-Rey, J. Casado-Pascual (Springer, Cham, 2018), p. 261
7. N. Boechler, G. Theocharis, S. Job, P.G. Kevrekidis, M.A. Porter, C. Daraio, Phys. Rev. Lett. **104**, 244302 (2010)

8. S. Shige, K. Miyasaka, W. Shi, Y. Soga, M. Sato, A.J. Sievers, Europhys. Lett. **121**, 30003 (2018)
9. M. Sato, B.E. Hubbard, A.J. Sievers, B. Ilic, H.G. Craighead, Europhys. Lett. **66**, 318 (2004)
10. S.J. Orfanidis, Phys. Rev. D **18**, 3822 (1978)
11. Yu.S. Kivshar, D.K. Campbell, Phys. Rev. E **48**, 3077 (1993)
12. J.M. Speight, R.S. Ward, Nonlinearity **7**, 475 (1994)
13. J.M. Speight, Nonlinearity **10**, 1615 (1997)
14. J.M. Speight, Nonlinearity **12**, 1373 (1999)
15. O.F. Oxtoby, D.E. Pelinovsky, I.V. Barashenkov, Nonlinearity **19**, 217 (2006)
16. M. Peyrard, M.D. Kruskal, Phys. D **14**, 88 (1984)
17. Y. Zolotaryuk, I.O. Starodub, in *Nonlinear Systems*, vol. 2, ed. by J.F.R. Archilla, F. Palmero, M.C. Lemos, B. Sánchez-Rey, J. Casado-Pascual (Springer, Cham, 2018), p. 315
18. I.V. Barashenkov, T.C. van Heerden, Phys. Rev. E **77**, 036601 (2008)
19. L.A. Cisneros, A.A. Minzoni, Phys. D **237**, 50 (2008)
20. A. Aigner, A. Champneys, V. Rothos, Phys. D **186**, 148 (2003)
21. C. Cattuto, G. Costantini, T. Guidi, F. Marchesoni, Phys. Rev. E **63**, 5 (2001)
22. A. Champneys, B. Malomed, J. Yang, D. Kaup, Phys. D **152–153**, 340 (2001)
23. P.G. Kevrekidis, M.I. Weinstein, Phys. D **142**, 113 (2000)
24. A.V. Savin, Y. Zolotaryuk, J.C. Eilbeck, Phys. D **138**, 265 (2000)
25. S. Flach, Y. Zolotaryuk, K. Kladko, Phys. Rev. E **59**, 6105 (1999)
26. O.M. Braun, Yu.S. Kivshar, M. Peyrard, Phys. Rev. E **56**, 6050 (1997)
27. H.S.J. van den Zant, T.P. Orlando, S. Watanabe, S.H. Strogatz, Phys. Rev. Lett. **74**, 174 (1995)
28. P.T. Dinda, C.R. Willis, Phys. Rev. E **51**, 4958 (1995)
29. R. Boesch, C.R. Willis, M. El-Batanouny, Phys. Rev. B **40**, 2284 (1989)
30. Y. Ishimori, T. Munakata, J. Phys. Soc. Jpn. **51**, 3367 (1982)
31. V.H. Schmidt, Phys. Rev. B **20**, 4397 (1979)
32. I.V. Barashenkov, O.F. Oxtoby, D.E. Pelinovsky, Phys. Rev. E **72**, R035602 (2005)
33. D.K. Campbell, J.F. Schonfeld, C.A. Wingate, Phys. D **9**, 1 (1983)
34. P.G. Kevrekidis, C.K.R.T. Jones, T. Kapitula, Phys. Lett. A **269**, 120 (2000)
35. P. Anninos, S. Oliveira, R.A. Matzner, Phys. Rev. D **44**, 1147 (1991)
36. D.K. Campbell, M. Peyrard, Phys. D **18**, 47 (1986)
37. D.K. Campbell, M. Peyrard, Phys. D **18**, 165 (1986)
38. R.H. Goodman, R. Haberman, SIAM J. Appl. Dyn. Syst. **4**, 1195 (2005)
39. K. Kladko, I. Mitkov, A.R. Bishop, Phys. Rev. Lett. **84**, 4505 (2000)
40. P.G. Kevrekidis, I.G. Kevrekidis, A.R. Bishop, Phys. Lett. A **279**, 361 (2001)
41. P.G. Kevrekidis, Phys. D **183**, 68 (2003)
42. E.B. Bogomol'nyi, Sov. J. Nucl. Phys. **24**, 449 (1976)
43. S.V. Dmitriev, P.G. Kevrekidis, N. Yoshikawa, J. Phys. A: Math. Gen. **39**, 7217 (2006)
44. P.G. Kevrekidis, V. Putkaradze, Z. Rapti, in *Dynamical Systems and Differential Equations*, ed. by M. de León et al.. 10th AIMS International Conference, Madrid (Spain), July 2014. AIMS Proceedings 2015 (American Institute of Mathematical Sciences, Springfield, Missouri, USA), p. 696
45. A.M. Bloch, P.S. Krishnaprasad, J.E. Marsden, R.M. Murray, Arch. Ration. Mech. Anal. **136**, 21 (1996)
46. J.E. Marsden, M. West, Acta Numer. **10**, 357 (2001)
47. S.V. Dmitriev, P.G. Kevrekidis, N. Yoshikawa, D.J. Frantzeskakis, Phys. Rev. E **74**, 046609 (2006)
48. F. Cooper, A. Khare, B. Mihaila, A. Saxena, Phys. Rev. E **72**, 036605 (2005)
49. S.V. Dmitriev, P.G. Kevrekidis, A. Khare, A. Saxena, J. Phys. A: Math. Theor. **40**, 6267 (2007)
50. A. Khare, S.V. Dmitriev, A. Saxena, J. Phys. A: Math. Theor. **42**, 145204 (2009)
51. C.M. Bender, A. Tovbis, J. Math. Phys. **38**, 3700 (1997)
52. S.V. Dmitriev, A. Khare, P.G. Kevrekidis, A. Saxena, L. Hadzievski, Phys. Rev. E **77**, 056603 (2008)
53. I. Roy, S.V. Dmitriev, P.G. Kevrekidis, A. Saxena, Phys. Rev. E **76**, 026601 (2007)
54. J.M. Speight, Y. Zolotaryuk, Nonlinearity **19**, 1365 (2006)
55. Y. Zhou, B.G.-G. Chen, N. Upadhyaya, V. Vitelli, Phys. Rev. E **95**, 022202 (2017)

Chapter 7
Discrete Breathers in ϕ^4 and Related Models

Jesús Cuevas–Maraver and Panayotis G. Kevrekidis

Abstract In this chapter, we touch upon the wide topic of discrete breathers with a special emphasis on the prototypical system of interest, namely the ϕ^4 model. We start by introducing the model and discussing some of the application areas/motivational aspects of exploring time periodic, spatially localized structures, such as the discrete breathers. Our main emphasis is on the existence, and especially on the stability features of such solutions. We explore their spectral stability numerically, as well as in special limits (such as the vicinity of the so-called anti-continuum limit of vanishing coupling) analytically. We also provide and explore a simple, yet powerful stability criterion involving the sign of the derivative of the energy versus frequency dependence of such solutions. We then turn our attention to nonlinear stability, bringing forth the importance of a topological notion, namely the Krein signature. Furthermore, we briefly touch upon linearly and nonlinearly unstable dynamics of such states. Some special aspects/extensions of such structures are only touched upon, including moving breathers and dissipative variations of the model and some possibilities for future work are highlighted. While this chapter by no means aspires to be comprehensive, we hope that it provides some recent developments (a large fraction of which is not included in time-honored discrete breather reviews) and associated future possibilities.

J. Cuevas–Maraver (✉)
Grupo de Física No Lineal, Universidad de Sevilla, Departamento de Física Aplicada I, Escuela Politécnica Superior, C/ Virgen de África, 7, 41011 Sevilla, Spain
e-mail: jcuevas@us.es

J. Cuevas–Maraver
Instituto de Matemáticas de la Universidad de Sevilla (IMUS), Edificio Celestino Mutis. Avda. Reina Mercedes s/n, 41012 Sevilla, Spain

P. G. Kevrekidis
Department of Mathematics and Statistics, University of Massachusetts, Amherst, MA 01003-4515, USA
e-mail: kevrekid@math.umass.edu

P. G. Kevrekidis and J. Cuevas-Maraver (eds.), *A Dynamical Perspective on the ϕ^4 Model*, Nonlinear Systems and Complexity 26,
https://doi.org/10.1007/978-3-030-11839-6_7

7.1 A Brief Description of Discrete Breathers: Definition, Historical Perspective and Applications

Dynamics of localized excitations is, undoubtedly, one of the most important topics within the realm of nonlinear science. In discrete systems, the localized excitations that can be argued to be most generic [1–3] are the so-called discrete breathers (DBs). These can be defined as time-periodic spatially-localized coherent structures emerging at coupled nonlinear oscillator lattices. This term was coined by Campbell and Peyrard [4] in order to distinguish them from the "continuous" breathers found as exact solutions through the inverse scattering machinery in the sG PDE [5]. [It is worthwhile to mention in passing that continuous breathers have a particularly interesting history associated with them in ϕ^4 models [6, 7], which is described in detail in Chap. 8 of this special volume.] Such DBs are also known as intrinsic localized modes; this name was introduced in order to emphasize that their origin was in the intrinsic nonlinearity of the system, contrary to the case of Anderson modes, where localization stems from lattice disorder [8] and can exist even in the linear limit.

A major development that springboarded the study of DBs took place in 1988, through the numerical study of DBs in some prototypical lattice models as reported in two pioneering works [9, 10] by Takeno, Sievers and Kisoda. In these papers DBs are calculated for the first time in the two basic kinds of lattices where they can emerge, namely Fermi–Pasta–Ulam–Tsingou (FPUT)[1] [9] and KG [10]. Additionally, some of their stability properties were also determined. FPUT lattices are characterized by the absence of substrate (on-site) potential and the nonlinearity of the inter-site forces, whereas KG lattices possess nonlinear on-site potential and (typically) linear inter-site forces. Interestingly, in those references, potentials of the ϕ^4 form and related models were considered. There are also mixed cases of lattices where nonlinearities are present in both the substrate and intersite potentials (see e.g. [12]); these lattices are sometimes denoted as KG/FPUT lattices.

While these early works provided credible numerical evidence and planted the seed for studying DBs, they did not rigorously prove their existence. The latter came in a celebrated 1994 paper by MacKay and Aubry [13] and was based on the concept of the so-called anti-continuous (AC) limit. They basically demonstrated in a rigorous fashion, by making use of the implicit function theorem, that DBs can generically exist in nonlinear KG lattices as a result of unobstructed (when some suitable resonance restrictions are avoided) continuation of periodic orbits of individual (uncoupled) oscillators. Mackay–Aubry's theorem, together with the rigorous proof of stability by Aubry in 1997 [14] and the numerical methods developed by Marín among others [15] led to an intense interest on DBs in the late 1990's and the beginning of the 21st Century, not only from the theoretical but also from the

---

[1]These lattices have been traditionally denoted as Fermi–Pasta–Ulam, forgetting the outstanding role of Mary Tsingou who was the person responsible for all the numerical simulations in these first computations of nonlinear lattice dynamics [11].

experimental point of view. The relevant activity has been very well summarized by now in a series of reviews; see, e.g., [1–3, 16, 17].

DBs have been sought for in many fields of Physics. They have been experimentally generated in arrays of Josephson junctions [18, 19], mechanical [20] and magnetic pendula [21] and microcantilevers [22], nonlinear electrical lattices [23, 24] and granular media [25, 26]. They have been observed in molecular and ionic crystals like the so-called PtCl [27] and antiferromagnets [28], and have been argued as plausible explanations of observational findings in systems such as α-Uranium [29] or NaI [30]. They have also been speculated to play an important role in DNA denaturation in an extensive literature reviewed, e.g., in [31], as well as to arise in reconstructive chemical reactions [32] — and reaction kinetics in general [33, 34] — and in the slow decay of luminiscence in Pb-doped alkali-halide crystals such as KBr [35], or in (ab initio) molecular dynamics in crystals of Niobium and Nickel [36] or carbon materials like graphene [37], carbon nanotubes, fullerenes or hydrocarbons (see [38] for a review). Under suitable conditions, DBs can move along the lattice and are known in that setting as moving breathers. As such, they have been proposed to be responsible for some of the observed tracks in muscovite mica sheets [16, 39], for the infinite charge mobility (also dubbed as *hyperconductivity*) experimentally evidenced in such crystals [40, 41], and they have also been suggested as the explanation of annealing and modification of defects in Ge brought about by low energy plasmas [42–44]. Admittedly, these are only some examples of an ever increasing list of applications which is by necessity, due to the limited scope of this Chapter, rather incomplete. Nevertheless, it serves to illustrate the generic nature and wide impact of such structures and their broad relevance of study.

The present Chapter is devoted to reviewing more concretely some results on the existence, stability and dynamics of DBs in KG lattices with ϕ^4 on-site potential, and some other miscellaneous topics related to DBs in such lattices; many of these results have been found for generic KG lattices, but we will focus here on the principal theme of this special volume, namely the ϕ^4 potential. Moreover, by choice, the emphasis will be on some recent results, not only due to their connections to the research interests and recent work of the authors, but also because, to the best of our knowledge, the associated findings and the resulting over-arching stability perspective have not been collected in such a summarizing body of work to date elsewhere.

7.2 The Klein-Gordon Lattice and the Anti-continuous Limit

After the brief presentation of the DB concept, in this Section we will introduce some definitions. First of all, we need to address the concept of a KG nonlinear dynamical lattice; from a mathematical point of view, it can be defined as a system of coupled second-order ODEs of the form:

$$F_n(u) \equiv \ddot{u}_n + V'(u_n) + \sum_m C_m(u_n - u_{n+m} - u_{n-m}) = 0 \tag{7.1}$$

where n and m are D-dimensional indices, $V(u_n)$ is the on-site potential, which is not necessarily homogeneous, and C_m is the coupling constant, depending on the distance from the neighbors. In most case examples, the inter-site force is nearest-neighbour with $C_m = C\delta_{m,1}$, in which case the previous equation can be written e.g. for the one-dimensional lattice as

$$F_n(u) \equiv \ddot{u}_n + V'(u_n) + C(2u_n - u_{n+1} - u_{n-1}) = 0. \tag{7.2}$$

This dynamical equation derives from the following Hamiltonian:

$$H = \sum_n h_n = \sum_n \frac{1}{2}\dot{u}_n^2 + V(u_n) + \frac{C}{4}\left[(u_n - u_{n-1})^2 + (u_n - u_{n+1})^2\right] \tag{7.3}$$

with h_n being the energy density. As mentioned above, the anti-continuous (AC) limit introduced by MacKay–Aubry's theorem [13] is of paramount importance in order not only to prove the existence of discrete breathers in nonlinear KG lattices, but also to give a hint on how to obtain them numerically. The AC limit is that of all the lattice oscillators being uncoupled (i.e. $C = 0$) and either oscillating with the same frequency ω_b or remaining at rest. By virtue of the implicit function theorem, MacKay and Aubry established that the solution can be continued from the AC limit to a finite value of $C \to 0$ whenever two conditions are fulfilled: (1) the potential of the isolated oscillators is anharmonic and (2) no integer multiples of the DB frequency ω_b resonate with the linear modes frequency (i.e. the so-called phonon band which we will quantify further below). If these conditions are fulfilled, a coherent (in the sense that all the sites oscillate with the same frequency) and exponentially localized structure of a DB form will exist. These discrete breathers are exact periodic orbit solutions (up to machine precision) of the nonlinear KG equation.

This theorem is not only of theoretical value, but also of practical usefulness as it provides a strategy on how to numerically calculate DBs that was exploited by Marín and collaborators [15]. Methods based on the AC limit are quite simple: as at the AC limit the oscillators are uncoupled, it suffices to get a periodic orbit of single oscillators subjected to potential $V(u)$ and continue this solution by means of fixed-point methods (like Newton–Raphson) up to the desired coupling. Among the numerical methods used for attaining DBs, we can highlight two: (1) Fourier space methods in which DBs are represented by a Galerkin truncation up to index k_m in a Fourier series expansion of the form:

$$u_n = \sum_{k=-k_m}^{k_m} z_k e^{ik\omega_b t} , \tag{7.4}$$

transforming the coupled ODE system (7.1) into a set of nonlinear algebraic equations; and (2) shooting methods, where DBs are fixed points of the map:

$$(\{u_n(0)\}, \{\dot{u}_n(0)\}) \to (\{u_n(T)\}, \{\dot{u}_n(T)\}) \ , \tag{7.5}$$

with $T = 2\pi/\omega_b$ being the breather period. Fourier methods have the advantage of using an analytical Jacobian but one has to pay the price of handling a larger number of equations. In addition, there are pathological potentials [23, 45] for which the convergence of Fourier series is very slow and this method cannot be used.

There is a great number of potentials that can be found in the DB literature. They can be classified as soft or hard, if the energy of a single oscillator decreases or increases, respectively, with the frequency; a simple way to discern if a potential is soft or hard is by making use of the hardening coefficient $h = 3V''''(0) - 5V'''(0)$ [46] so that the potential is soft (hard) when $h < 0$ ($h > 0$). Moreover, a soft (hard) oscillator vibrates with a frequency ω_b which is smaller (greater) than its natural frequency $\omega_o = \sqrt{V''(0)}$ (notice that in most cases, $\omega_o = 1$). Typical examples of soft potentials include the Morse, Lennard-Jones, cubic (ϕ^3), sG and double-well potentials, many of which arise in applications [1, 2]. Hard potentials are usually particular cases of polynomials which, depending on parameters, can be either soft or hard, namely cubic-quartic or purely quartic (ϕ^4) anharmonicities, where the quartic term arises with a positive sign; see below. This latter potential, on which we will focus in the present Chapter, is given by

$$V(u) = \frac{1}{2}u^2 + \frac{1}{4}su^4 \tag{7.6}$$

When $s = 1$ ($s = -1$), the potential is hard (soft). The orbits $u(t)$ of an isolated oscillator can be expressed in terms of Jacobi elliptic functions and the parameter m is related to the oscillation frequency ω_b through a transcendental equation. Thus, if $s = 1$,

$$u(t) = \sqrt{\frac{2m}{1-2m}}\,\mathrm{cn}\left(\frac{2K(m)\omega_b t}{\pi}, m\right), \qquad \omega_b = \frac{\pi}{2\sqrt{1-2m}\,K(m)} \tag{7.7}$$

and, if $s = -1$

$$u(t) = \sqrt{\frac{2m}{1+m^2}}\,\mathrm{cd}\left(\frac{2K(m)\omega_b t}{\pi}, m\right), \qquad \omega_b = \frac{\pi}{2\sqrt{1+m^2}\,K(m)} \tag{7.8}$$

In the equations above, $K(m)$ is the complete elliptic integral of the first kind of parameter m.

Notice that in the soft case, there are heteroclinic orbits separating oscillating states from unbounded ones representing escape from the potential [47]. Another

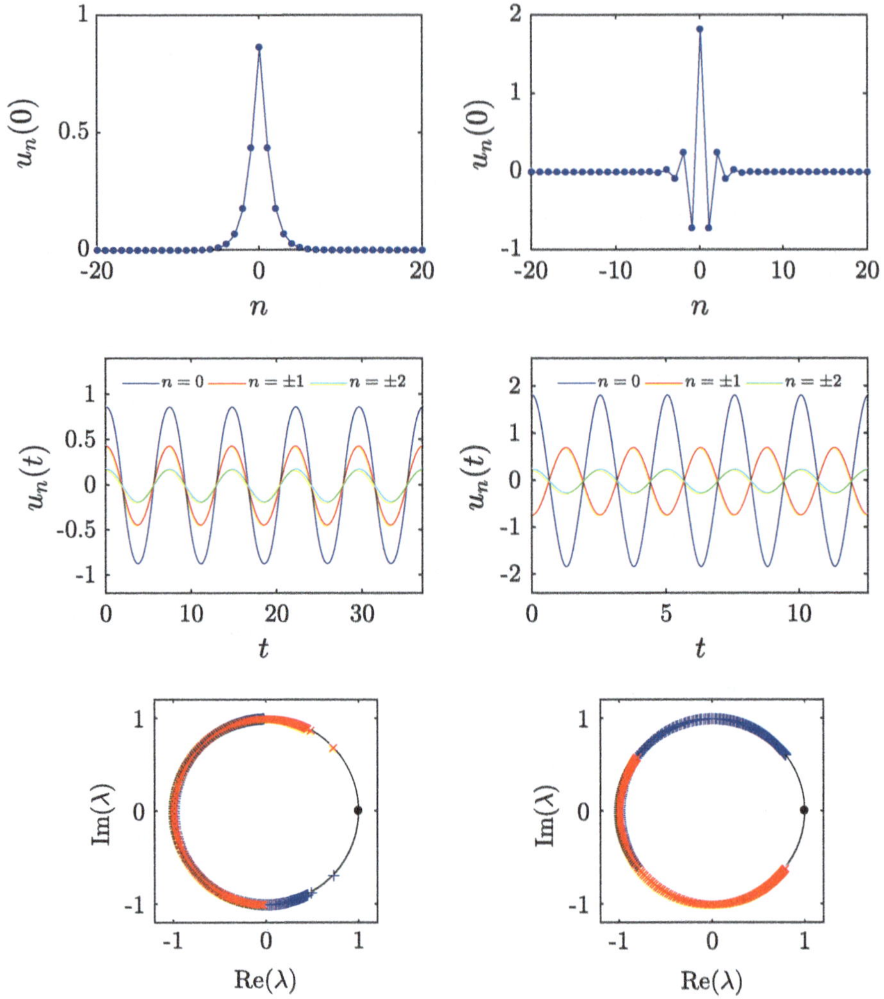

Fig. 7.1 Profile (top panels), time-dependence of the central sites (middle panels) and Floquet multiplier spectrum (bottom panels) of a 1-site breather in a soft (left panels) and in a hard (right panels) potential. In the former case, parameters are $\omega_b = 0.85$ and $C = 0.3$, whereas in the latter case, $\omega_b = 2.5$ and $C = 1$. In the bottom panels, multipliers with positive (negative) Krein signatures are depicted with red crosses $\times$ (blue pluses $+$); see the associated discussion around (7.15) below

interesting feature that distinguishes soft and hard potentials is related to the oscillation pattern of the tails. When the potential is hard and $C > 0$, the tails are staggered and they are unstaggered in the soft case [48]. An example of discrete breathers in soft and hard potentials is shown in Fig. 7.1 where the profile and the time-evolution is displayed.

At the AC limit, it is possible to construct DB-like solutions with more than one excited sites, dubbed as *multibreathers*. When suitably constructed (typically with the "excited" oscillators in- or out-of-phase as summarized, e.g., in [49]), these are also found to persist when the coupling is switched on. Over the years, some of the DB structures have been endowed with distinguishing names. The Sievers-Takeno mode corresponds to a DB with only one excited site (it can also be denoted as single site or 1-site breather). Also, the Page mode is a multibreather where two adjacent sites are excited; notice that the former is a site-centered breather whereas the latter corresponds to a bond-centered (inter-site-centered) one. When all the sites at the AC limit are excited, we are dealing with a nonlinear phonon or *phonobreather*; a dark breather [50] is a phonobreather with one (or a few) non-excited site(s), resembling the functional form of a dark soliton of the NLS equation.

Multibreathers are usually formed by time-reversible oscillators. In these cases, they can be characterized by a coding sequence $\sigma \equiv \{\sigma_n\}$ indicating the phase and the excitation state of the relevant sites at the AC limit. This code is $\sigma_n = 0$ if the n-th oscillator is at rest, $\sigma_n = 1$ if it oscillates with a frequency ω_b and phase 0 (i.e. $u_n(0) > 0$) and $\sigma_n = -1$ if it oscillates with initial phase π (i.e. $u_n(0) < 0$). For instance, the Sievers-Takeno and Page modes are represented by $\sigma = \{1\}$ and $\sigma = \{1, 1\}$ (for soft potentials) or $\sigma = \{1, -1\}$ (for hard potentials), respectively. There are some cases where the code can be more complex as in sG potentials, where rotors can coexist with oscillators; in such cases, structures called *rotobreathers* (see [14, 51]) can emerge if at least one of the sites corresponds to a rotor; these have also been experimentally generated in Josephson junction arrays [18, 19].

There are also, however, some special cases where multibreathers are constituted by non-time-reversible oscillators, as demonstrated in [14]. One can find such kind of solutions in 1D chains with periodic boundary conditions in the form of phonobreathers with phase torsion [52, 53], in 2D percolating clusters or so-called discrete vortices [49, 54, 55]. They can also emerge in 1D lattices with long-range interactions [56, 57]. Additionally, they constitute a potential attractor in periodically forced and damped oscillator networks [58].

The non-resonance condition of MacKay-Aubry's theorem clearly establishes that the breather frequency ω_b must be either above the phonon band or in the gaps between different phonon bands; additionally, harmonics of this frequency must avoid resonances with the band to ensure the absence of energy dispersing mechanisms affecting the DB. In the case of 1D KG lattices, there is an optical band of phonons given by

$$\omega_{\text{ph}}^2 = \omega_{\text{o}}^2 + 4C \sin^2 \frac{q}{2} \tag{7.9}$$

where q is the phonon wavenumber and ω_{ph} the frequency of the associated effectively plane wave excitations $\sim e^{i(qn-\omega_{\text{ph}}t)}$. In finite lattices (which are needed for numerical computations), the value of q is quantized with the quantization being determined by the nature of the imposed boundary conditions and the number of lattice nodes N. Consequently, breathers in hard potentials, can be continued towards higher frequencies without resonances and to lower frequencies until ω_b collides with the upper edge of the phonon band ($q \approx \pi$), an event occurring when $C \approx (\omega_b^2 - \omega_o^2)/4$ (the approximation symbol is used to highlight that, depending on the quantization of q, it could happen that $q = \pi$ is not in the band). Resonances in soft potentials are caused by integer multiples of the breather frequency colliding with the frequencies of the phonon band. The relevant critical point emerges when the second harmonic in asymmetric potentials and the third one in symmetric ones (like the purely quartic ϕ^4 analyzed in this chapter) collides once again with the upper edge of the phonon band; that is, when $C \approx \omega_b^2 - \omega_o^2/4$ or $C \approx (9\omega_b^2 - \omega_o^2)/4$, respectively. In addition, in a finite lattice and soft potentials, there are gaps in the phonon band and breathers can "bypass" the resonance frequency, by existing within these finite gaps. In this case, the breather hybridizes with the bifurcating phonon creating structures called *phantom breathers*; strictly speaking, they are non-exponentially-localized breathers as they have a tail oscillating with a frequency $n\omega_b$ with n depending on the multiple of the breather frequency that resonates [59]. These structures are the discrete analogue of the nanoptera observed at the continuum limit of KG lattices with the ϕ^4 double well potential [7]. As shown in [60] for sG and ϕ^4 double well potentials, continuation up to the continuum limit is similar to a Wannier–Stark ladder. This phenomenon (i.e. breather-phonon hybridization) takes place because the staggering character of the phonons is different than that of the breather; if the breather tails had the same staggering character of the bifurcating phonon, the breather would be smoothly continued from the phonon and its amplitude would be zero at the phonon frequency. Notice that gaps in the phonon band appear naturally in disordered lattices because of Anderson localization; breathers in such systems can either delocalize [61] or remain localized [62, 63]. The above scenario is typical of Sievers-Takeno and Page modes. When other multibreathers are considered, breathers cannot reach the phonon band and the bifurcation scenario is more complex (see e.g. [60]).

Let us also mention that decay of breathers is not necessarily exponential. In lattices with long-range intersite interactions, the decay can be algebraic [64], possibly featuring a transition from exponential to algebraic; for a recent experimental realization of this transition in a system based on magnets, see, e.g., [65]. In addition, in some KG/FPUT lattices with ϕ^4 intersite potential, breathers can decay superexponentially being dubbed as (nearly) compact DBs [66, 67].

Finally, we must remark that there are several more existence proofs. For instance, a variational proof was introduced in [68]; this is valid only for hard potentials. A proof based on the center manifold theorem was introduced in [46] and is valid for DBs whose frequency is close to the phonon band edge.

7.3 Stability of Discrete Breathers

This section can be considered as the core of the present chapter. The treatment of the topic will be as follows: first of all, we will present an introduction to the Floquet theory applied to the linear stability of discrete breathers; then, we will show different approaches to the linear stability of multibreathers in the vicinity of the AC limit and introduce the energy-vs-frequency monotonicity criterion for the linear stability of breathers and multibreathers at arbitrary coupling; afterwards, nonlinear stability criteria will be summarized together with the dynamical evolution of some examples of unstable solutions.

7.3.1 Floquet Analysis

In order to assess the dynamical robustness of the identified DB solutions and their potential accessibility in physical experiments, a key step is the determination of their stability. In the paper where MacKay and Aubry demonstrate the existence of breathers [13], it is speculated that 1-site breathers are likely to be stable. This was finally proven by Aubry in [14] for finite lattices and MacKay and Sepulchre in [69] for infinite lattices.

As breathers are time-periodic solutions of the equation system (7.1), their spectral stability can be determined by means of a Floquet analysis. To this aim, we need to evaluate the evolution of a perturbation $\xi_n(t)$ to a solution $v_n(t)$. We thus introduce— in e.g. the 1D equation (7.2)— the solution $u_n(t) = v_n(t) + \epsilon\xi_n(t)$ with ϵ being a small constant. Then, the equation that the perturbation satisfies to $O(\epsilon)$ is

$$\ddot{\xi}_n + V''(v_n)\xi_n + C(2\xi_n - \xi_{n+1} - \xi_{n-1}) = 0. \tag{7.10}$$

This equation can be written in a more compact form as

$$\mathcal{N}(v(t))\xi = 0 \tag{7.11}$$

where $\xi \equiv \{\xi_n(t)\}$ and $v(t) \equiv \{v_n(t)\}$. $\mathcal{N}$ is known as the linearization operator. If $\xi \in C^2$, the study of this operator spectrum offers information about stability. If, moreover, ξ is a time-reversible function of frequency ω_b, this operator can be identified as the Jacobian (Fréchet derivative) of the dynamical equations, i.e. $\partial_v F(v(t)) = \mathcal{N}(v(t))$. Equation (7.11) can be viewed as the particular case $E = 0$ of the eigenvalue equation for the Newton operator

$$\mathcal{N}(v(t))\xi = E\xi \tag{7.12}$$

The study of the spectrum of $\mathcal{N}$ is related to Aubry's band theory [14] as we will see further in the present section. In Hamiltonian dynamical systems, this linearization operator is time-symmetric, real, symplectic and Hermitian. In addition, it is invariant under time translations of period T, so by virtue of Bloch's theorem, the corresponding eigenfunctions $\xi(t)$ can be expressed as Bloch functions:

$$\xi(t) = e^{i\theta t/T}\upsilon(t). \tag{7.13}$$

To perform Floquet analysis, we need to study the spectrum of the Floquet operator $\mathcal{F}$, defined from the following map:

$$\Omega(T) = \mathcal{F}_o\Omega(0), \qquad \text{with } \Omega(t) = [\xi(t), \dot{\xi}(t)] \tag{7.14}$$

The representation of $\mathcal{F}_o$ in $\mathbb{R}^{2N}$ is denoted as the so-called monodromy matrix. The eigenfunctions of $\mathcal{F}_o$ are those of the Newton operator with $E = 0$. Thus, as a consequence of Bloch's theorem (7.13), $\Omega(T) = \exp(i\theta)\Omega(0)$. In other words, monodromy eigenvalues (also known as Floquet multipliers) are of the form $\lambda = \exp(i\theta)$ with $\theta \in \mathbb{C}$. θ is dubbed as Floquet argument.

As the linearization operator is real and symplectic, the Floquet operator also possesses both properties. From the fact that $\mathcal{F}$ is real, one can deduce that if λ is a Floquet multiplier, λ^* is also a multiplier. Because of the symplecticity of $\mathcal{F}$, $1/\lambda$ is also a multiplier. In other words, Floquet multipliers always come in quadruplets $(\lambda, \lambda^*, 1/\lambda, 1/\lambda^*)$ if $\lambda \notin \mathbb{R}$ and in pairs $(\lambda, 1/\lambda)$ if $\lambda \in \mathbb{R}$. Consequently, a necessary and sufficient condition for a Hamiltonian dynamical system to be linearly stable is that $\theta \in \mathbb{R}$ (i.e. that the multipliers lie at the unit circle of the complex plane). If the system is not Hamiltonian, the symplecticity can be broken and the previous property does not hold. Then, the condition for stability is that $|\lambda| \leq 1$.

An important concept in Floquet analysis is the Krein signature, which, for a given eigenvalue λ is defined as

$$\kappa(\lambda) = \operatorname{sgn}\left(\sum_{n=1}^{N} \operatorname{Im}\left[\xi_n^*(t)\dot{\xi}_n(t)\right]\right) \tag{7.15}$$

Given the symplectic nature of $\mathcal{N}$, the total Krein signature (i.e. the sum of the Krein signature of every Floquet multipliers) is conserved in time. If λ is real, $\xi_n \in \mathbb{R}\ \forall n$ and $\kappa(\lambda) = 0$. The Krein signature is helpful for predicting bifurcations when system parameters vary: due to the properties of Floquet multipliers in Hamiltonian systems, the only way that an instability takes place is that two eigenvalues coincide at $\theta = 0$ or $\theta = \pi$ or a quartet of eigenvalues coincides at a θ different than 0 or π. However, the coincidence of eigenvalues is not a necessary condition for an instability to take place. Krein's criterion establishes that a necessary condition for the instability emergence is that the coincident multipliers have different Krein signature. Nevertheless, this criterion is not a sufficient condition as discussed, e.g., in the context of Aubry's band theory in [14].

The starting point of the band theory is (7.12), from which Bloch's theorem establishes that eigenvalues come in a set of non-overlapping bands $E_\nu(\theta)$, which is a continuous function, non constant and 2π-periodic. From (7.13), one can write $\xi(T) = \exp(i\theta_\nu(E))\xi(0)$. θ_ν can be chosen in the first Brillouin zone, i.e. $\theta_\nu \in (-\pi, \pi]$. Then, each set of eigenvalues with θ in the first Brillouin zone is denoted as band ν. Bands are associated with stable solutions if $\theta_\nu \in \mathbb{R}$. Moreover, due to the properties of the linearization operator, bands are symmetric with respect to $\theta_\nu = 0$. The values $\theta_\nu(E)$ are found by diagonalizing matrix $\mathcal{F}_E$, which is attained by applying (7.14) to the integration of (7.12). The monodromy coincides with $\mathcal{F}_0$ and, consequently, the Floquet arguments are $\theta_\nu(0)$. That is, in order for the solution to be stable, there must exist $2N$ bands that cut or are tangent to the $E = 0$-axis. As mentioned above, the band analysis allows to improve Krein's criterion: as demonstrated in [14], the Krein signature of a Floquet multiplier is minus the sign of the slope at $E = 0$ of the band corresponding to this multiplier; i.e.

$$\kappa(\theta_\nu) = -\text{sgn}\left(\frac{\mathrm{d}E_\nu(\theta)}{\mathrm{d}\theta}|_{\theta=\theta_\nu}\right) \tag{7.16}$$

As discussed in [14, Sect. 4.4] a necessary and sufficient condition for a bifurcation to occur is that the coincident eigenvalues belong to the same band, as in that case the slopes at the bifurcation point will have opposite signs.

Floquet analysis can also help to identify what kind of bifurcations a breather experiences. If a pair of eigenvalues collides at $\theta = 0$, the breather undergoes an exponential (tangent) bifurcation, whereas collisions at $\theta = \pi$ correspond to a period-doubling bifurcation. In the rest of cases, there is a quartet of multipliers colliding out of the real axis and the breather experiences a (Hamiltonian) Hopf bifurcation.[2]

Prior to showing the stability properties of breathers, we want to indicate that when the dynamical equations are invariant under time translation, there is a pair of eigenvalues that always remain at $\theta = 0$, known as the phase and growth mode, which come from the fact that $\xi_n = \dot{u}_n$ and $\xi_n = \partial u_n/\partial\omega_b$ are solutions of (7.10). The growth mode is, in fact, a marginal mode, as perturbations along its direction grow linearly. This mode is associated to a generalized eigenvector along this eigendirection. These modes will be particularly important when considering stability criteria below.

7.3.2 Linear Stability Near the Anti-continuous Limit

Prior to introducing the existing approaches for the stability in the vicinity of the AC limit, we need to understand the structure of the monodromy spectrum at that limit. Let us suppose that the lattice is formed by N oscillators and we have a multibreather with p excited sites at $C = 0$. Then, it is easy to deduce from (7.10) that there are p pairs of degenerated phase/growth modes at $\theta = 0$ corresponding to excited sites, and $N - p$ pairs of multipliers at $\theta = \pm\omega_o T \bmod 2\pi$ corresponding to the oscillators at rest. The Krein signature of the multipliers at $\theta = 0$ is zero, but for the oscillators at rest, it depends on the hardness/softness of the substrate potential, the breather frequency and the sign of the coupling constant. From now on, let us suppose that $C > 0$. In that case, the multiplier in the upper half-circle will have $\kappa = 1$ (and those of the lower half-circle will have $\kappa = -1$) when $\omega_b > 2\omega_o$ if the on-site potential

[2]Notice that, strictly speaking, we are dealing with a Neimark-Sacker bifurcation (Hopf bifurcation of periodic orbits). However, through a slight abuse of the relevant terminology, we denote it simply as a Hopf bifurcation.

is hard and $\frac{2}{2k+1}\omega_o < \omega_b < \frac{1}{k}\omega_o$ with $k \in \mathbb{N}$ if the potential is soft. On the contrary, if $\omega_o < \omega_b < 2\omega_o$ for hard potentials and $\frac{1}{k+1}\omega_o < \omega_b < \frac{2}{2k+1}\omega_o$ for soft ones, the multipliers at the upper half-circle will have $\kappa = -1$ [69, 70].

When the coupling is switched on, we need to analyze the fate of the multipliers at $\theta \neq 0$, that will correspond to the linear modes, and those at $\theta = 0$. The former will expand so that the isolated eigenvalue pair at $\theta \neq 0$ leads to a couple of arcs with $\theta \approx \pm\omega_{ph}T \bmod 2\pi$, being ω_{ph} the phonon frequencies (7.9). Notice that the Krein signature of the eigenvalues of an arc is the same as the multiplier from which they are born. With this in mind, it is easy to deduce that a 1-site breather must remain stable when the coupling is switched on because, on the one hand, there are no bifurcations stemming from the linear mode arcs and, on the other hand, $p = 1$ and the only pair at $\theta = 0$ corresponds to the phase/growth modes and must remain there for every coupling. Figure 7.1 shows a couple of examples of Floquet multipliers spectrum for a 1-site breather in soft and hard potentials.

When dealing with a multibreather, $p > 1$ and there are $p - 1$ eigenvalues whose evolution is not clear (i.e. if they move along the circle or they abandon it through the real axis). The first attempt to give a generic explanation of the emerging possibilities was based on Aubry's band theory [71], while a second approach, which makes use of the effective Hamiltonian methodology [72], is able to prove the existence of breathers at small coupling and also to give an expression of the individual multipliers at low coupling. The work of [49] demonstrated that the two approaches are equivalent if one assumes that Aubry's bands are parabolic. The main drawback of these findings is that they can only predict the stability or instability of multibreathers at low coupling when all the excited sites are adjacent. The latter limitation was overcome in the work of [73] by considering higher order perturbations. However, for simplicity, we will hereafter focus on the description of multibreathers whose excited sites are adjacent, as per the earlier studies by the authors in [49, 71, 72].

As a general result, we proved that, for $C \to 0^+$, multibreathers are stable when all the excited sites at the AC limit are in phase (anti-phase) if the potential is hard (soft). Two sites i and j are said to be in phase (anti-phase) if $\sigma_i\sigma_j = +1$ ($\sigma_i\sigma_j = -1$). This property is reversed if $C < 0$; such a case is relevant when considering e.g. dipole-dipole interactions in DNA models [74]. If the excitation pattern is neither uniformly in phase, nor uniformly in anti-phase (e.g. $\sigma = \{1, -1, -1, 1\}$) the multibreather is unstable, independently of the hardness of the potential or sign of coupling constant, as demonstrated in [75]. The work of [72] provided a detailed count of the unstable eigendirections in such a case and an estimate of the associated multipliers in the small C limit.

As an instructive special case example, let us focus on the simplest multibreather case, namely the 2-site DB in the hard and soft ϕ^4 potential (7.6). Among these multibreathers, the ones that are unstable close to the AC limit, by virtue of the theorems of [71, 72], are the Page modes, whose codes are $\sigma = \{1, 1\}$ (if the potential is soft) and $\sigma = \{1, -1\}$ (if the potential is hard). The theorems also predict that one of the multiplier pairs lying at $\theta = 0$ at the AC limit moves along the real axis with its (imaginary) argument given by:

$$\theta = \pm \frac{2i\pi}{\omega_b} \sqrt{2sC \frac{J}{\omega_b} \frac{\partial \omega_b}{\partial J}}, \tag{7.17}$$

with

$$J = \frac{1}{2\pi} \int_0^T [\dot{u}(t)]^2 \mathrm{d}t = \frac{\omega_b}{2} \sum_{k \geq 1} k^2 z_k^2, \tag{7.18}$$

being the action of an isolated oscillator. In the case of the ϕ^4 potential, finding an analytical form of $\partial\omega_b/\partial J$ by using (7.7) and (7.8) is a cumbersome task. An alternative is to use the rotating wave approximation (RWA) $u(t) \approx z_1 \cos(\omega t)$, obtaining that:

$$z_1 = 2\sqrt{\frac{\omega_b^2 - 1}{3s}} \tag{7.19}$$

Then, the Floquet argument of the analyzed multibreathers can be written as:

$$\theta = \pm \frac{2i\pi}{\omega_b} \sqrt{2sC \frac{\omega_b^2 - 1}{3\omega_b^2 - 1}}. \tag{7.20}$$

In Fig. 7.2, we compare the Floquet arguments of the stable 2-site multibreathers found by integrating equation (7.10) with those of the predictions of [49, 72], i.e. (7.20) for $C \to 0^+$. Notice the good agreement between all the curves.

The stability theorems have also been useful towards investigating the stability or instability (at low coupling) of dark breathers [71] and multibreathers in the presence of inhomogeneities [75]. They have also been applied to multibreathers with broken time-reversibility, as phonobreathers with phase torsion [53], multibreathers in 1D lattices with long-range interactions [56, 57] and vortices in triangular [55] and rectangular latices [49]. In the latter case, the stability theorems are unable to get analytical predictions when the on-site potential is even (as e.g. in the hard and soft ϕ^4 potentials), as the perturbations are of order higher than $\sim \sqrt{C}$. This type of scenario has been explored in the case of the simpler class of DNLS models, as summarized in [76]. In that context, it has been dubbed as the super-symmetric case, and requires higher order perturbative analysis in order to identify the leading order contributions to the multipliers for which the corresponding discrete vortices arise at $O(C)$ — although for some of them, they arise at much higher order (such as $O(C^3)$). A systematic consideration of such stability findings for DBs is a worthwhile investigation for future work. It should be noted that in the present contribution we do not expand further on the specifics of both the Aubry band and the effective Hamiltonian methods and their analytical predictions and comparisons with full

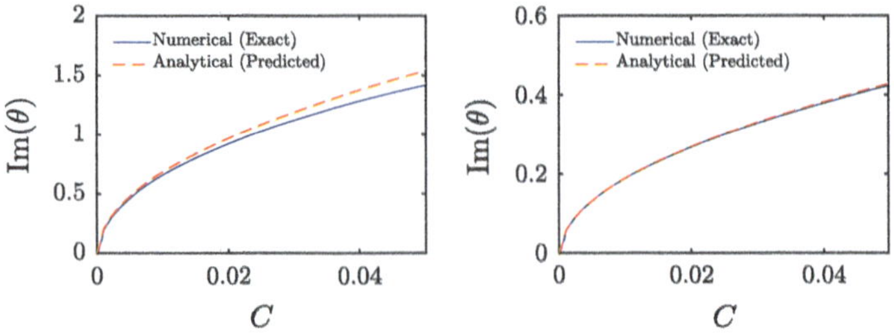

Fig. 7.2 Imaginary part of the argument of the Floquet multiplier responsible for the instability of Page modes with respect to C in soft potentials with $\omega_b = 0.80$ (left panel) and hard potentials with $\omega_b = 2.5$ (right panel). The figure, which highlights the region of small coupling, also shows the analytical prediction of (7.20)

numerical results. A reasonably recent and relatively up-to-date summary of these methods can be found in more detail in [49].

7.3.3 Linear Stability at Arbitrary Coupling. An Energy-Based Stability Criterion

When the coupling is increased from the situation described in the previous Subsection, the phonon arcs will be expanding. The mapping between angles and phonon frequencies is not exact because there exists a hybridization between the linear modes and the breather itself, which is more pronounced when the coupling is progressively higher. In fact, in many cases, beyond a threshold coupling, exponentially localized so-called "internal modes" can bifurcate from the phonon band. These modes localize in the rightmost (leftmost) edge of the arc if the Krein signature of the phonons of the upper half-circle is positive (negative). In addition, in the case of multibreathers, there can also exist localized multipliers that were at $\theta = 0$ at the AC limit. These modes may move on the unit circle for finite coupling and may lead to various scenarios of collision between Floquet multipliers. For instance, two pairs of localized modes can collide at $\theta \neq 0$ or a localized mode pair can collide with the phonon arcs. Recall that the Krein signature of the colliding multipliers must be opposite in order for a (Hamiltonian Hopf) bifurcation to take place (otherwise the pairs are led to a so-called avoided crossing scenario without producing an instability). In the opposite Krein multiplier collision case, we are dealing with a Hopf bifurcation (also called Krein crunch) that generically persists when the lattice is enlarged. Another situation occurs when the phonon arcs themselves collide. As they have opposite Krein signature, more Krein crunches can take place. If this bifurcation takes place at $\theta = 0$, it is a signal that a resonance of the breather with the phonon band has taken place and, consequently, the existence condition of the MacKay-Aubry theorem has been violated. If the arcs' overlap takes place at $\theta \neq 0$, the bifurcations are generally ruled out at infinite lattices [77].

There are other interesting cases. For instance, when localized modes collide at $\theta = \pi$, they bring about a period-doubling bifurcation; such a bifurcation was found for ϕ^4 Hamiltonian lattices at low coupling in [73]. Another scenario corresponds to the case when a multiplier, which escapes from the phonon arc, corresponds to an anti-symmetric mode and moves along the unit circle, eventually colliding with its conjugate at $\theta = 0$; in this case, the mode is structurally similar to the translational one and a perturbation of a breather along its direction increases linearly (as the model is marginal), giving rise to a moving breather [78]. We will deal with this topic in more detail in Sect. 7.4.1.

Figure 7.3 shows examples of the Floquet multiplier dependence with respect to C for Page (unstable two-site) modes and also for a stable configuration close to the AC limit, which corresponds to the $\sigma = \{1, 1\}$ multibreather in the hard potential. In the case of the Page modes one can see that the only instability is the one caused by the Floquet multiplier that abandons the circle at the AC limit, and that, in the case of hard potentials, this mode comes back to the circle when the breather frequency resonates with the linear modes band. For the $\sigma = \{1, 1\}$ multibreather in the hard potential, it is observed that the instability is caused by the collision with the phonon arc of the localized mode that departed from $\theta = 0$ at the AC limit. Notice that the $\sigma = \{1, -1\}$ multibreather, which is stable for small coupling when the on-site potential is soft, undergoes a Hopf bifurcation caused by the collision of two localized modes in KG lattices with Morse on-site potential (see e.g. [79]); however, in soft ϕ^4 potentials, this bifurcation does not take place and, instead, there is a Hopf bifurcation similar to the one of the $\sigma = \{1, 1\}$ multibreather in the hard potential.

Tangent bifurcations can also occur when there exists an extremum in the energy dependence with respect to the breather frequency. In KG lattices, such extrema seem only to appear in two-dimensional lattices and, in several cases, are related to energy thresholds for the existence of breathers [32, 80, 81]. The stability criterion that we analyzed in [82] is related to such bifurcations. This criterion is a generalization to KG and FPUT lattices (and their DB solutions) of the well known Vakhitov–Kolokolov (VK) criterion defined for the NLS equation [83]. In fact, as shown in [82], upon the multiscale expansions that lead from the former equations to the latter one, the VK criterion is retrieved as a special case example. This establishes a sufficient condition for the instability of breathers. In KG lattices, a family of breathers is exponentially unstable if its energy increases with the frequency, if the on-site potential is soft, or if the energy decreases with the frequency if the on-site potential is hard. In fact, if the dependence of the energy versus frequency presents an extremum, there is a tangent bifurcation at that point. This energy criterion is unable to predict oscillatory instabilities (in our case, caused by Hopf bifurcations). In addition, it is also unable to predict the bifurcations from translational modes, i.e. the ones which produce moving breathers, or those bifurcations causing blow-up.

In the case of ϕ^4 potentials, one-dimensional 1-site breathers do not present energy thresholds. But, as mentioned above, the threshold exists for 2D and 3D lattices. This implies an extremum in the energy-frequency dependence and, consequently,

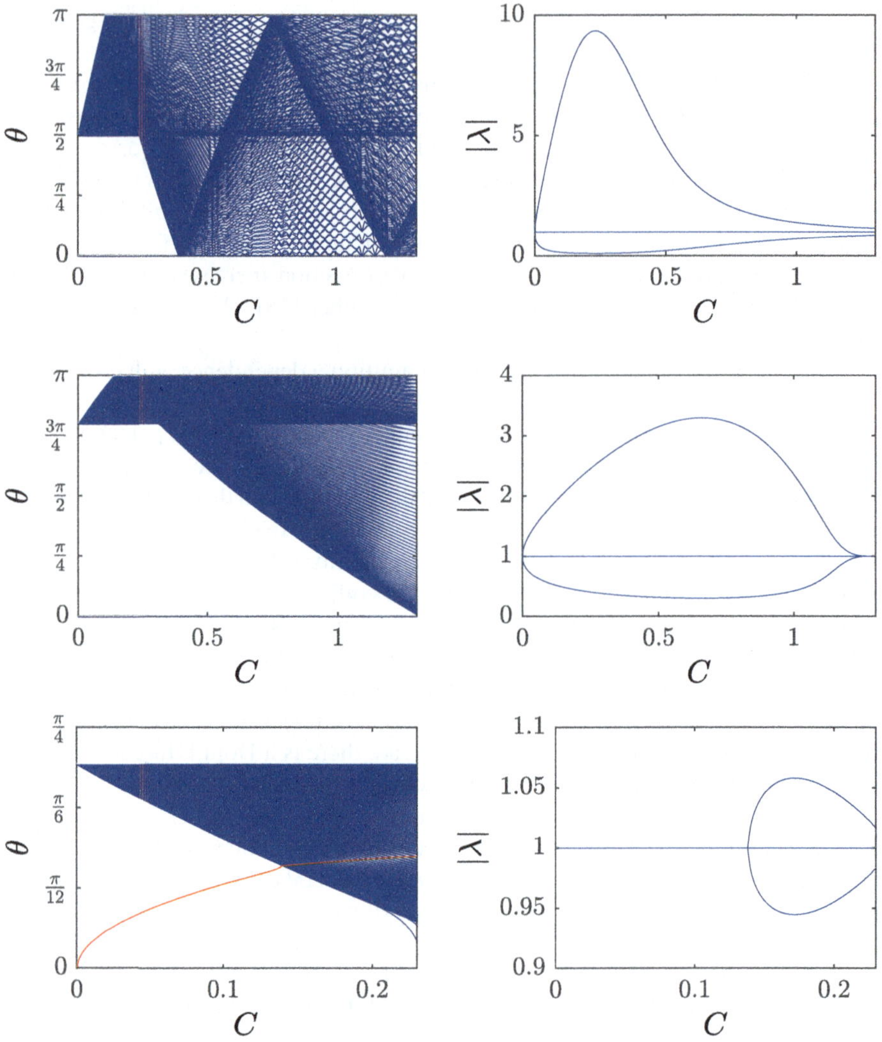

Fig. 7.3 Dependence with respect to the coupling constant C of the argument (left panels) and modulus (right panels) of the Floquet multipliers corresponding to the following configurations: (top panels) $\sigma = \{1, 1\}$ multibreather with $\omega_b = 0.8$ in soft potentials; (middle panels) $\sigma = \{1, -1\}$ multibreather with $\omega_b = 2.5$ in hard potentials; (bottom panels) $\sigma = \{1, 1\}$ multibreather with $\omega_b = 1.5$ in hard potentials. Notice that in the bottom panels, the mode with positive (negative) Krein signature is depicted in red (blue)

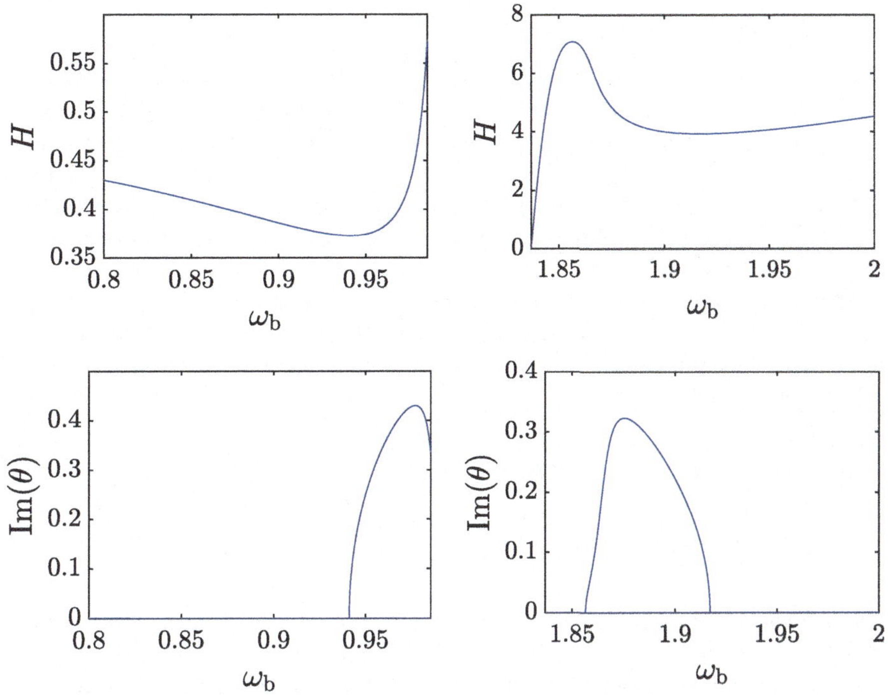

Fig. 7.4 The top panels show the energy-frequency dependence of discrete breathers in 2D lattices with a soft potential and $C = 0.1$ (left panels) and a hard potential and $C = 0.3$ (right panels). Bottom panels show the imaginary part of the argument of the Floquet multipliers of these solution families

a tangent bifurcation that can be predicted by the energy-based criterion of [82]. This is shown in Fig. 7.4. There, it is clear that for the 2D lattices the changes of monotonicity lead to the corresponding changes of stability.

The energy criterion has been successfully applied for predicting stability changes in travelling waves (supersonic lattice solitons) in FPUT lattices with soft potentials [84, 85]. In particular, the connection between breathers and traveling waves is an intriguing one: traveling waves on a lattice typically return to themselves upon translation by a single lattice site. As such, if one considers the operation including traveling by one lattice site and then back-shift to the original starting point, then it has a fixed point (if the original lattice has a genuine traveling wave). The period of the associated periodic orbit is directly associated with the speed of the traveling wave (TW) since $\omega_b = 2\pi s/h$ where s is the speed of the wave and h the spacing of the lattice. We thus give a very short proof of the theorem for the case of traveling waves and, by direct analogy, for DBs.

Consider the dynamical system: $U_t = \begin{pmatrix} u \\ p \end{pmatrix}_t = J\nabla\mathcal{H}(U)$ where $J = \begin{pmatrix} 0 & I \\ -I & 0 \end{pmatrix}$. We seek a TW in the form: $u = u_0(n - st) = u_0(\xi)$ [solving $-sU_{0,\xi} = J\nabla\mathcal{H}(U_0)$] and linearize around it according to: $u(\xi, t) = u_0(\xi) + \epsilon W(\xi, t)$ and $p(\xi, t) = p_0(\xi) + \epsilon P(\xi, t)$. Then, the linearization operator and its adjoint can be found explicitly as: $\mathcal{M} := s\partial_\xi + J\nabla^2\mathcal{H}(U_0), \quad \mathcal{M}^* = (-\nabla^2\mathcal{H}(U_0)J - s\partial_\xi) = J\mathcal{M}J$. Similarly to the breather problem, the time translation invariance induces an eigenvector and a generalized eigenvector with 0 eigenvalue of the form: $\partial_\xi U_0$ (the eigenvector) and $\mathcal{M}(-\partial_s U_0) = \partial_\xi U_0$ (the equation for the generalized eigenvector).

Then the straightforward proof of the stability theorem is as follows: If an extra eigenvector $\tilde{Y}$ with $\lambda = 0$ exists, then it must satisfy $\mathcal{M}\tilde{Y} = \partial_s U_0$ (yielding an additional generalized eigenvector). This, however, imposes the following symplectic orthogonality condition:

$$0 = \langle J\partial_\xi U_0, \mathcal{M}\tilde{Y}\rangle = \int (J\partial_\xi U_0) \cdot (\partial_s U_0) \mathrm{d}\xi = \int \frac{1}{s}\nabla\mathcal{H}(U_0) \cdot \frac{\partial U_0}{\partial s}\mathrm{d}\xi$$
$$= \frac{1}{s}\int \frac{\partial\mathcal{H}(U_0)}{\partial s}\mathrm{d}\xi = \frac{1}{s}H'(s).$$

Consequently, at this critical point the derivative $H'(s)$ must vanish, and similarly for DBs $H'(\omega_b) = 0$. As explained in [82] for DBs and [84, 85] for TWs, one can take the calculation further providing estimates of the bifurcating instability-inducing multipliers on the two sides of the critical point.

7.3.4 Nonlinear Stability

As illustrated in [86] for a wide range of NLS type models, linear stability is not the last word, as there are cases of e.g. spatially-antisymmetric solitons, which are linearly stable but the dynamics can be nonlinearly unstable, if it is evolved over sufficiently long time scales. In both lattice and continuum NLS modes, it was shown in that work that spectral stability may be inconclusive if modes of negative Krein signature exist in the spectrum and if their (nonlinearity induced) harmonics are in resonance with the continuous spectrum. In particular, an ODE was derived suggesting that for positive energy modes, their energetic content is depleted due to dispersive wave radiation. However, for negative Krein signature modes, the reverse process occurs, eventually pumping (over a slow, power-law in time procedure) the internal mode and ultimately leading to the demise of the coherent structure due to this genuinely nonlinear mechanism.

Based in such findings, in [79] we explored the nonlinear instability of linearly stable multibreathers. We found that nonlinear instability is possible when an integer multiple of the frequency of an internal (i.e. localized) eigenmode resonates with the phonon band and the Krein signature of such a mode and the phonon arc are opposite to each other. Notice that the frequency Ω of the internal eigenmode and its Floquet argument follows the relation $\theta = \pm\Omega T \mod 2\pi$; consequently, if the coupling constant is small enough, the internal mode has not collided with the phonon arc (in fact, this condition is necessary in order for the breather to be stable against potential Hamiltonian Hopf bifurcations), and then $\Omega = \theta/T = \omega_b\theta/(2\pi)$. In this case, however, a harmonic of such a mode can be located inside the arcs. If it is the 2nd harmonic, then the slow growth follows a $t^{-1/2}$ law, if the 3rd, a $t^{-1/4}$ law etc., as derived by the corresponding ODE [79, 86].

Having in mind the stability properties of multibreathers near the AC limit, the only stable solutions among them are those whose codes σ are in phase if the on-site potential is hard and in anti-phase if the potential is soft. As demonstrated in [79], the internal modes that detach from $\theta = 0$ have, in the upper half-circle (i.e. for $\theta \in [0, \pi]$), Krein signature $\kappa = 1$ if the potential is soft and $\kappa = -1$ if it is hard. On the other hand, in order to have nonlinear instability, it is needed that the Krein signature of the phonon arcs in the upper half-circle are the opposite to the internal modes. Because of this, as explained at the beginning of Sect. 7.3.2, nonlinear instability can only be possible if $\omega_o < \omega_b \leq 2\omega_o$ if the potential is hard and $\frac{2}{2k+1}\omega_o < \omega_b < \frac{1}{k}\omega_o$ with $k \in \mathbb{N}$ if the potential is soft. These conditions were corroborated in the detailed numerical computations of [79]. In the simpler case of the NLS and DNLS models, no such conditions need to be imposed, however numerical computations verified the existence of the instability in [86]. Here, we only provide a prototypical case example of the associated phenomenology in the bottom panels of Fig. 7.3. This shows the particular case of the $\sigma = \{1, 1\}$ multibreather in a hard potential. This solution is linearly stable before the Hopf bifurcation takes place, although it presents a nonlinear instability as the conditions of the paragraph above are fulfilled. The dynamics eventually manifests this instability although the latter only arises over extremely long times, much longer than the linearly unstable cases of the top and bottom panels.

7.3.5 Dynamics

In this subsection we will further discuss some prototypical examples of unstable dynamics stemming from linear and nonlinear instabilities of the (2-site) multibreathers shown previously (especially in Fig. 7.3).

First of all, we consider linearly unstable multibreathers in the soft ϕ^4 potential. In that case, the main behavior is a blow-up caused by a phenomenon similar to the escape described in [47]. The left panels of Fig. 7.5 illustrate the blowing-up dynamics observed in an unstable $\sigma = \{1, 1\}$ multibreather. Notice that blow-up is caused by both exponential and oscillatory instabilities. I.e., the linearly unstable growth leads the oscillators to exit the finite height potential barrier and hence tend to $\pm\infty$.

In the case of hard potentials, the main dynamical features consist of the transformation of the multibreather into a less energetic 1-site breather. In this transformation, some energy is shed from the breather and, in some cases, it can be localized. The

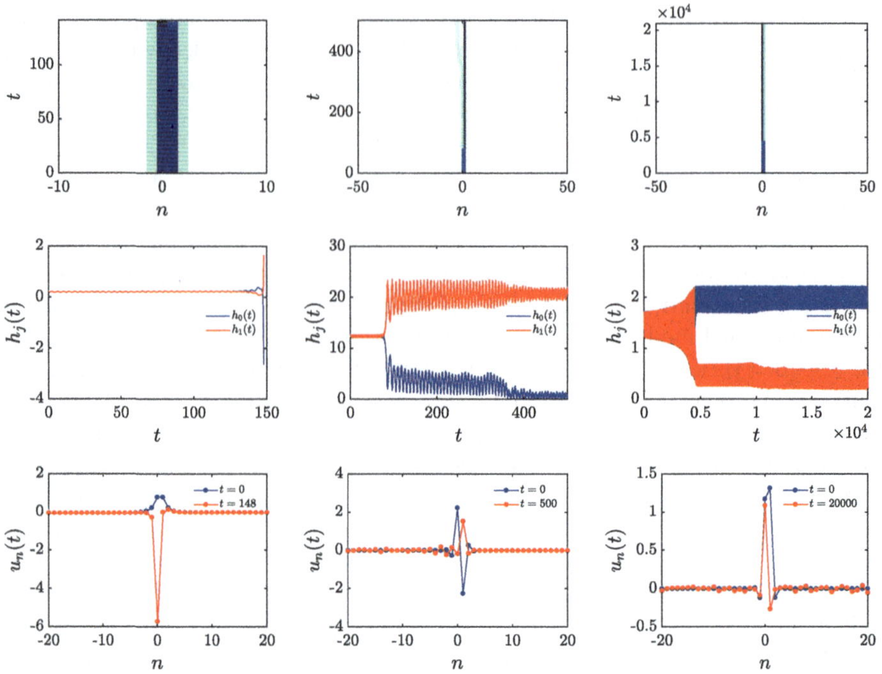

Fig. 7.5 Dynamical evolution of the unstable configurations displayed in Fig. 7.3. Top panels show the space-time dependence of the energy density, middle panels depict the energy density evolution of the central sites and bottom panels compare to the profiles of the perturbed breather at the beginning and at the end of the simulation. The multibreathers considered at the figures are: (left panels) $\sigma = \{1, 1\}$ in a soft potential with $\omega_b = 0.8$ and $C = 0.2$; (central panels) $\sigma = \{1, -1\}$ in a hard potential with $\omega_b = 2.5$ and $C = 0.5$; (right panels) $\sigma = \{1, 1\}$ in a hard potential with $\omega_b = 1.5$ and $C = 0.1$

central panels of Fig. 7.5 shows the dynamics of the $\sigma = \{1, -1\}$ multibreather. It is natural for the configuration to tend to the site-centered variant as that is generically stable, as we discussed above.

Finally, the right panels of the figure consider a linearly stable but nonlinearly unstable multibreather with $\sigma = \{1, 1\}$, displaying a similar behaviour. The nonlinearly unstable dynamics fulfills the conditions of the theorem of [79] as the frequency of the internal mode is $\Omega = 0.2073$ and the spectral bands expands in $[0.32, 0.5]$. As a result, the second harmonic of the internal mode lies in the phonon band, ultimately (at very long times) slowly feeding the nonlinear growth and eventually leading to the destruction of the two-site configuration. Notice, however, as also indicated above, the characteristically longer (by at least an order of magnitude) time needed for the manifestation of this (nonlinear) instability.

7.4 Some Glimpses on Other Breather Features

Although the core of the present chapter focuses on the stability properties of discrete breathers in ϕ^4 lattices, we mention in passing some developments related to a judicious selection of aside topics, namely moving breathers and the generation of discrete breathers in lattices with dissipation.

7.4.1 Moving Breathers

As mentioned in the previous section, when an anti-symmetric eigenmode detaches from the phonon arc, it can reach $\theta = 0$ bringing about a tangent bifurcation. Just at $\theta = 0$, the mode is marginal and resembles a translational mode, so that a perturbation along it can set the breather into motion. Contrary to continuous breathers, moving discrete breathers generically radiate phonons (see e.g. [78]) and they eventually stop.

Due to the special features of moving breathers, there does not exist a systematic underlying mathematical theory that can clearly characterize them. Some attempts of defining a Peierls–Nabarro barrier similar to kinks have been performed, but they only seem to work in FPUT lattices close to the continuum limit [87]. In any case, there must be a mechanism similar to the Peierls–Nabarro barrier related to the existence of tangent bifurcations of the translational mode that detaches from the phonon arc. Because of this, moving breathers can only be observed in lattices for which the breathers experience such bifurcations. There are only a few reported cases of one-dimensional KG lattices, namely, with Morse, sG and double-well potentials [48]. We have also been able to generate moving breathers in two-dimensional KG lattices with Morse potential, a result that has not been published yet. Moving breathers (with high mobility) have been observed in two-dimensional lattices with in-plane degrees of freedom modeling e.g. muscovite mica [88, 89] which are also known as quodons and, as mentioned in the Introduction, are speculated to play an important role in charge transport properties in such materials.

Moving breathers do not exist in KG lattices with the ϕ^4 potential, as the tangent bifurcation of the translational mode does not take place. However, such a bifurcation was observed for KG/FPUT lattices with on-site and interaction potentials of the hard ϕ^4 form [70]. Such a lattice has been used for modeling micromechanical cantilever arrays [22]. In [90] we generated moving breathers in such a model and analyzed their interaction with geometrical defects.

7.4.2 Dissipative Lattices

Most of the experimental findings of discrete breathers have been achieved on lattices with dissipation and external driving, such as micromechanical cantilever [22], pendulum [20] and Josephson junctions [18, 19] arrays, nonlinear electrical lattices [23, 24] or granular media [25, 26]. Such classes of systems remain quite popular to this day with numerous variations continuously arising including, e.g., piecewise-linear systems emulating the β-form of the celebrated FPUT lattice [91], or electrical systems involving beyond-nearest-neighbor interactions [92].

As demonstrated in [93], discrete breathers in dissipative lattices can also exist away from the anticontinuum limit. Contrary to Hamiltonian lattices, there are no resonances with phonons (the spectrum of plane wave excitations is pushed to the left half of the complex spectral plane); as a result such states are now potential attractors of the system. The work of [58] shows the complex phenomenology that is observed in driven and damped Frenkel–Kontorova lattices. If the system is driven with a frequency ω_b, discrete breather solutions acquire the same frequency as the driving force. In general, all the lattice sites oscillate with the same frequency; however, we have found an electrical lattice where subharmonic resonance emerges (that is, the excited sites of the breather oscillate with half of the frequency of the low amplitude sites) [94]. Recently, multistable variations of pendula, potentially applicable to SQUID metamaterials, have been manifested as potential sources of more complex breathing patterns such as the celebrated chimera states [95].

Discrete breathers have been studied in driven and damped one-dimensional KG lattices with hard ϕ^4 potentials in [96]. Such a lattice is defined by equation:

$$\ddot{u}_n + \alpha\dot{u}_n + V'(u_n) + C(2u_n - u_{n+1} - u_{n-1}) = F_n(t) + \eta_n(t) \tag{7.21}$$

with $F_n(t)$ being a periodic function of frequency ω_b and $\eta_n(t)$ is a Gaussian white noise with zero mean and autocorrelation $<\eta_n(t)\eta_m(t')>= 2D\delta_{nm}\delta(t-t')$. In the deterministic case ($D = 0$) and for a staggered driving of the form $F_n(t) = (-1)^n f\sin(\omega_\text{b}t)$, the phenomenology is quite simple: at given damping and frequency, discrete breathers only exist above a threshold f_th. However, if the noise is introduced in the lattice, there are two interesting phenomena: if the driving amplitude is suprathreshold ($f > f_\text{th}$), breathers with frequency ω_b can be generated even if the initial condition is uniform; if the driving amplitude is subthreshold ($f < f_\text{th}$), breathers are still produced by the concerted action of noise and the driving force, in a way that noise, on the one hand, enables system transitions between the uniform and the coherent localized states and, on the other hand, destroys any degree of order of the system if its amplitude is large enough: in other words, we are dealing with a stochastic resonance phenomenon.

7.5 Outlook and Future Directions

From the above discussion, it is clear that the theme of discrete breathers is one that is increasingly maturing and emerging in a wide range of applications and a diverse array of systems. In particular, we are gradually departing from the simplest nearest-neighbor scenarios of either just KG or just FPUT types and moving on to a new, more elaborate phase where systems can be designed with beyond-nearest [92] and even long-range interactions [65] and also with multiple and potentially competing [12] interactions, or with ones that are progressively more amenable to analytical considerations [91]. This suggests that there is a significant need for further theoretical and computational developments to support the corresponding emerging experimental platforms.

While the early stages of development of DBs favored analytical proofs of existence and associated techniques of numerical existence and stability, subsequent ones favored a more systematic exploration of spectral properties and an attempt to classify the different multibreathers and offer systematic guidelines about when they may be expected to be dynamically robust. In this Chapter, we summarized some of this systematic effort in the previous decade and some of its crystallized results and convergence of different methods over the past few years. More recently, further tools have arisen in probing spectral and dynamical features of DBs. Among others, we have explored and summarized here the energy-versus-frequency monotonicity criteria and how they relate to linear instabilities and given connections between these and the stability of traveling waves in lattices. Additionally, we have warned the reader against the naive expectation that spectral stability is the full story, presenting case examples where this fails to be true due to the nonlinear instability of internal modes with opposite Krein signature than that of the phonon arcs. The slow, power-law nature of the latter instabilities, as opposed to the exponential growth of linear instabilities was highlighted. Lastly, some possibilities of further developments towards moving breathers or dissipative lattices were briefly touched upon.

Clearly, there is need for further theoretical systematics. Many of the relevant points were raised in parts of our discussion. Understanding the stability of phase-shift multibreathers and vortex breathers in higher-dimensional systems is an important open topic. Carrying out the associated stability computations to higher order is particularly relevant. Recent work, in fact, brings up the possibility that relevant solutions may fail to exist at higher orders in some important case examples [97]. Studying also long-range interactions may bring about surprises and produce gaps in the spectrum where novel DBs may exist, as per the recent work of [98]. Again, this is a topic meriting further exploration. The study of systems with nontrivial tails (nanoptera) and the examination of whether the stability features/criteria presented herein apply to them is also an open topic. The same holds true for systems with external drive and damping: can we offer some guidelines to characterize their stability characteristics, suitably adapting what we know in the more structured Hamiltonian cases or perhaps not ? Plus then there are topics which, while touched upon, still seem fairly poorly understood or wide open for new insights: among them moving breathers, or quasi-periodic solutions and their existence and stability, as well as the role of DBs in asymptotic dynamics and thermalization (see, e.g., [99] for a recent summary in the discrete NLS case) are only some that come to mind. In summary, discrete breathers may have matured but have many more challenges to offer for the years to come both at the fundamental, at the computational and at the experimental level...

Acknowledgements This material is based upon work supported by the National Science Foundation under Grant No. DMS-1809074 (P.G.K.). J.C.-M. thanks financial support from MAT2016-79866-R project (AEI/FEDER, UE). P.G.K. also gratefully acknowledges support from the US-AFOSR under Grant No. FA9550-17-1-0114. We are grateful to J.F.R. Archilla for his useful comments.

References

1. S. Flach, C.R. Willis, Phys. Rep. **295**, 181 (1998)
2. S. Flach, A.V. Gorbach, Phys. Rep. **467**, 1 (2008)
3. S. Aubry, Phys. D **216**, 1 (2006)
4. D.K. Campbell, M. Peyrard, in *CHAOS/XAOC: Soviet-American Perspectives on Nonlinear Science*, ed. by D.K. Campbell (American Institute of Physics, New York, 1990), p. 305
5. M. Ablowitz, D.J. Kaup, A.C. Newell, H. Segur, Stud. App. Math. **53**, 249 (1974)
6. H. Segur, M. Kruskal, Phys. Rev. Lett. **58**, 747 (1987)
7. J.P. Boyd, Nonlinearity **3**, 177 (1990)
8. P.W. Anderson, Phys. Rev. **109**, 1492 (1958)
9. A.J. Sievers, S. Takeno, Phys. Rev. Lett. **61**, 970 (1988)
10. S. Takeno, K. Kisoda, A.J. Sievers, Prog. Theor. Phys. Suppl. **94**, 242 (1988)
11. T. Dauxois, Phys. Today **61**, 55 (2008)
12. M. Kimura, T. Hikihara, Chaos **19**, 013138 (2009)
13. R.S. MacKay, S. Aubry, Nonlinearity **7**, 1623 (1994)
14. S. Aubry, Phys. D **103**, 201 (1997)
15. J.L. Marín, S. Aubry, Nonlinearity **9**, 1501 (1996)
16. J.F.R. Archilla, N. Jiménez, V.J. Sánchez-Morcillo, L.M. García-Raffi (eds.), *Quodons in Mica. Nonlinear Travelling Excitations in Crystals* (Springer, Cham, 2015)
17. S.V. Dmitriev, E.A. Korznikova, Yu.A. Baimova, M.G. Velarde, Phys. Uspekhi **59**, 446 (2016)
18. E. Trías, J.J. Mazo, T.P. Orlando, Phys. Rev. Lett. **84**, 741 (2000)
19. P. Binder, D. Abraimov, A.V. Ustinov, S. Flach, Y. Zolotaryuk, Phys. Rev. Lett. **84**, 745 (2000)
20. J. Cuevas, L.Q. English, P.G. Kevrekidis, M. Anderson, Phys. Rev. Lett. **102**, 224101 (2009)
21. F.M. Russell, Y. Zolotaryuk, J.C. Eilbeck, T. Dauxois, Phys. Rev. B **55**, 6304 (1997)
22. M. Sato, B.E. Hubbard, A.J. Sievers, Rev. Mod. Phys. **78**, 137 (2006)
23. F. Palmero, L.Q. English, J. Cuevas-Maraver, R. Carretero-González, P.G. Kevrekidis, Phys. Rev. E **82**, 026605 (2011)
24. L.Q. English, F. Palmero, J.F. Stormes, J. Cuevas-Maraver, R. Carretero-González, P.G. Kevrekidis, Phys. Rev. E **88**, 022912 (2013)
25. N. Boechler, G. Theocharis, S. Job, P.G. Kevrekidis, M.A. Porter, C. Daraio, Phys. Rev. Lett. **104**, 244302 (2010)
26. C. Chong, F. Li, J. Yang, M.O. Williams, I.G. Kevrekidis, P.G. Kevrekidis, C. Daraio, Phys. Rev. E **89**, 032924 (2014)
27. B.I. Swanson, J.A. Brozik, S.P. Love, G.F. Strouse, A.P. Shreve, A.R. Bishop, W.-Z. Wang, M.I. Salkola, Phys. Rev. Lett. **82**, 3288 (1999)
28. M. Sato, A.J. Sievers, Nature **432**, 486 (2004)

29. M.E. Manley, M. Yethiraj, H. Sinn, H.M. Volz, A. Alatas, J.C. Lashley, W.L. Hults, G.H. Lander, J.L. Smith, Phys. Rev. Lett. **96**, 125501 (2006)
30. M.E. Manley, A.J. Sievers, J.W. Lynn, S.A. Kiselev, N.I. Agladze, Y. Chen, A. Llobet, A. Alatas, Phys. Rev. B **79**, 134304 (2009)
31. M. Peyrard, Nonlinearity **17**, R1 (2004)
32. J.F.R. Archilla, J. Cuevas, M.D. Alba, M. Naranjo, J.M. Trillo, J. Phys. Chem. B **110**, 24112 (2006)
33. V.I. Dubinko, P.A. Selyshchev, J.F.R. Archilla, Phys. Rev. E **83**, 041124 (2011)
34. V.I. Dubinko, J.F.R. Archilla, S.V. Dmitriev, V. Hizhnyakov, in *Quodons in Mica. Nonlinear Travelling Excitations in Crystals*, ed. by J.F.R. Archilla, N. Jiménez, V.J. Sánchez-Morcillo, L.M. García-Raffi (Springer, Cham, 2015), p. 381
35. E. Mihóková, L.S. Schulman, J. Phys. A: Math. Theor. **43**, 183001 (2010)
36. M. Haas, V. Hizhnyakov, A. Shelkan, M. Klopov, A.J. Sievers, Phys. Rev. B **84**, 144303 (2011)
37. I.P. Lobzenko, G.M. Chechin, S. Bezuglova, Yu.A. Baimova, A. Korznikova, S.V. Dmitriev, Phys. Solid State **58**, 633 (2016)
38. J.A. Baimova, E.A. Korznikova, I.P. Lobzenko, S.V. Dmitriev, Rev. Adv. Mater. Sci. **42**, 68 (2015)
39. J.F.R. Archilla, F.M. Russell, Lett. Mater. **6**, 3 (2016)
40. F.M. Russell, in *Nonlinear Systems*, vol. 2, ed. by J.F.R. Archilla, F. Palmero, M.C. Lemos, B. Sánchez-Rey, J. Casado-Pascual (Springer, Cham 2018), p. 241
41. F.M. Russell, J.F.R. Archilla, F. Frutos, S. Medina-Carrasco, Europhys. Lett. **120**, 46001 (2017)
42. J.F.R. Archilla, S.M.M. Coelho, F.D. Auret, V.I. Dubinko, V. Hizhnyakov, Phys. D **297**, 56 (2015)
43. J.F.R. Archilla, S.M.M. Coelho, F.D. Auret, C. Nyamhere, V.I. Dubinko, V. Hizhnyakov, in *Quodons in Mica. Nonlinear Travelling Excitations in Crystals*, ed. by J.F.R. Archilla, N. Jiménez, V.J. Sánchez-Morcillo, L.M. García-Raffi (Springer, Cham, 2015), p. 343
44. S.M.M. Coelho, J.F.R. Archilla, F.D. Auret, J.M. Nel, in *Quodons in Mica. Nonlinear Travelling Excitations in Crystals*, ed. by J.F.R. Archilla, N. Jiménez, V.J. Sánchez-Morcillo, L.M. García-Raffi (Springer, Cham, 2015), p. 363
45. G. James, P.G. Kevrekidis, J. Cuevas, Phys. D **251**, 39 (2013)
46. G. James, B. Sánchez-Rey, J. Cuevas, Rev. Math. Phys. **21**, 1 (2009)
47. V. Achilleos, A. Álvarez, J. Cuevas, D.J. Frantzeskakis, N.I. Karachalios, P.G. Kevrekidis, B. Sánchez-Rey, Phys. D **244**, 1 (2013)
48. D. Chen, S. Aubry, G. Tsironis, Phys. Rev. Lett. **77**, 4776 (1996)
49. J. Cuevas, V. Koukouloyannis, P.G. Kevrekidis, J.F.R. Archilla, Int. J. Bif. Chaos **21**, 2161 (2011)
50. A. Álvarez, J.F.R. Archilla, J. Cuevas, F.R. Romero, New J. Phys. **4**, 72 (2002)
51. S. Takeno, M. Peyrard, Phys. D **92**, 140 (1996)
52. T. Cretegny, S. Aubry, Phys. D **113**, 162 (1998)
53. J. Cuevas, J.F.R. Archilla, F.R. Romero, J. Phys. A: Math. Theor. **44**, 035102 (2011)
54. T. Cretegny, S. Aubry, Phys. Rev. B **55**, R11929 (1997)
55. V. Koukouloyannis, R.S. MacKay, J. Phys. A: Math. Gen. **38**, 1021 (2005)
56. V. Koukouloyannis, P.G. Kevrekidis, J. Cuevas, V. Rothos, Phys. D **242**, 16 (2013)
57. Z. Rapti, Phys. Lett. A **377**, 1543 (2013)
58. J.L. Marín, F. Falo, P.J. Martínez, L.M. Floría, Phys. Rev. E **63**, 066603 (2001)
59. A.M. Morgante, M. Johansson, S. Aubry, G. Kopidakis, J. Phys. A: Math. Gen. **35**, 4999 (2002)
60. M. Chirilus-Bruckner, C. Chong, J. Cuevas-Maraver, P.G. Kevrekidis, in *The Sine-Gordon Model and Its Applications*, ed. by J. Cuevas-Maraver, P.G. Kevrekidis, F.L. Williams (Springer, Heidelberg, 2014), p. 31
61. G. Kopidakis, S. Aubry, Phys. D **130**, 155 (1999)
62. G. Kopidakis, S. Aubry, Phys. D **139**, 247 (2000)
63. J.F.R. Archilla, R.S. MacKay, J.L. Marín, Phys. D **134**, 406 (1999)
64. S. Flach, Phys. Rev. E **58**, R4116 (1998)

65. M. Molerón, C. Chong, A.J. Martínez, M.A. Porter, P.G. Kevrekidis, C. Daraio, arXiv:1801.09560
66. B. Dey, M. Eleftheriou, S. Flach, G.P. Tsironis, Phys. Rev. E **65**, 017601 (2001)
67. J.C. Comte, Phys. Rev. E **65**, 067601 (2002)
68. S. Aubry, G. Kopidakis, V. Kadelburg, Discret. Contin. Dyn. Syst. Ser. B **1**, 271 (2001)
69. R.S. MacKay, J.-A. Sepulchre, Phys. D **119**, 148 (1998)
70. J.L. Marín, S. Aubry, M. Floría, Phys. D **113**, 283 (1998)
71. J.F.R. Archilla, J. Cuevas, B. Sánchez-Rey, A. Álvarez, Phys. D **180**, 235 (2003)
72. V. Koukouloyannis, P.G. Kevrekidis, Nonlinearity **22**, 2269 (2009)
73. D.E. Pelinovsky, A. Sakovich, Nonlinearity **25**, 3423 (2012)
74. J. Cuevas, J.F.R. Archilla, Yu.B. Gaididei, F.R. Romero, Phys. D **163**, 106 (2002)
75. J. Cuevas, J.F.R. Archilla, J.F.R. Romero, Nonlinearity **18**, 769 (2005)
76. P.G. Kevrekidis, *The Discrete Nonlinear Schrödinger Equation: Mathematical Analysis, Numerical Computations and Physical Perspectives* (Springer, Heidelberg, 2009)
77. J.L. Marín, S. Aubry, Phys. D **119**, 163 (1998)
78. S. Aubry, T. Cretegny, Phys. D **119**, 34 (1998)
79. J. Cuevas-Maraver, P.G. Kevrekidis, D.E. Pelinovsky, Stud. Appl. Math. **137**, 214 (2016)
80. S. Flach, K. Kladko, R.S. MacKay, Phys. Rev. Lett. **78**, 1207 (1997)
81. M. Kastner, Nonlinearity **17**, 1923 (2004)
82. P.G. Kevrekidis, J. Cuevas-Maraver, D.E. Pelinovsky, Phys. Rev. Lett. **119**, 094101 (2016)
83. N.G. Vakhitov, A.A. Kolokolov, Radiophys. Quantum Electron. **16**, 783 (1973)
84. J. Cuevas-Maraver, P.G. Kevrekidis, A. Vainchtein, H. Xu, Phys. Rev. E **90**, 042903 (2017)
85. H. Xu, J. Cuevas-Maraver, P.G. Kevrekidis, A. Vainchtein, Phil. Trans. R. Soc. A **376**, 2017092 (2018)
86. P.G. Kevrekidis, D.E. Pelinovsky, A. Saxena, Phys. Rev. Lett. **114**, 214101 (2015)
87. M. Kastner, J.-A. Sepulchre, Discret. Contin. Dyn. Syst. Ser. B **5**, 719 (2005)
88. J. Bajars, J.C. Eilbeck, B. Leimkuhler, Phys. D **301–302**, 8–20 (2015)
89. J. Bajars, J.C. Eilbeck, B. Leimkuhler, in *Quodons in Mica. Nonlinear Travelling Excitations in Crystals*, ed. by J.F.R. Archilla, N. Jiménez, V.J. Sánchez-Morcillo, L.M. García-Raffi (Springer, Cham, 2015), p. 35
90. J. Cuevas, P.G. Kevrekidis, Phys. Rev. E **69**, 056609 (2004)
91. Y. Watanabe, T. Nishida, Y. Doi, N. Sugimoto, Phys. Lett. A **382**, 1957 (2018)
92. X.-L. Chen, S. Abdoulkary, P.G. Kevrekidis, L.Q. English, arXiv:1806.07494
93. J.-A. Sepulchre, R.S. MacKay, Nonlinearity **10**, 679 (1997)
94. L.Q. English, F. Palmero, P. Candiani, J. Cuevas, R. Carretero-González, P.G. Kevrekidis, A.J. Sievers, Phys. Rev. Lett. **108**, 084101 (2012)
95. J. Hizanidis, N. Lazarides, G.P. Tsironis, Phys. Rev. E **94**, 032219 (2016)
96. D. Cubero, J. Cuevas, P.G. Kevrekidis, Phys. Rev. Lett. **102**, 205505 (2009)
97. T. Penati, M. Sansottera, S. Paleari, V. Koukouloyannis, P.G. Kevrekidis, Phys. D **370**, 1 (2018)
98. Y.Y. Yamaguchi, Y. Doi, Phys. Rev. E **97**, 062218 (2018)
99. T. Mithun, Y. Kati, C. Danieli, S. Flach, Phys. Rev. Lett. **120**, 184101 (2018)

Chapter 8
ϕ^4 Continuum Breathers: Forty Years After

John P. Boyd

Abstract More than forty years ago, a perturbation series in powers of a small parameter ϵ, proportional to amplitude, was found as a description of breathers solving the ϕ^4 wave equation. These breathers are periodic in time and localized in space. Thirty-five years ago, Segur and Kruskal showed that such breathers were weakly nonlocal in space because of emission of radiation with an amplitude exponentially small in $1/\epsilon$. In this chapter we discuss existence-to-all-orders and beyond-all-orders nonexistence. We also describe progress in numerical algorithms for weakly nonlocal breathers. We also explain the slow decay of breathers on multiple logarithmic time scales.

8.1 Introduction: The ϕ^4 Breather

A "breather" is a solitary wave in which a steady translation, or a lack of translation, is accompanied by the rising and falling of a standing wave, or more poetically, the compression of lungs during respiration as illustrated in Fig. 8.1. Such solitary waves have been given a wide variety of other names in the literature as cataloged in Table 8.1.

In this chapter, we shall concentrate on direct calculations of the breather. It is possible to readily generate breathers from very non-breather-like initial conditions by colliding a kink soliton with an antikink soliton as shown by Kudryavtsev [5], Sugiyama [9], Campbell, Schonfeld and Wingate [10] and Campbell and Peyrard [11] and others. The concept of breather-as-kink-antisoliton bound state is responsible for many of the variant names in Table 8.1.

The ϕ^4 breather is both pedestrian and paradox. The breather is approximated by a small amplitude perturbation series which, to all orders, satisfies the criteria of a

J. P. Boyd (✉)
Department of Climate and Space Sciences and Engineering, University of Michigan, 2455 Hayward Avenue, Ann Arbor, MI 48109, USA
e-mail: jpboyd@umich.edu

P. G. Kevrekidis and J. Cuevas-Maraver (eds.), *A Dynamical Perspective on the ϕ^4 Model*, Nonlinear Systems and Complexity 26,
https://doi.org/10.1007/978-3-030-11839-6_8

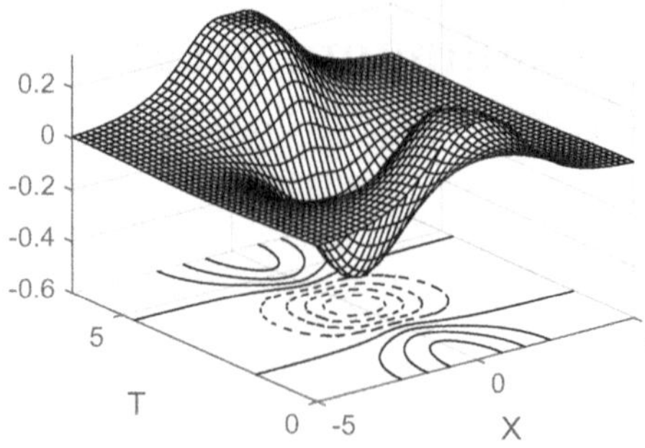

Fig. 8.1 Schematic of a breather over one complete cycle, $T \in [0, 2\pi]$, where $T = \omega t$ and ω is the breather frequency. Negative-valued isopleths in the contour plot are dashed. The infinite spatial domain has been truncated to $X \in [-5, 5]$ where $X = \epsilon x$ is the spatial coordinate scaled by the amplitude parameter ϵ

Table 8.1 Synonyms for the breather

Name	Reference
Breather	This chapter, Boyd (1998,1987,1990,1995) [1–4]
Double soliton	Kudryavtsev (1975) [5]
Soliton-antisoliton bound state	Geicke (1983) [6]
Pulsons	Geicke (1983) [6]
Bions	Getmanov (1976) [7] Geicke (1983) [6]
Oscillons	Hertzberg (2010) [8]

classical solitary wave, but Segur and Kruskal [12] have proved that the soliton does not exist!

The ϕ^4 continuum model is the PDE

$$\phi_{xx} - \phi_{tt} + \phi - \phi^3 = 0 \tag{8.1}$$

where subscript x or t denotes differentiation with respect to that coordinate. Because breathers oscillate about either $\phi = 1$ or $\phi = -1$, it is convenient to define the new unknown

$$u = \phi \pm 1 \tag{8.2}$$

which gives

$$u_{xx} - u_{tt} - 2u \mp 3u^2 - u^3 = 0 \tag{8.3}$$

Our choice is the upper sign [negative]. Segur and Kruskal make the opposite choice, but this difference is insignificant because their breather is just the negative of the "plus" breather.

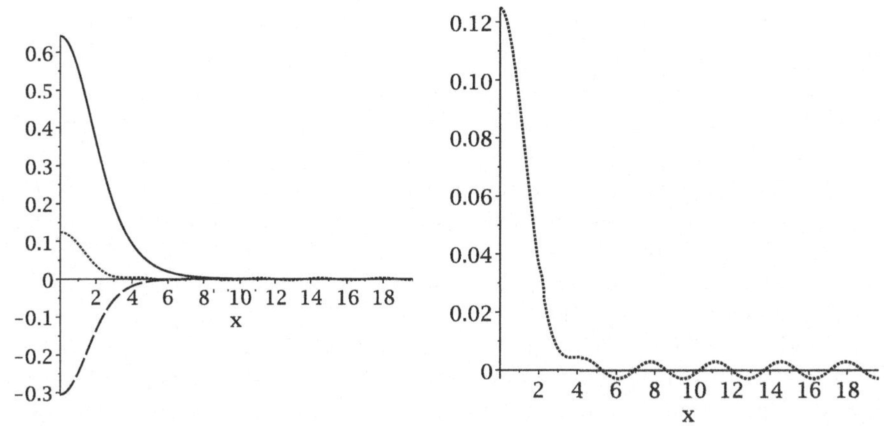

Fig. 8.2 Left: The first three time-harmonics at $t = 0$ for the breather with $\epsilon = 4/5$.The time coordinate is T, which is time rescaled so that the temporal period in the new coordinate is 2π. Only positive x is graphed because the breather and all its individual time-harmonics $A_n(x)$ are invariant to reflection with respect to the spatial origin. Solid: coefficient $A_1(x)$ of $\cos(T)$. Dashed: mean (time-independent component) $A_0(x)$. Dotted: second harmonic A_2. Right panel: Same as the left panel but showing the second harmonic only where $A_2(x)$ is the coefficient of $\cos(2T)$ in the Fourier cosine series in time

Section 8.4 reviews the multiple scales/small amplitude perturbation theory. This is divergent but asymptotic and gives a simple and explicit description of the *core* of the breather.

Unfortunately, the series in powers of an amplitude parameter ϵ completely misses an important feature: the breather steadily radiates quasi-sinusoidal waves of very small amplitude. Figure 8.2 (left) shows the first three time-harmonics for a moderate value of ϵ. An important part of the classical concept of a solitary wave or any type of coherent structure is *spatial localization*, that is, the solution decays rapidly with distance from the maximum (usually exponentially fast although Benjamin-Ono solitons decay proportionally to x^{-2}). The figure seems to suggest that "localization" is a property of each time-harmonic, but this is an illusion as is obvious when the second harmonic is graphed by itself (right).

Segur and Kruskal showed that the amplitude of this radiation is

$$\alpha(\epsilon) \sim 0.018 \exp(-3.85/\epsilon), \qquad \epsilon \ll 1 \tag{8.4}$$

where ϵ is the amplitude of the core $A_1(x)$ of breather.

The power series of $\exp(-q/\epsilon)$ for any positive constant q is the trivial series

$$\exp(-q/\epsilon) \sim 0 + 0\epsilon + 0\epsilon^2 + 0\epsilon^3 + \cdots \tag{8.5}$$

because $\lim_{\epsilon\to 0} d^k \exp(-q/\epsilon)/d\epsilon^k = 0$ for all orders of differentiation k. This is a consequence of the identity that

$$\lim_{\epsilon \to 0} \exp(-q/\epsilon)/\epsilon^{j} = 0 \quad \text{for all finite } j \tag{8.6}$$

This same identity implies that the exponential-of-a-reciprocal goes to zero more rapidly than *any finite power* of ϵ as $\epsilon \to 0$ and therefore is said to lie "beyond-all-orders" in ϵ.

Thus, the steadily emitted radiation is completely missed by the power series, even if carried to trillions of terms. It can only be calculated by a collection of methods variously called "beyond-all-orders asymptotics", "exponential asymptotics" or "hyperasymptotics" [1, 13]. Later, we discuss the subtleties of "all-orders existence" and "hyperasymptotic nonexistence".

Numerical methods are also challenged by beyond-all-orders effects. Below we review the derivation and application of special "radiation basis functions" to augment standard rational Chebyshev basis functions and compute the radiation when ϵ is not small.

The author has previously discussed the ϕ^4 breather in [3, 14] and a chapter in his book [1], a related ordinary differential in time in [15], and the closely related breathers and traveling solitons of the KG equation in [4], and these works cite many innovative papers by other authors. We shall therefore review many topics only in summary, and focus on unsolved problems and new perspectives.

8.2 Generalized Solitary Waves: The Existence of Solitons and the Nonexistence of Nanopterons

The ϕ^4 breather is but one example of a large bestiary of "weakly nonlocal" solitary waves. "Nonlocal" because the spatial localization intrinsic to the classical definition of a "solitary" wave is violated by quasi-sinusoidal radiation that is unbounded in space. "Weakly" because the amplitude of the radiation is very small. The "core" is the big central mountain. Everywhere outside the core is the "wings", dominated by the tiny sinusoidal ripples on the right of Fig. 8.2.

The cause of nonlocalization is *resonance* between the solitary wave and sine waves of infinitesimal amplitude and a particular wavenumber [1, 13, 16–19]. Equatorially-trapped Rossby waves in the ocean (with mean currents ignored) is an illuminating example. *Linear, small amplitude* Rossby waves have nondimensional phase speeds in the range $c \in [-1/3, 0]$. Solitons associated with latitudinal mode one have nonlinear phase speeds always more negative than $-1/3$; the $m = 1$ mode is a true solitary wave, moving too fast to be resonant with any linear Rossby waves (Chap. 16 of [20]). However, the $m = 3$ soliton has a nonlinear phase speed slightly more negative than $-1/7$, matching the phase speed of linear Rossby waves of nondimensional wavenumber two in the first latitudinal mode. Williams and Wilson's numerical simulations confirm the predicted radiative decay [16, 20, 21].

When the space-filling ripples are exponentially small in $1/\epsilon$, the "weakly nonlocal solitary wave" is a "nanopteron". Some solitary wave species (not discussed here) radiate with amplitude proportional to a power of ϵ; such "power-law-oscillations" are "micropterons". A "nanopteroidal wave" is a spatially-periodic generalization of a nanopteron. The former is often used to approximate the latter because when the "cores" of the endlessly-repeating wavetrain are tall and narrow relative to the spatial period, each peak of the nanopteroidal wave is dynamically isolated from its infinitely numerous identical siblings, and is therefore indistinguishable from a nanopteron, which is a single core "alone on a wide, wide sea", to borrow from the poet Coleridge.

The jargon is not terribly important, but the ubiquity of nonlocal solitary waves and of "beyond-all-orders" phenomena is. Good reviews may be found in the books [20, 22, 23].

The most familiar example of nonlocal solitary waves are surface gravity waves. When the nondimensional parameter known as the "Bond number" is larger than $1/3$, the waves are classical solitons, but the Bond number is this large only when the wavelength is a few millimeters or less. The much longer waves that corrugate the air-ocean interface are weakly nonlocal because the gravity wave solitons are resonant with small amplitude capillary waves, known in this context as "Wilton ripples" [20].

8.2.1 Asymptotic Existence and Beyond-All-Orders Nonexistence

To describe the subtleties of nanopterons, we need all of the following definitions.

Definition 1 (*Beyond-All-Orders Nonexistent*) A function which is "asymptotically existent" is "beyond-all-orders" nonexistent if there are hyperasymptotic corrections to the asymptotic power series which violate the required properties.

Segur and Kruskal, who entitled their 1987 paper "On the nonexistence of small amplitude breather solutions in ϕ^4 theory" [12] mean "nonexistence" in this "beyond-all-orders" sense.

Definition 2 (*Asymptotically Existent*) A function with a certain set of properties, such as spatial localization or the absence of high-frequency wiggles, is "asymptotically existent" if there is an asymptotic but divergent series with the required properties.

It is a frequent occurrence that hyperasymptotic corrections, proportional to $\exp(-q/\epsilon)$ for some positive constant q and the small positive perturbation parameter ϵ, drastically change the *qualitative* properties of a solution. It is surely wise to know that the soliton core is radiating, or that the electron wavefunction of a hydrogen atom is decaying exponentially with time due to ionization caused by an external magnetic or electric field.

However, because the hyperasymptotic correction is an *exponential* function of $1/\epsilon$, the beyond-all-orders effect may be *qualitatively negligible*. For the ϕ^4 breather, the ratio of the amplitude of the core to the amplitude of the sinusoidal wings is, when $\epsilon = 1/10$, 2.4×10^{24}!

Evagrius Ponticus (345–399), the great theologian who originated the concept of the Seven Deadly Sins, distinguished between *gnostike*, which is abstract, contemplative theory taught only to his best students and *praktike*, which is concrete, practical knowledge shared with all.

Definition 3 (*Praktike Existence*) An entity that violates a certain classification only through hyperasymptotic corrections too small to be of any significance in an observational or engineering sense exists in a "practical" or *praktike* sense.

8.2.2 Hyperasymptotically Nonexistent But Asymptotically and Praktike Existent: The Slow Manifold of Atmospheric Dynamics

Large-scale atmospheric flow is in a state of so-called quasi-geostrophic balance in which wind and temperature are inextricably connected through the thermal wind equation. The errors in anemometers, which measure wind, and themometers, which measure temperature, are uncorrelated and therefore unbalanced. Initializing a weather forecast with raw data is a disaster wrapped in a catastrophe. The imbalance excites spurious gravity waves which trigger vast amounts of wholly imaginary precipitation on the scale of the computational grid. Therefore, all operational weather forecasting systems preprocess the data by slightly adjusting the initial point in the phase space of the model from the observed point that will give slow-with-a-little-bit-of-fast dynamics to the nearest point on the "slow manifold", the subspace of balanced dynamics. Baer and Tribbia showed that initialization is equivalent to a singular perturbation expansion in powers of the Rossby number [24, 25]. Although Baer-Tribbia and equivalent initializations are very effective in practice, the Rossby number power series is asymptotic but divergent and the slow manifold doesn't exist in the same sense that nonlocal solitons don't exist! More precisely, the slow manifold is asymptotically existent and *praktike* existent and only nonexistent hyperasymptotically [4, 26–30]. The beyond-all-orders nonexistence of the slow manifold is, to a meteorologist, completely irrelevant.

8.2.3 The Three Deaths of Solitary Waves

An essential part of the classical definition of a solitary wave is that a soliton is non-decaying: the breather oscillates forever with undiminished amplitude. There are, however, at least three different terminal diseases for solitary waves.

First, fluids are always viscous; other wave media are always dissipative, so a non-decaying structure with no time dependence except a steady translation or oscillation is impossible.

To a fluid dynamicist, the non-dissipative, non-decaying solitary wave is an *idealization* and an *approximation*, often a superb approximation. A soliton is always fiction because of viscosity.

Second, for some species of soliton and some parameter ranges, the decay can be accelerated by resonant radiation.

Third, waves decay because of propagation through a varying medium. It has long been known that when the Korteweg–deVries (KdV) equation has variable coefficients, modeling water waves in a channel with spatially-varying depth, the core of the solitary wave is stable and persistent. However, the soliton trails a shelf behind it as the core adjusts its width and height to stay in nonlinear-dispersive balance as the environment changes [31].

Surface water waves propagating from the ocean interior onto a beach are not channel-confined nor well-modeled by the KdV equation, but a similar, more violent evolution unfolds. As the water depth shallows, dispersion decreases and the waves evolve to a balance between nonlinearity and dispersion and the narrowing, self-focused pulses become dynamically independent. Such luminaries as Joseph Keller, Walter Munk and D. Howell Peregrine found it useful to describe this as "Trains of Solitary Waves" [32]. However, the solitons are not steady except for translation; as the water continues to shallow, nonlinearity overwhelms the weakening dispersion, the wave steepens and then breaks, dissolving into hissing foam.

Nevertheless, the literature conventionally regards a solitary wave as nonexistent if it is weakly nonlocal, even though there is always viscosity and usually also a spatially-varying environment. One is reminded of the joke about two cowboys who decide to resolve their argument with a gunfight. One cowboy is very tall and wide; the other small. The large cowboy says, "It's not fair! I'm twice the target you are!" The small cowboy replies, "I'll draw an outline of myself with chalk on your clothes, and any bullets that hit outside the chalk won't count."

It is equally arbitrary and preposterous to speak of the "nonexistence" of the breather because of resonant radiation (perhaps on a timescale longer than the lifetime of the universe) while ignoring viscous decay and spatial inhomogeneity decay. Viscosity and variations in the wave environment are bullets that hit and kill the soliton, no matter how much one may pretend "they don't count".

When a nonlocal soliton depends on multiple parameters, it may happen that the radiation coefficient is zero for some discrete parameter values. This is regarded as so significant that new jargon was invented: the non-radiating special solitary waves are called "embedded solitons" or "gap solitons".

8.2.4 Rectification of Names

An important concept in Confucian thought is the Rectification of Names, which may be briefly summarized as: Call everything by precise, accurately-descriptive names instead of sloppy, impulsively-chosen labels that mislead or confuse.

In this spirit, Segur and Kruskal should have replaced "nonexistence" by "radiatively-decaying".

"Embedded-soliton" and "gap soliton" are more precisely "radiation-free soliton".

8.3 Far Field Analysis

Kudryavtsev, who first created the ϕ^4 breather in an initial value simulation, suspected that the breather leaked rather more radiation to infinity than could be attributed to numerical error [5]. Eleonskii, Kulagin, Novozhilova and Silin [33] observed that if the breather were simply periodic-in-time, then it could be expanded as the Fourier series

$$u(x,t) = \sum_{n=0}^{\infty} A_n(X;\epsilon)\ \cos(n\omega t) \tag{8.7}$$

where we assume that the breather is symmetric in time, as true of the multiple scales series described later, by restricting the Fourier series to cosines. Substitution of the Fourier series into the ϕ^4 equation followed by Galerkin projection [34] gives an infinite set of coupled ODEs in x only.

In the far field, $|x| >> 1$, the breather has decayed to such small amplitude that the differential equations for the time-harmonics are *linear* and *uncoupled*:

$$A_{n,xx} + \left(n^2\,\omega^2 - 2\right) A_n = 0 \tag{8.8}$$

The bounded solutions of these are

$$A_0 \sim a_0\ \exp(\pm\sqrt{2}\,x) \qquad x \to \infty \tag{8.9}$$

$$A_1 \sim a_1\ \exp\left(\pm\sqrt{2-\omega^2}\,|x|\right) \tag{8.10}$$

$$A_n \sim \alpha_n\ \sin\left(\sqrt{n^2\omega^2 - 2}\,|x| + \Phi_n\right), \quad n \geq 2 \tag{8.11}$$

Define $\epsilon(\omega)$ to be the e-folding scale of the fundamental, that is, A_1 decays as $\exp(-\epsilon|x|)$. This is consistent with the differential equation for $A_1(x)$ if and only if $\epsilon = (2-\omega^2)^{1/2}$, and this relationship between ω and ϵ must be true to all orders in perturbation theory.

However, all the higher harmonics are *oscillatory* for large $|x|$.

$$k_n = \sqrt{n^2\,\omega^2 - 2} = \sqrt{2n^2 - 2 - n^2\epsilon^2} \qquad n \geq 2 \tag{8.12}$$

Neglecting $O(\epsilon)$ spatially variable coefficients, the system of ODEs can be written *everywhere*, and not merely for large $\epsilon|x| \gg 1$, as

$$A_{n,xx} + \left(n^2\,\omega^2 - 2\right) A_n = f_n(\epsilon x;\, \epsilon) \tag{8.13}$$

where the precise form of $f_n(\epsilon x;\, \epsilon)$ is for the moment irrelevant except to note that it is a function of the harmonics $A_n(\epsilon x)$. By analog with the *linear* inhomogeneous ODE $u_{xx} + u = f(\epsilon x)$ discussed in [35, 36] and Sect. 3.8 of [1] and specifically in conjunction with nonlinear problems in Sect. 9.2 of [1], the amplitude of the far field oscillation for the nth harmonic $A_n(\epsilon x)$ will be proportional to the spatial Fourier Transform $\tilde{A}_1(k)$ of the fundamental $A_1(\epsilon\, x)$, evaluated at $k = k_n$, yielding

$$\alpha_n \sim \beta_n(\epsilon)\, \exp\left(-\frac{k_n\,\pi}{2\epsilon}\right), \qquad \epsilon << 1 \tag{8.14}$$

where $\beta_n(\epsilon)$ are slower-than-exponential functions such as a polynomial. For $\epsilon \ll 1$ ($\leftrightarrow \omega \approx 2^{1/2}$), $k_2 \approx 6^{1/2}$ whereas $k_3 \approx 4$ and the wavenumber k_n increases roughly linearly with n for the higher modes. This implies that α_2 will be exponentially large in comparison to the radiation coefficients of all the higher harmonics.

Therefore, a consistent lowest order approximation to the far field is to ignore all time-harmonics except the second. All the numerical calculations in [3] used a single radiation basis function so that the far field is approximated by a sine wave of wavenumber $k_2(\epsilon)$.

8.4 Multiple Scales/Weakly Nonlinear Power Series for Soliton Core

8.4.1 Overview

The breather has a power series in ϵ. To *all orders*,

$$\epsilon = \sqrt{2 - \omega^2} \qquad \Leftrightarrow \qquad \omega^2 = 2 - \epsilon^2 \tag{8.15}$$

so the breather can be parameterized either by ω or ϵ as convenient. The leading terms are, defining $T \equiv \omega t$ and $X \equiv \epsilon x$,

$$
\begin{aligned}
u(x,t) \sim\ & \sqrt{2/3}\,\epsilon\ \text{sech}(X)\cos(T) \\
&+ \epsilon^2\ \text{sech}^2(X)\left\{\frac{1}{6}\cos(2T) - \frac{1}{2}\right\} \\
&+ \epsilon^3\sqrt{\frac{2}{3}}\left\{\frac{1}{24}\text{sech}^3(X)\cos(3T) + \left[\frac{49}{36}\text{sech}(X) - \frac{103}{72}\text{sech}^3(X)\right]\cos(T)\right\} \\
&+ O(\epsilon^4)
\end{aligned}
\tag{8.16}
$$

Since a detailed derivation is given in Chap. 12 of the author's book [1] and in [3], and a pseudocode to compute the series is given for the closely-related KG breather in [37] in addition to Maple and Matlab codes here, and a table of up to fifth order is presented on p. 313 of [1], we shall omit the details in favor of brief comments.

1. This series (with notational and normalization differences) was first derived by Dashen, Hasslacher and Neveu (1975) [38].
2. The series gives a good description of the *core* of the breather.
3. The series *completely misses* the *far field oscillations* whose amplitude is proportional to $\exp(-3.85/\epsilon)$ and therefore lies "beyond-all-orders" of any ϵ-power series.
4. The series is a weakly nonlinear *amplitude* expansion because the magnitude of the core is $O(\epsilon)$.
5. The series is also a *multiple scales* series because the spatial dependence is through the "slow variable" ϵx.
6. The series is factorially divergent like most multiple scales series [35, 36].
7. The ϵ^n solution is a trigonometric cosine polynomial of degree n in time whose spatial dependence is an nth degree polynomial in powers of $\text{sech}(\epsilon x)$.
8. Even orders, proportional to ϵ^{2m}, involve only *even* time harmonics $\{1, \cos(2\omega t), \ldots \cos(2m\omega t)\}$ and only positive *even* powers of $\text{sech}^2(\epsilon x)$, $\text{sech}^4(\epsilon x) \ldots \text{sech}^{2m}(\epsilon x)\}$.
9. Odd orders, proportional to ϵ^{2m+1}, are sums only of *odd* time harmonics $\{\cos(\omega t), \cos(3\omega t), \ldots \cos((2m+1)\omega t)\}$ and only positive *odd* powers of $\text{sech}(\epsilon x)$, $\text{sech}^3(\epsilon x) \ldots \text{sech}^{2m+1}(\epsilon x)\}$.
10. The numerical coefficients at all orders are *rational numbers*.
11. If $u(X, T; \epsilon)$ is a solution, then $-u(X, T; -\epsilon)$, $u(-X, T; \epsilon)$ and $u(X, -T; \epsilon)$ are solutions, too.

8.4.2 Chirugery and Computer Symbolurgy: Paper-and-Pencil Convolutions and the Limits of Computer Algebra

In the days before digital computers, voluminous tables of functions evaluated at closely-spaced points (implicitly followed by linear interpolation and hand-drawn, professionally-engraved graphs) were the only way to capture special functions. The early Digital Age replaced the value tables by much shorter tables of Chebyshev polynomial coefficients or the numerical coefficients of the cosine-and-sech polynomials of the ϕ^4 multiple scales series on p. 313 of [1]. Now the shift is increasingly to tables not of numbers but instead of computer code, the programming to generate those formerly-tabulated numerical coefficients. This is partly because computers are now fast enough to generate many, many pages of coefficients from a half page of code. This is partly also because language systems like Matlab and Maple have built-in libraries that greatly shorten the codes. In keeping with this evolution,

`Dense Value Tables` $\Rightarrow$ `Chebyshev Polynomial Tables` $\Rightarrow$ `Maple/Matlab Code`

we present no tables of values or of perturbation series coefficients, but only in the Appendix the Maple and Matlab codes to generate the multiple scales series.

Our motive in presenting the same task in two languages is not merely to cater to different language preferences, but rather to make some important points of very broad applicability far beyond breather or nonlinear wave theory.

Computer algebra programs like Maple, Mathematica and the now fading REDUCE and MACSYMA allow one to manipulate symbols and not merely numbers. One can replace a lot of chirugery, that is, paper-and-pencil algebra, by computational symbolurgy. Hurrah! But substituting brute-force silicon cycles is not wholly satisfactory as a substitute for a box of pencils and a good eraser. Computer algebra pioneer Richard Fateman wrote in pp. 515–516 of [39]

> As in many other computational matters, the devil is in the details. A totally brute-force method may not work because an explosive growth of terms can force the computational effort beyond available resources. [After describing a series of optimizations that delete or simplify intermediate expressions, he continues] Without such optimizations the program bogs down rather spectacularly; we tried removing all such speedups and wrote a totally naive program directly mimicking the mathematics. **We found it could not complete in one hour what the optimized program could do in two seconds.**

Many decades ago, the physicist Edward Teller formulated the following:

Assertion Teller's Law A state of the art computation is a hundred-hour run on the state of the art computer. □

What he meant was that there is an almost irresistable temptation to push computing resources to the limit. As decades pass, the state of the art escalates in complexity — from two-dimensional flow over the cross-section of an airfoil to three-dimensional flow over a full wing to supersonic flow around an entire space shuttle — but the ambitious are always making runs that take several days of processor time whatever the century.

Perturbation theorists are no less prone to hero-computing. With an abundant plenitude of coefficients, one can make a Domb–Sykes plot to estimate the radius of convergence, extend the convergence domain through Padé approximants or conformal mapping of the perturbation parameter and so on.

Unfortunately, symbolic manipulation languages have considerable overhead. A number is always and only a number, but a symbol must be tagged with a boatload

Table 8.2 Computer algebra versus FORTRAN for calculating perturbation series. The small numbers in the third column and the large numbers in the fourth column are the maximum order calculated by the authors cited in the second column. The calculations in the third column used algebraic manipulation languages (MACSYMA, REDUCE) while the fourth column used languages restricted to purely numerical languages (FORTRAN, BASIC)

Application	Reference	$N_{\text{computer algebra}}$ /Symbolic language	$N_{\text{numerical}}$ /Numerical language
Unforced van der Pol limit cycle	Andersen and Geer [40]	22 MACSYMA	162 FORTRAN
Kelvin–Helmholtz roll-up	Meiron, Baker and Orszag [41]	7 MACSYMA	38 FORTRAN
Periodic nonlinear water waves	Hui and Tenti [42] Schwartz [43, 44]	11 MACSYMA	117 FORTRAN
Standing water waves	Aoki [45] Schwartz and Whitney [46]	8 REDUCE	25 FORTRAN
Fifth-order KdV cnoidal waves	Boyd [47]	4 REDUCE	11 BASIC
Taylor-green turbulence	Fateman [39] Brachet et al. [48]	14 MACSYMA	80 FORTRAN

of auxiliary information beginning with "number or variable"? The Maple code in the Appendix is only a little shorter than its Matlab work-alike, but required only half as long to program and debug because much algebra devolved on Maple, and the ability to show symbolic dependencies is advantageous in debugging. However, to compute the multiple scales series to twentieth order, the Maple program required 4200 s, the Matlab code only 6 s, a ratio of 700!

This not atypical. The running time to calculate a series to $O(\epsilon^N)$ is usually a cubic or higher power of N. Nevertheless, Table 8.2 shows that for a given problem, the maximum N calculated in a numerical language is usually an order of magnitude larger than with a symbolic language, a difference of at least three orders of magnitude in operation count.

John Fitch, a pioneer in applied computer algebra, described the slow growth of symbolurgy as a "creeping flow". Maple and Mathematica are widely used for routine tasks, but research articles where lengthy computer algebra is the heart of discovery are indeed growing at a "creeping" pace. *Numerical* perturbation theory is the only path to high order N.

8.5 Hyperasymptotic Perturbation Theory

Segur and Kruskal went "beyond-all-orders" by applying the method of matched asymptotic approximations in the complex plane [12]. This is but one of many "hyper-asymptotic" strategies as reviewed in [1, 13, 22, 23, 36]. Segur and Kruskal found that the amplitude α of the far field radiation is

$$\alpha(\epsilon) \sim 0.018 \exp\left(-\sqrt{\frac{3}{2}}\pi \frac{1}{\epsilon}\right), \qquad \epsilon \ll 1 \tag{8.17}$$

In this and other nonlocal solitary wave problems, the *exponential* coefficient, here approximately 3.847, is *easy* because the exponential is simply the exponential of the Fourier Transform of the core evaluated at the resonant wavenumber, here $k_2 = \sqrt{6 - 4\epsilon^2}$. The Segur-Kruskal formula can be significantly improved by replacing the limiting wavenumber, $k_2(\epsilon = 0) = \sqrt{6}$ by the exact wavenumber $k_2(\epsilon)$ to obtain, as a special case of (8.14),

$$\alpha_{\text{Boyd}}(\epsilon) \sim 0.018 \exp\left(-\sqrt{6 - 4\epsilon^2}\frac{\pi}{2}\frac{1}{\epsilon}\right), \qquad \epsilon \ll 1 \tag{8.18}$$

Figure 8.3 compares the numerically-computed radiation coefficient with the Boyd and Segur-Kruskal formulas. Both approximations converge towards the numerically computed α as $\epsilon \to 0$ or equivalently as $1/\epsilon \to \infty$. The Boyd formula is a significant improvement, but both approximations are an order of magnitude too small for $\epsilon \sim O(1)$. The situation could be improved by calculating corrections to the constant multiplying the exponential as done by Joshi and Grimshaw for the

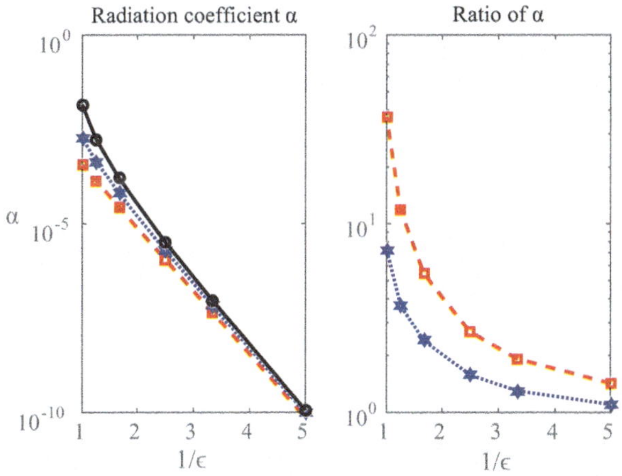

Fig. 8.3 Left: A comparison of the far field radiation $|\alpha|$ versus $1/\epsilon$. Black-with-circles: numerical computations by double Fourier numerical method. Red-dashed-with-squares: Segur-Kruskal formula. Blue-dotted-with-hexagrams: Boyd approximation. Right: the ratio of the numerically-computed α to asymptotic formulas. Red-dashed-squares: $\alpha_{\text{num}}(\epsilon)/\alpha_{\text{Segur-Kruskal}}$. Blue-dotted-with-hexagrams: $\alpha_{\text{num}}(\epsilon)/\alpha_{\text{Boyd}}(\epsilon)$

FKdV equation [1], but obtaining even the lowest order constant is *difficult*, requiring matched asymptotics in the complex-plane or Borel Summation tricks. No such extension has been made for the ϕ^4 breather.

8.6 Numerical Calculations of Breathers, I: The General Initial-Value Problem

The ϕ^4 equation is usually solved by *truncating* the spatial domain from the entire real axis to a *finite interval* with *periodic boundary conditions*. For large times, radiation may spuriously exit one endpoint and enter from the other; such artifacts can be controlled by imposing a spatially-varying dissipation which is large at the endpoints, creating so-called "sponge layers" to absorb the transients (p. 341 of [34]). This is only needed if the integration duration is very large.

With periodic boundary conditions, a Fourier pseudospectral spatial discretization combined with any of a wide variety of time-marching schemes offers an unbeatable combination of high accuracy with simplicity of programming as explained in Yang's treatise on nonlinear waves. He discusses Fourier methods at length with many brief Matlab codes on pp. 327–358 of [49]. Rigorous convergence theorems and numerical results for periodic boundary conditions and a Fourier basis are given in [50] and similarly for a finite interval with Dirichlet or Neuman boundary conditions with a Legendre polynomial expansion in [51]. (Their generalized KG equation includes the ϕ^4 equation as a special case.)

In summary, the initial value problem is easy.

In contrast, direct computation of the breather and nothing but the breather is hard as explained in the next section.

8.7 Numerical Calculations of Breathers, II: Nanopterons and Nanopteroidal Waves

8.7.1 Overview

If a Korteweg–deVries traveling wave is assumed to depend upon x and t only as a univariate function of $X = x - ct$, the time-dependent KdV equation can be reduced to an ODE in x. If the amplitude is fixed, one must compute c as an eigenvalue simultaneously with $u(X)$. Breathers can be similarly be computed in "pure form" as the solution to a nonlinear eigenvalue problem with ω as the eigenvalue by substituting the bivariate series (8.7) (a Fourier series in time) into the ϕ^4 PDE.

For arbitrary choices of the spatial period P or its equivalent for the nanopteron, which is the phase parameter Φ, it is straightforward to compute symmetric but spatially nonlocalized solutions [3] which asymptote to

$$u(x,t) \sim \alpha_2(x;\epsilon,\Phi)\sin(k|x|+\Phi)\cos(2\omega t), \qquad P/2 \gg \epsilon|x| \gg 1 \qquad (8.19)$$

where we have dropped the subscripts “2” since only the dominant far field oscillation, that of the second harmonic, will be considered further. Neglecting terms of $O(\alpha_3, \alpha^2)$, this asymptotic approximation applies to both the nanopteron and nanopteroidal wave, provided the period for the latter is sufficiently large in comparison to $1/\epsilon$.

Standard infinite interval spectral methods fail for nonlocal solitary waves; a finite number of rational Chebyshev functions (or whatever) cannot faithfully mimic a never-ending oscillation over all of space. To obtain a solution which is uniformly accurate everywhere with an error smaller than $O(\alpha)$ where α is (as always) the amplitude of the wings, one has but two options.

One is to use a large but finite periodic interval and apply the Fourier pseudospectral method. The other is to compute directly on the infinite interval using rational Chebyshev functions or other standard basis set augmented by a special *radiation basis function* to approximate the oscillatory wings of the nanopteron. Both methods give similar accuracy.

There are two technical complications.

The first is the asymptotic phase Φ. The condition of boundedness at infinity is not sufficient to determine a unique nanopteron. Instead, the asymptotic phase factor Φ is an extra degree of freedom which must be explicitly incorporated into the radiation basis function. For the spatially periodic formulation, this phase factor is implicitly determined by the ratio of the spatial period P to the wavelength W of the wing oscillations.

The second technical complication is that the far field oscillations are linear sine waves only to a first approximation. The self-interaction of the far field oscillations generates a second harmonic of amplitude $O(\alpha^2)$ which can be incorporated into the radiation basis function by “cnoidal matching” as explained in Sect. 9.5 of [1]. However, this refinement has not yet been applied to nonlocal breathers. Note that since $\alpha_3 \gg \alpha^2$, one must include the radiation in the third harmonic before correcting for the self-interaction of the far field of the second time-harmonic.

8.7.2 Fourier Pseudospectral Method

To calculate spatially periodic solutions (nanopteroidal waves) both for their own intrinsic interest and as approximations to nanopterons, we must first choose a spatial period P. The pseudospectral method is then applied in a completely orthodox way as described in [3, 34]. The multiple scales theory shows that the breather is symmetric in both time and space, which simplifies a general Fourier series (with sines and cosines) to the bivariate cosine series

$$u(x,t) = \sum_{m=0}^{M} \sum_{n=0}^{N} a_{mn} \cos(m\kappa x) \cos(n\omega t) \tag{8.20}$$

where $\kappa \equiv 2\pi/P$.

The second harmonic in the far field has the general form $\alpha \sin(k|x| + \Phi)$. Spatial periodicity with period P demands that $\Phi = \pi(1/2 - P/W) + \pi m$ where m is an arbitrary integer and W is the far field wavelength

$$W \equiv 2\pi/k_2 \tag{8.21}$$

Both the phase Φ and radiation coefficient α are functions of period P.

Through arguments in Chaps. 9 and 12 of [1] are too complicated to repeat here, the far field radiation is *minimized* by choosing Φ as

$$\Phi_{\min} = \frac{\sqrt{6}}{5}\epsilon \tag{8.22}$$

or equivalently in the periodic case

$$P = (m + 1/2)W - \Phi_{\min}(W/P) \tag{8.23}$$

where m is an arbitrary positive integer (a *large* positive integer when approximating a nanopteron using a periodic domain).

The radiation coefficient α has a simple pole whenever P is an integral multiple of W (Sect. 1.5 of [1]). The resonance arises because the jth Fourier component of $f_P(x)$, where $j = P/W$, has a wavenumber which exactly matches that of the homogeneous solution of $A_{2,xx} + k_2^2 A_2 = f_2(\epsilon x)$. The resonance poles are confirmed by Fig. 4 of [37]. However, otherwise, the parametric neighborhood of *large* α has never been explored.

8.7.3 Rational Chebyshev Functions with Radiation Basis Function

The idea of the "radiation function" method is to augment the standard infinite interval basis set with a special function chosen to match the asymptotic behavior of the wings.

There is some arbitrariness in the choice of the non-sinusoidal factors of the basis function. The hyperbolic tangent of ϵx used in [3] and here varies on the same scale as the breather core and thus does not introduce a new unphysical length scale into the numerical solution.

The infinite interval computations in [37] employed the basis

$$u(x,t) = \sum_{m=0}^{M} \sum_{n=0}^{N} a_{mn} \left[T B_{2m}(x) - 1 \right] \cos(n\omega t) + \alpha \tanh(\epsilon |x|) \sin(k_2 |x| + \Phi) \sin(2\omega t) \tag{8.24}$$

where the $TB_m(x; L)$ are the rational Chebyshev functions defined in [1, 3, 14, 34]. The choice of map parameter $L = 2/\epsilon$ was effective. The coefficient of the special basis function, α, asymptotes to the true radiation coefficient α_2 in the limit $M \to \infty$, $N \to \infty$.

Unfortunately, the radiation coefficient is very sensitive to small changes in ϵ, the forcing function, the coefficients of the PDE and so on as explained in Sect. 6 of Chap. 9 of [1]. However, this is a *physics* problem rather than a numerical issue.

When $\epsilon < 0.8$, Newton's iteration with Armijo line search ([52–54] and Sect. 8.9 of [1]) converges monotonically from an initial guess which is the multiple scales series up to and including the $O(\epsilon^2)$ terms. However, it is difficult, even when applying the continuation strategy with extremely small steps in ϵ, to compute solutions with $\epsilon > 1$, and this parameter range remains *terra incognita*.

8.7.4 Typical Numerical Results

Figures 8.4 and 8.5 show two different visualizations for ϵ near one where the multiple scales perturbation fails, at any order, in successfully initializing Newton's iteration. (The continuation method, marching in small steps in ϵ from small ϵ where the ϵ-power series is accurate, was used to compute nanopteroidal waves for $\epsilon > 0.9$.)

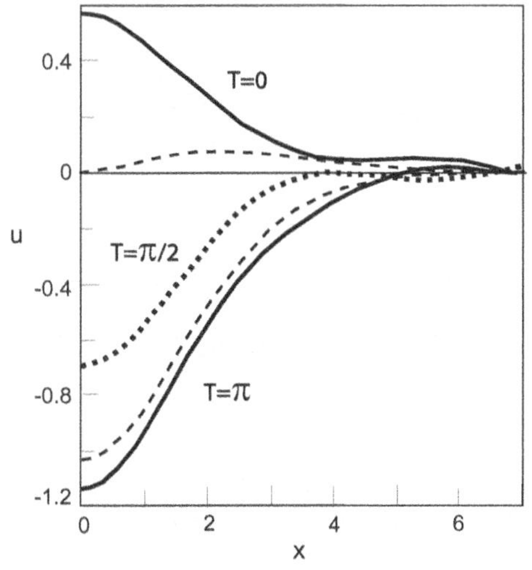

Fig. 8.4 $u(x,t)$ at five different times for the ϕ^4 breather: $\epsilon = \omega = 1$ and $\Phi = 0$. Because the oscillation is symmetric with respect to both $x = 0$ and $t = 0$, it suffices to illustrate only positive x for one-half of a temporal cycle

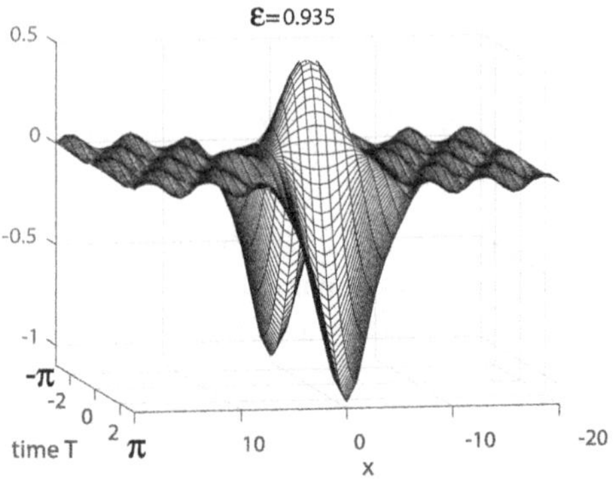

Fig. 8.5 $u(x, t)$ for $\epsilon = 0.935$, shown over a full period in $T = \omega t$

The symmetry of the positive and negative peaks of the breather for small ϵ has been drastically modified: The valley is twice as deep as the peak. Figure 8.4 shows that the breather is negative for two-thirds of the period. Nevertheless, even for this large amplitude where the ϵ power series is not very accurate, the far field oscillations are still two orders of magnitude smaller than the core.

8.8 Radiative Decay: The ODE $\mathrm{d}w/\mathrm{d}t = -\exp(-1/w)$

8.8.1 Early Work

The amplitude of the core of the breather, which is $O(\epsilon)$, decays through radiation whose amplitude is proportional to $p \exp(-q/\epsilon)$ where p and q are positive constants. This suggests that a useful heuristic model for understanding the decay of the breather of the ϕ^4 PDE is the ODE in time $\mathrm{d}u/\mathrm{d}t = -p \exp(-q/u)$ as suggested by Segur and Kruskal (their (16)) [12]. They noted that lowest order approximation is $u \sim \text{constant}/\log(t)$. Theorem 1 of [15] shows that $u(t)$ with $u(0) = \varpi$, an arbitrary positive constant, can be expressed in terms of the parameter-free function $w(t)$ as $u(t; p, q, \varpi) = q w((p/q)(t - \varpi))$ where w solves the canonical problem that will occupy the rest of this section:

$$\frac{\mathrm{d}w}{\mathrm{d}t} = -\exp\left(-\frac{1}{w}\right), \qquad w(0) = 1 \tag{8.25}$$

Through arguments in [15] too complicated to repeat here, Boyd showed that for *extremely* large times $(\log(\log(t)) \gg 1)$

$$t^* \equiv t + \exp(1) \tag{8.26}$$

$$\tau \equiv \log(\log(t^*)) \tag{8.27}$$

$$B(\tau) \equiv 1 + (4\tau - 2)\exp(-\tau) + 4\,\tau^2\,\exp(-2\tau) \tag{8.28}$$

$$w \sim \frac{1}{\log\left\{t^* \log^2(t^*)B(\tau)\right\}} \tag{8.29}$$

Geicke carried out a very long time integration beginning from an initial condition that was the superposition of a kink and an antikink [55]. If the amplitude of these solitary waves is not too large, their collision generates a breather. Integrating to $t = 2 \times 10^6$, he applied two different empirical curve fits to try to understand the long-term behavior. Unfortunately, he imposed boundary conditions to allow the emitted radiation to pass harmlessly out of the numerical domain which assumed that the radiation is traveling at unit phase speed. The actual phase speed of the resonant radiation is $\sqrt{3}/2$. This generated spurious wave reflection from the computational boundaries. Furthermore, although he was careful to check conservation of energy as his very long integration proceeded, he used a rather long time step of 0.1. Furthermore, the forms assumed in his empirical curve fits had no theoretical justification. He was successful in confirming the inverse logarithmic decay predicted by Segur and Kruskal. However, because of the development of extremely long time scales, $O(\log(\log(t)))$ behavior, initial value studies even today are inadequate to correct the leading $1/\log(t)$ behavior.

8.8.2 Implicit Solution

Although neither Boyd nor Geicke attempted this approach, Boyd's problem (8.25) can be solved using the method of separation of variables (pp. 296–298 of [56] and pp. 37–42 of [57]) to obtain the exact but *implicit* solution

$$t = \int_w^1 \exp\left(\frac{1}{y}\right) dy \tag{8.30}$$

$$= \Upsilon(w) \tag{8.31}$$

where

$$\Upsilon(w) = \text{Ei}(1, -1) - \text{Ei}\left(1, -\frac{1}{w}\right) + \exp(1) - w\exp\left(\frac{1}{w}\right) \tag{8.32}$$

$$= -w\exp(1/w) + \exp(1) + \log(1/w) - \text{Ein}(-1/w) + \text{Ein}(-1) \tag{8.33}$$

$$= -\log(w) + \exp(1) - w\exp\left(\frac{1}{w}\right) + \sum_{n=1}^{\infty} \frac{w^{-n} - 1}{n\,n!} \tag{8.34}$$

and

$$\mathrm{Ei}(1, -x) = -\gamma - \log(x) + \mathrm{Ein}(-x) \tag{8.35}$$

$$\mathrm{Ein}(x) = -\sum_{n=1}^{\infty} \frac{(-x)^n}{n\, n!} \tag{8.36}$$

$$= x - \frac{1}{4}x^2 + \frac{1}{18}x^3 - \frac{1}{96}x^4 + \frac{1}{600}x^5 + O\left(x^6\right)) \tag{8.37}$$

$$= \int_0^x \frac{1 - \exp(-t)}{t} dt \tag{8.38}$$

with $\gamma = 0.577\ldots$ as the Euler–Mascheroni constant.

The exact solution of $dw/dt = -\exp(-1/w)$ with $w(0) = 1$ is

$$w(t) = \Upsilon^{-1}(t) \tag{8.39}$$

where Υ^{-1} is the inverse of $\Upsilon(w)$. Further analysis of this implicit solution is an open problem.

8.9 Summary

In broad detail, the ϕ^4 breather is well understood. The amplitude of the wings of the nanopteron has been calculated both by numerical and perturbative means. The infinite interval solutions generalize to periodic solutions, and these, too, have been computed. The ϵ-power series gives a good approximation to the core of the soliton for small ϵ. It has not been rigorously proved that the series is divergent, as typical of multiple scales expansions, but the rapid growth of the numerical coefficients evident in the table on p. 313 of [1] leave little doubt. In addition, much work has been done on non-breather solutions of the ϕ^4 equation as reviewed by Campbell et al. (1983).

However, many subtleties and open problems remain including the following.

1. Theory predicts that the breather should be nonlocal through an *infinite* number of time harmonics, but the role of all but $A_2(x)$, the second harmonic, has been ignored in all calculations to date.
2. The radiation basis function has not been improved, as for the FKdV computations described in Sect. 5 of Chap. 9 of [1], by inclusion of perturbative nonlinear corrections ("cnoidal matching"). This will be much complicated than for the FKdV case because the lowest order in α_3, the radiation coefficient of the third harmonic, is exponentially large compared to α_2^2.
3. Segur-Kruskal hyperasymptotics has not been extended beyond lowest order.

4. There is no theory analogous to the inverse scattering method for predicting the amplitude of the breather that will evolve from an arbitrary initial condition or even forecasting when a breather will form.
5. There have been no computations of the breather for $\epsilon > 1$. It is not known whether breathers exist in this regime.
6. It seems likely that there are resonances, that is, choices of the phase factor $\Phi(\epsilon)$ or period P such that $\alpha \to \infty$ for fixed ϵ but these have not been studied.
7. Each time-harmonic $A_n(x)$ for $n \geq 2$ has its own phase parameter Φ_n—an infinite number of free parameters. There may be a complex system of bifurcations and branches for α_n and/or ϵ not small, but this has not been investigated.

Acknowledgements This work was supported by the U. S. National Science Foundation through grant DMS-1521158. I thank Panos Kevrekidis and Jesús Cuevas-Maraver for their invitation to write this chapter and their editorial labors on the book as a whole. I thank Joseph Tribbia whose comments on [2] directly led to my work on the slow manifold of atmospheric dynamics [4, 26].

Appendix: Codes for Multiple Scales Perturbation Theory

Maple code

```
restart; with(numapprox); with(orthopoly);
interface(rtablesize=infinity) ; with(LinearAlgebra):
Nmax:=6; # highest power of epsilon computed;
nF:= Nmax + 3:
for m from 0 to nF do for N from 0 to maxpertorder do
vv[m, N]:=0;od: od: vv\cite{bib0,bib1}:= sqrt(2/3)* sech(X);
wsq:= 2 - epsilon*epsilon; # square of frequency omega;
# START OF LOOP OVER N (PERTURBATION ORDER);
for N from 1 to Nmax do v:=0;
# special treatment of terms in cos(T), which are determined;
# at O(epsilon**2) later than their own order;
vv[1, N]:=0;
if modp( N,2) = 0 then
for kkk from 1 to N do vv[1, N]:=vv[1, N]
    + Delaya[kkk, N] * (sech(X))**(2*kkk-1); od: fi:
for jorder from 0 to ( N-1) do for n from 0 to nF do
v:=v + vv[n,jorder]*cos(n*T)* epsilon**jorder; od: od:
resid:=collect(combine(epsilon*epsilon*diff(v,X,X)-wsq* diff(v,T,T)
-2*v -3*epsilon*v*v - epsilon*epsilon*v*v*v , trig),epsilon):
resid:=subs(tanh(X)*tanh(X)=1-sech(X)*sech(X),resid);
ra[ N]:= coeff(resid,epsilon, N); rum:= combine( ra[ N],trig);
# Fourier decomposition;
for jj from 1 to nF do uFou[jj]:= coeff( rum, cos(jj*T), 1);
rum:= rum - cos(jj*T)*uFou[jj];
od: uFou\cite{bib0}:= rum;
# delayed calculation of terms proportional to cos(T) and of;
# O(epsilon**( N-2)) by zeroing residual at O(N);
```

```
# All nonzero terms proportional to cos(T) are;
# for even powers of epsilon;
       if modp( N,2)=0 then
       if  N > 3 then print(' special solve for cos(T), order=',N);
       uFouone:= subs(sech(X)=A/sqrt(2/3), uFou\cite{bib1});  icounter:=0;
         for kkk from 1 to  N do resone:=coeff(uFouone,A,2*kkk-1);
       if resone=0 then # do nothing;
          else icounter:=icounter+1;  residone[icounter]:= resone;
Delaya[icounter,N-2]:=solve(residone[icounter],Delaya[icounter,N-2]);
        fi:  od:   fi:   fi:
for n from 2 to nF do vv[n, N]:= uFou[n]/(2 - 2*n*n); od:
vv[0,N]:= uFou\cite{bib0}/2;
od: # END of N LOOP (PERTURBATION ORDER);
   v:=0: for jorder from 0 to  Nmax do
for n from 0 to nF do v:=v+vv[n,jorder]*cos(n*T)*epsilon**jorder;
od: od:
print(' final answer, u=',collect(epsilon*v,epsilon));
```

Matlab code for multiple scales perturbation theory

```
clear, tic, Nmax=10, % "Nmax" specifies maximum order in epsilon
DD=zeros(Nmax,Nmax+1,Nmax);  R=zeros(Nmax+1,Nmax);
OM=zeros(1,Nmax+1);    DTILDE=zeros(1,Nmax);
OM(1)=1;    OM(3)=- 1/4;  DD(1,2,1)=1;   DD(2,3,2)=1/4;
DD(2,1,2)=-3/4;   DD(3,4,3)=1/16;
% End of initialization. Start of loop over order N
for N=4:Nmax,  R=zeros(Nmax+1,Nmax); OMSQUARED=0;
for m=1:(N-2), OMSQUARED=OMSQUARED + OM(N-m)*OM(m+1); end % m
R(2,1)=2*OMSQUARED;
for k=1:(N-2), for m=0:k, for ii=0:(N-k), for j=1:(N-k)
R(ii+1,j)=R(ii+1,j) + 2*ii*ii*OM(k-m+1)*OM(m+1)*DD(N-k,ii+1,j);
end, end, end, end
% ******* QUADRATIC NONLINEARITY block *********
for m=1:(N-1), for i1=0:(N-m), for j1=1:(N-m),
for i2=0:m, for j2=1:m
jsum=j1+j2; pp=i1+i2; pm=abs(i1-i2);
SUBFACT=1.5*DD(N-m,i1+1,j1)*DD(m,i2+1,j2);
R(pp+1,jsum)=R(pp+1,jsum)-1.5*DD(N-m,i1+1,j1)*DD(m,i2+1,j2);
R(pm+1,jsum)=R(pm+1,jsum)-1.5*DD(N-m,i1+1,j1)*DD(m,i2+1,j2);
end, end,  end,  end,  end
% ******* CUBIC NONLINEARITY ***********
for q=2:(N-1), for p=1:(q-1), for i1=0:(N-q), for j1=1:(N-q)
for i2=0:(q-p), for j2=1:(q-p), for i3=0:p, for j3=1:p,
jsum=j1+j2+j3;
COD=0.25*DD(N-q,i1+1,j1)*DD(q-p,i2+1,j2)*DD(p,i3+1,j3);
pp=i1+i2+i3; pm=abs(i1+i2-i3); mp=abs(i1-i2)+i3;
mm=abs(abs(i1-i2)-i3);
R(pp+1,jsum)=R(pp+1,jsum) - COEFFPROD;
R(pm+1,jsum)=R(pm+1,jsum) - COD;
R(mp+1,jsum)=R(mp+1,jsum) - COEFFPROD;
R(mm+1,jsum)=R(mm+1,jsum) - COD;
end, end, end, end, end, end, end, end
% 2d x-derivative, omitting terms proportional to cos(t), which
```

```
% are determined as part of the eigenvalue block.
for ii=0:(N-2), if ii ~= 1, for j=1:(N-2)
     R(ii+1,j)=R(ii+1,j) + DD(N-2,ii+1,j)*j*j;
     R(ii+1,j+2)=R(ii+1,j+2) - 1.5*j*(j+1)*DD(N-2,ii+1,j);
     end, end, end
% Convert the residual factors in the third order coefficients
for ii=0:N, if ii ~= 1, for j=1:N,
DD(N,ii+1,j)=R(ii+1,j) / (2 - 2*ii*ii);
end, end, end
% **** EIGENCONDITIONS ***************
if rem(N,2) == 1, k=N-2;
DD(N-2,2,N-2)=R(2,N) / (1.5 * (-6 + (N-1)*(N-2)) );
DTILDE(N-2) = DD(N-2,2,N-2);
while (k > 1),
 DD(N-2,2,k-2)=(R(2,k)+(k*k-1)*DD(N-2,2,k))/(1.5*(-6+(k-1)*(k-2)));
 DTILDE(k-2)=DD(N-2,2,k-2); k=k-2; end, % while
OM(N) = - 0.25 * R(2,1); % FIXUP of order N-1
for m=2:(N-1),
DD(N-1,1,m)=DD(N-1,1,m) -1.5*DTILDE(m-1);
DD(N-1,3,m)=DD(N-1,3,m) + 0.5*DTILDE(m-1);
end, % FIXUP of order N
for m=3:N, DD(N,4,m)=DD(N,4, m) + (3/16)*DTILDE(m-2); end % k
end % if % END OF NONSECULARITY BLOCK
end % N, end of Loop over epsilon orde
% printing loop
for N=1:Nmax, for k=0:Nmax, for m=1:Nmax, if abs(DD(N,k+1,m))>0,
  display(['j=',int2str(N),' k=',int2str(k),...
  ' m=',int2str(m),' D(j,k,m)=',num2str(DD(N,k+1,m))]);
  end, end, end, end, elapsedtime=toc
```

References

1. J.P. Boyd, *Weakly Nonlocal Solitary Waves and Beyond- All-Orders Asymptotics: Generalized Solitons and Hyperasymptotic Perturbation Theory* (Kluwer, Amsterdam, 1998)
2. J.P. Boyd, in *Dynamics of Low-Frequency Phenomena in the Atmosphere*, ed. by G. Brantstator, J.J. Tribbia, R. Madden (National Center for Atmospheric Research, Boulder, Colorado, 1987), p. 717
3. J.P. Boyd, Nonlinearity **3**, 177 (1990) [The eigenfunction calculation (5.15, etc.) has some typographical errors corrected in Chapter 12 of [1]]
4. J.P. Boyd, Dyn. Atmos. Oceans **22**, 49 (1995)
5. A.E. Kudryavtsev, JETP Lett. **22**, 82 (1975)
6. J. Geicke, Phys. Lett. B **133**, 337 (1983)
7. B.S. Getmanov, JETP Lett. **24**, 291 (1976)
8. M.P. Hertzberg, Phys. Rev. D **82**, 045022 (2010)
9. T. Sugiyama, Prog. Theor. Phys. **61**, 1550 (1979)
10. D.K. Campbell, J.F. Schonfeld, C.A. Wingate, Phys. D **9**, 1 (1983)
11. D.K. Campbell, M. Peyrard, Phys. D **18**, 47 (1986)
12. H. Segur, M.D. Kruskal, Phys. Rev. Lett. **58**, 747 (1987) [Erratum, Phys. Rev. Lett. **58**, 1158 (1987)]
13. J.P. Boyd, Acta Appl. Math. **56**, 1 (1999)
14. J.P. Boyd, J. Comput. Phys. **69**, 112 (1987)
15. J.P. Boyd, Adv. Appl. Math. **9**, 358 (1988)

16. J.P. Boyd, in *Mesoscale/Synoptic Coherent Structures in Geophysical Turbulence*, ed. by J.C.J. Nihoul, B.M. Jamart. *Proceedings of the 20th Liege Colloquium on Hydrodynamics* (Elsevier, Amsterdam, 1989), p. 103
17. R. Grimshaw, in *Asymptotic Methods in Fluid Mechanics: Survey and Recent Advances*, ed. by H. Steinruck (Springer, Udine, 2010), p. 71
18. J.P. Boyd, in *Nonlinear Topics of Ocean Physics*, ed. by A.R. Osborne. Fermi Summer School, Course LIX (North-Holland, Amsterdam, 1991), p. 527
19. J.P. Boyd, in *Advances in Applied Mechanics*, vol. 27, ed. by T.Y. Wu, J.W. Hutchinson (Academic Press, New York, 1989), p. 1
20. J.P. Boyd, *Dynamics of the Equatorial Ocean* (Springer, Heidelberg, 2018)
21. G.P. Williams, R.J. Wilson, J. Atmos. Sci. **45**, 207 (1988)
22. H. Steinrück (ed.), *Asymptotic Methods in Fluid Mechanics: Survey and Recent Advance* (Springer, Udine, 2010)
23. H. Segur, S. Tanveer, H. Levine (eds.), *Asymptotics Beyond All Orders* (Plenum, New York, 1991)
24. F. Baer, Contrib. Atmos. Phys. **50**, 350 (1977)
25. F. Baer, J.J. Tribbia, Mon. Weather Rev. **105**, 1536 (1977)
26. J.P. Boyd, J. Atmos. Sci. **51**, 1057 (1994)
27. J. Vanneste, Ann. Rev. Fluid Mech. **45**, 147 (2013)
28. J. Vanneste, J. Atmos. Sci. **61**, 224 (2004)
29. J. Vanneste, SIAM J. Appl. Dyn. Syst. **5**, 783 (2006)
30. J. Vanneste, SIAM J. Appl. Dyn. Syst. **7**, 1163 (2008)
31. B. Long, P. Chang, J. Phys. Oceanogr. **20**, 1826 (1990)
32. J.P. Boyd, Geophys. Astrophys. Fluid Dyn. (2018). (Submitted)
33. V.M. Eleonskii, E. Kulagin, S. Novozhilova, P. Silin, Theor. Math. Phys. **60**, 896 (1984)
34. J.P. Boyd, *Chebyshev and Fourier Spectral Methods* (Dover, New York, 2001)
35. J.P. Boyd, Appl. Numer. Math. **7**, 453 (1991)
36. J.P. Boyd, SIAM Rev. **47**, 553 (2005)
37. J.P. Boyd, Wave Motion **21**, 311 (1995)
38. R.F. Dashen, B. Hasslacher, A. Neveu, Phys. Rev. D **11**, 3424 (1975)
39. R.J. Fateman, Int. J. Mod. Phys. C **9**, 509 (1998)
40. C.M. Andersen, J.G. Geer, SIAM J. Appl. Math. **42**, 678 (1982)
41. D.I. Meiron, G.R. Baker, S.A. Orszag, J. Fluid Mech. **114**, 283 (1982)
42. W.H. Hui, G. Tenti, Z. Angew, Math. Phys. **33**, 569 (1982)
43. L.W. Schwartz, Ph.D. dissertation, Stanford University, 1972
44. L.W. Schwartz, J. Fluid Mech. **62**, 553 (1974)
45. H. Aoki, J. Phys. Soc. Jpn **49**, 1598 (1980)
46. L.W. Schwartz, A.K. Whitney, J. Fluid Mech. **107**, 147 (1981)
47. J.P. Boyd, Phys. D **21**, 227 (1986)
48. M.E. Brachet, D.I. Meiron, S.A. Orszag, B.G. Nickel, R.H. Morf, U. Frisch, J. Fluid Mech. **130**, 411 (1983)
49. J. Yang, *Nonlinear Waves in Integrable and Nonintegrable Systems* (SIAM, Philadelphia, 2010)
50. W. Cao, B. Guo, J. Comput. Phys. **108**, 296 (1993)
51. B.Y. Guo, X. Li, L. Vazquez, Comput. Appl. Math. **15**, 19 (1996)
52. J.P. Boyd, *Solving Transcendental Equations: The Chebyshev Polynomial Proxy and Other Numerical Rootfinders, Perturbation Series and Oracles* (SIAM, Philadelphia, 2014)
53. J.P. Boyd, Math. Comput. Simul. **74**, 72 (2007)
54. J.P. Boyd, J. Comput. Phys. **179**, 216 (2002)
55. J. Geicke, Phys. Rev. E **49**, 3539 (1994)
56. D. Zwillinger, *Handbook of Differential Equations* (Academic Press, New York, 1988)
57. W.E. Boyce, R.C. DiPrima, *Elementary Differential Equations and Boundary Value Problems*, 4th edn. (Wiley, New York, 1986)

Chapter 9
The Continuing Story of the Wobbling Kink

Igor Barashenkov

Abstract The wobbling kink is the soliton of the ϕ^4 model with an excited internal mode. We outline an asymptotic construction of this particle-like solution that takes into account the coexistence of several space and time scales. The breakdown of the asymptotic expansion at large distances is prevented by introducing the long-range variables "untied" from the short-range oscillations. We formulate a quantitative theory for the fading of the kink's wobbling due to the second-harmonic radiation, explain the wobbling mode's longevity and discuss ways to compensate the radiation losses. The compensation is achieved by the spatially uniform driving of the kink, external or parametric, at a variety of resonant frequencies. For the given value of the driving strength, the largest amplitude of the kink's oscillations is sustained by the *parametric* pumping — at its natural wobbling frequency. This type of forcing also produces the widest Arnold tongue in the "driving strength versus driving frequency" parameter plane. As for the *external* driver with the same frequency, it brings about an interesting rack and pinion mechanism that converts the energy of external oscillation to the translational motion of the kink.

9.1 Prologue

An unlikely insomniac wandering into the Dubna computer centre on one of those freezing nights in the winter of 1975, would invariably see the same slim figure rushing among mainframe dashboards, magnetic tape recorders and card perforation devices. A houndstooth blazer favoured by the jazzmen of the time, a ginger beard and a trademark cigarette holder — with some amazement, the passer-by would recognise Boris ("Bob") Getmanov, a pianist and a popular character of the local music scene. This time, experimenting with harmonies of nonlinear waves.

I. Barashenkov (✉)
University of Cape Town and National Institute for Theoretical Physics, Western Cape, South Africa
e-mail: Igor.Barashenkov@uct.ac.za

I. Barashenkov
Joint Institute for Nuclear Research, Dubna, Russia

P. G. Kevrekidis and J. Cuevas-Maraver (eds.), *A Dynamical Perspective on the ϕ^4 Model*, Nonlinear Systems and Complexity 26,
https://doi.org/10.1007/978-3-030-11839-6_9

The object that inspired Bob's syncopations, was a ϕ^4 kink with a nonlinearly excited internal mode — something he interpreted as a bound state of three kinks and called a musical term *tritone* [1]. In this chapter we review the continuation of the *tritone* story — the line of research started with blue sky experiments of an artist captivated by mysteries of the nonlinear world.

9.2 Wobbling Mode

Getmanov's numerical experiments were motivated by similarities between the ϕ^4-theory and the sine-Gordon model — arguably, the two simplest Lorentz-invariant nonlinear PDEs. While the ϕ^4 and sine-Gordon have so much in common, there is an important difference between the two systems. The sine-Gordon is completely integrable whereas the ϕ^4 model is not. Many a theorist tried to detect at least some remnants of integrability among the properties of the ϕ^4 kinks — and Getmanov with his computer codes was part of that gold rush — but only to find more and more deviations from the exact rules set by the sine-Gordon template.

One attribute that makes the ϕ^4 kink particularly different from its sine-Gordon twin, is the availability of an internal, or shape, mode. (See Fig. 9.1). This oscillatory degree of freedom may store energy and release it periodically — giving rise to resonances in the kink-antikink [2–11] and kink-impurity interactions [11, 12], as well as stimulating kink-antikink pair production [13–15]. The internal mode serves as the cause of the kink's counter-intuitive responses to spatially-uniform time-periodic

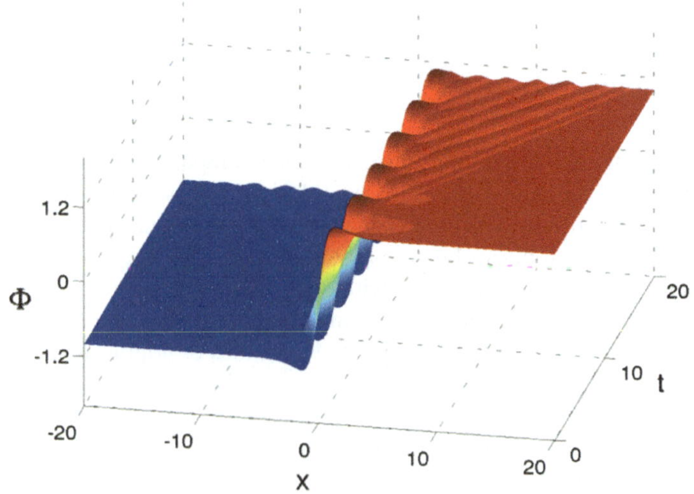

Fig. 9.1 The wobbling kink: a kink with an activated internal mode. This solution was obtained by the numerical simulation of (9.1) with the initial conditions $\phi_t(x, 0) = 0$ and $\phi(x, 0) = \tanh x + 2a \tanh x \operatorname{sech} x$, where $a = 0.3$

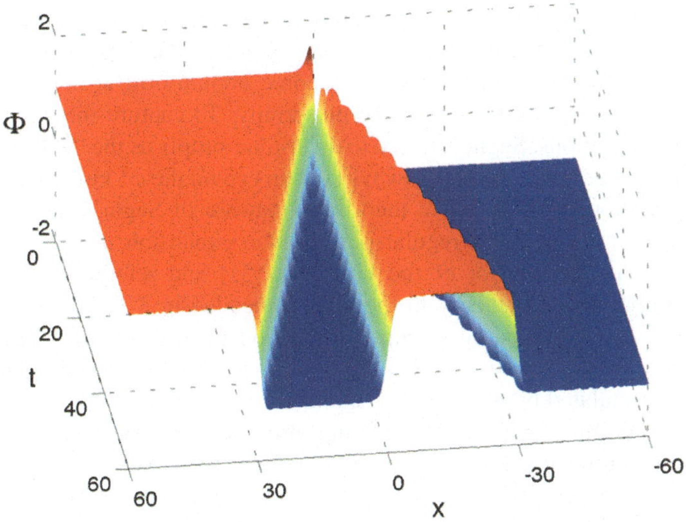

Fig. 9.2 Evolution of the same initial condition as in Fig. 9.1 but with $a = 0.6$. An attempt to excite the wobbling mode with a large amplitude results in the emission of a kink-antikink pair. (Note that time flows back to front in this figure.)

forcing [16, 17] and brings about its quasiperiodic velocity oscillations when the kink is set to propagate in a periodic substrate potential [18]. The excitation of the shape mode also provides a mechanism for the loss of energy and deceleration of the kink moving in a random medium [19].

In his simulations, Getmanov observed an amazing longevity of the kink's large-amplitude oscillations [1]. The kink seemed to be a comfortable location for the nonlinearity and dispersion to remain balanced in a high-energy excitation. Getmanov regarded the nonlinear excitation of the kink as a metastable bound state of *three* kinks — for this object was a product of a symmetric collision of two kinks and an antikink. (For more recent simulations of the bound-state formation, see [10].) This interpretation is consistent with the spontaneous production of a kink-antikink pair when the amplitude of excitation exceeds a certain threshold; see Fig. 9.2.

Getmanov's report of his *tritone* [1] was eclipsed by the storm of hype around the concurrent discovery of three-dimensional *pulsons* — in the same equation [20–24]. A more sustainable wave of interest starts forming when the one-dimensional ϕ^4 theory,

$$\frac{1}{2}\phi_{tt} - \frac{1}{2}\phi_{xx} - \phi + \phi^3 = 0, \tag{9.1}$$

was put forward as a model for the charge-density wave materials [25–27]. It has become clear that tritones should contribute to all characteristics of the material — alongside kinks and breathers. Rice and Mele described the tritone variationally, using the kink's width as the dynamical variable [28, 29]. Segur tried to construct this particle-like solution as a regular perturbation expansion in powers of the oscillation amplitude [30]. He determined the first two orders of the expansion (linear and quadratic), noted that expansion should become nonuniform at the order ϵ^3 and suggested a possible way to restore the uniformity. To capture the antisymmetric character of oscillations, Segur referred to the tritone simply as the "wobbling kink".

Another interest group that has always kept an eye on the ϕ^4 kink as the simplest topological soliton, is the particle theorists. Unaware of Segur's analysis, Arodź and his students developed a regular perturbation expansion using a polynomial approximation in the interior of the kink (see [31] and references therein) and extended the perturbation theory to the kink embedded in $3+1$ dimensions [32]. Romańczukiewicz obtained an asymptotic solution for the radiation wave emitted by the wobbling mode that is initially at rest [33] and discovered kink-antiknk pair productions stimulated by the wobbling-radiation coupling [14, 15].

Segur's suggestion for the circumventing of the perturbation breakdown, was to recognise the nonlinear nature of the wobbling mode by expanding its frequency in powers of its amplitude. This recipe constitutes the Lindstedt method in the theory of nonlinear oscillations; in the wobbling kink context this approach was later implemented by Manton and Merabet [13]. The Manton-Merabet analysis was successful in reproducing the $t^{-1/2}$ wobbling decay law that had been predicted by Malomed [19], on the basis of energy considerations. (See also [33].)

The Lindstedt method is known to be limited even when it is applied to solutions of ODEs. It proves to be a powerful tool for the calculation of anharmonic corrections to periodic orbits —but fails when the motion ceases to be periodic. More importantly, the method is not well suited for the analysis of PDEs— for it cannot handle the nonperiodic *spatial* degrees of freedom.

One alternative to the Lindstedt method is the Krylov–Bogolyubov collective-coordinate technique. This was used by Kiselev to obtain solutions of the ϕ^4 equation with a perturbed-kink initial condition posed on a characteristic line [34]. Although the introduction of collective coordinates allows one to recognise the hierarchy of coexisting space and time scales in the system, the resulting solutions lack the explicitness and transparency of the perturbation expansions in [13, 30, 32, 33]. The physical interpretation of their constituents is not straightforward either.

To preserve the lucidity of the regular expansion and, at the same time, take into account the coexistence of multiple space and time scales, Oxtoby and the present author have designed a singular perturbation expansion treating different space and time variables as independent [17, 35]. This asymptotic construction is reviewed in what follows. We also discuss the effects of the resonant forcing of the internal mode by a variety of direct and parametric driving agents.

The emphasis of the present note is on the fundamentals of our method, including the treatment of radiation, and the phenomenology of the wobbling kink's responses to driving. The reader interested in mathematical detail is welcome to consult the original publications [17, 35].

The outline of this chapter is as follows. In the next section we explain the basics of our approach as applied to the freely wobbling kink. Section 9.4 considers the equation for the wobbling amplitude and draws conclusions on the lifetime of this particle-like excitation. In Sect. 9.5 we formulate the asymptotic formalism for the consistent treatment of the long-range radiation. Section 9.6 is devoted to the effect of spatially-uniform temporally-resonant driving force. Some concluding remarks are made in Sect. 9.7.

9.3 Multiple Scales: Slow Times and Long Distances

In this and the next two sections we follow [35]. Instead of studying the kink travelling with the velocity v, we consider a motionless kink centred at the origin of the reference frame that moves with the velocity v itself. This is accomplished by the change of variables $(x,t) \to (\xi,\tau)$, where $\xi = x - \int_0^t v(t')\mathrm{d}t'$, $\tau = t$. The above transformation takes (9.1) to

$$\frac{1}{2}\phi_{\tau\tau} - v\phi_{\xi\tau} - \frac{v_\tau}{2}\phi_\xi - \frac{1-v^2}{2}\phi_{\xi\xi} - \phi + \phi^3 = 0. \tag{9.2}$$

An explicit occurrence of the soliton velocity in a relativistically-invariant equation is justifiable when there are factors that can induce time-dependence of v (e.g. radiation losses) or when the Lorentz invariance is broken by damping and driving terms. (These are discussed in Sect. 9.6.)

We expand ϕ about the kink $\phi_0 \equiv \tanh\xi$:

$$\phi = \phi_0 + \epsilon\phi_1 + \epsilon^2\phi_2 + \cdots. \tag{9.3}$$

Here ϵ is not pegged to any small parameter of the system (e.g. distance to some critical value) and has the meaning of the amplitude of the kink's perturbation. It can be chosen arbitrarily.

We also introduce a sequence of "long" space and "slow" time coordinates:

$$X_n \equiv \epsilon^n\xi, \quad T_n \equiv \epsilon^n\tau, \quad n = 0, 1, 2, \ldots.$$

In the limit $\epsilon \to 0$, the X_n and T_n are not coupled and can be treated as independent variables. Consequently, the ξ- and τ-derivatives are expressible using the chain rule:

$$\frac{\partial}{\partial\xi} = \partial_0 + \epsilon\partial_1 + \epsilon^2\partial_2 + \cdots, \quad \frac{\partial}{\partial\tau} = D_0 + \epsilon D_1 + \epsilon^2 D_2 + \cdots, \tag{9.4}$$

where $\partial_n \equiv \partial/\partial X_n$ and $D_n \equiv \partial/\partial T_n$.

We limit our analysis to the situation where the velocity is small. Hence we write $v = \epsilon V$ where V is of order 1. Furthermore, when the wobbling amplitude ϵ is small, it is natural to expect the velocity of the kink to vary slowly. Accordingly, V is taken to be a function of slow times only: $V = V(T_1, T_2, \ldots)$.

Substituting the above expansions into the ϕ^4 equation (9.2), we equate coefficients of like powers of ϵ. At the order ϵ^1, we obtain the linearised equation

$$\frac{1}{2}D_0^2\phi_1 + \mathcal{L}\phi_1 = 0,$$

where

$$\mathcal{L} = -\frac{1}{2}\partial_0^2 - 1 + 3\phi_0^2.$$

We choose a particular solution of this equation:

$$\phi_1 = A(X_1, X_2, \ldots; T_1, T_2, \ldots) \operatorname{sech} X_0 \tanh X_0 e^{i\omega_0 T_0} + \text{c.c.}, \tag{9.5}$$

where $\omega_0 = \sqrt{3}$ and c.c. stands for the complex conjugate of the preceding term. This is a wobbling mode with an undetermined amplitude A. The amplitude is a constant with respect to X_0 and T_0 but will generally depend on the stretched variables X_n and T_n $(n = 1, 2, \ldots)$.

At the second order we obtain

$$\frac{1}{2}D_0^2\phi_2 + \mathcal{L}\phi_2 = F_2, \tag{9.6}$$

where F_2 includes terms proportional to $e^{\pm i\omega_0 T_0}$, $e^{\pm 2i\omega_0 T_0}$ and e^0:

$$F_2 = (\partial_0\partial_1 - D_0 D_1)\phi_1 - 3\phi_0\phi_1^2 + V D_0\partial_0\phi_1 + \frac{1}{2}D_1 V \partial_0\phi_0 - \frac{1}{2}V^2\partial_0^2\phi_0.$$

Accordingly, the solution of (9.6) consists of a sum of three harmonics:

$$\phi_2 = \varphi_2^{(0)} + \varphi_2^{(1)} e^{i\omega_0 T_0} + \text{c.c.} + \varphi_2^{(2)} e^{2i\omega_0 T_0} + \text{c.c.}, \tag{9.7}$$

where the coefficients $\varphi_2^{(0)}$, $\varphi_2^{(1)}$ and $\varphi_2^{(2)}$ are determined to be

$$\varphi_2^{(0)} = 2|A|^2 \operatorname{sech}^2 X_0 \tanh X_0 + \left(\frac{V^2}{2} - 3|A|^2\right) X_0 \operatorname{sech}^2 X_0, \tag{9.8}$$

$$\varphi_2^{(1)} = -(\partial_1 A + i\omega_0 V A) X_0 \operatorname{sech} X_0 \tanh X_0, \tag{9.9}$$

and

$$\varphi_2^{(2)} = A^2 f_1(X_0), \tag{9.10}$$

with

$$f_1(X_0) = \frac{1}{8}\left\{6\tanh X_0 \operatorname{sech}^2 X_0 + (2 + ik_0 \tanh X_0 + \operatorname{sech}^2 X_0)\right.$$
$$\left.\times[J^*(X_0) - J_\infty]e^{ik_0X_0} + (2 - ik_0 \tanh X_0 + \operatorname{sech}^2 X_0)J(X_0)e^{-ik_0X_0}\right\}. \quad (9.11)$$

In the expression for f_1, $k_0 = \sqrt{8}$ and we have introduced the notations

$$J(X_0) = \int_{-\infty}^{X_0} e^{ik_0\xi} \operatorname{sech}^2\xi \, d\xi, \quad J_\infty = J(\infty). \quad (9.12)$$

The odd function $f_1(X_0)$ gives the short-range structure of the second-harmonic radiation. The particular solution (9.10)–(9.11) corresponds to the *outgoing* radiation:

$$\varphi_2^{(2)} e^{2i\omega_0 T_0} \to \frac{2 - ik_0}{8} J_\infty A^2 e^{i(2\omega_0 T_0 - k_0 X_0)} \text{ as } X_0 \to \infty;$$
$$\varphi_2^{(2)} e^{2i\omega_0 T_0} \to -\frac{2 - ik_0}{8} J_\infty A^2 e^{i(2\omega_0 T_0 + k_0 X_0)} \text{ as } X_0 \to -\infty. \quad (9.13)$$

To obtain the coefficients (9.8) and (9.9) we had to impose the constraints $D_1 V = 0$ and $D_1 A = 0$. These were necessary to make sure that $\varphi_2^{(0)}$ and $\varphi_2^{(2)}$ remain bounded as $|X_0| \to \infty$.

The ϵ^2-correction to the first-harmonic coefficient, function (9.9), decays to zero as $|X_0| \to \infty$; hence the term $\varphi_2^{(1)} e^{i\omega_0 T_0}$ in (9.7) is not secular. However, the function (9.9) becomes greater than the ϵ^1-coefficient (9.5) once X_0 has grown large enough. This contradicts our implicit assumption that all coefficients ϕ_n remain of the same order: $\phi_{n+1}/\phi_n = O(1)$ for all X_0. Accordingly, the term $\varphi_2^{(1)} e^{i\omega_0 T_0}$ with $\varphi_2^{(1)}$ in (9.9) is referred to as *quasi*-secular. To eliminate the resulting nonuniformity in the asymptotic expansion, we have to set

$$\partial_1 A + i\omega_0 V A = 0. \quad (9.14)$$

This equation will play a role in what follows.

Another quasi-secular term is the term proportional to $X_0 \operatorname{sech}^2 X_0$ in (9.8). This function is nothing but the derivative of $\tanh(\kappa X_0)$ with respect to κ. Hence we can eliminate the quasi-secular term simply by incorporating the coefficient $\frac{1}{2}V^2 - 3|A|^2$ in the variable width of the kink [35].

9.4 The Wobbling Mode's Lifetime

Proceeding to the order ϵ^3 we obtain an equation

$$\frac{1}{2}D_0^2\phi_3 + \mathcal{L}\phi_3 = F_3,$$

where

$$F_3 = (\partial_0\partial_1 - D_0 D_1)\phi_2 + (\partial_0\partial_2 - D_0 D_2)\phi_1 + \frac{1}{2}(\partial_1^2 - D_1^2)\phi_1 - \phi_1^3$$
$$-6\phi_0\phi_1\phi_2 + V D_0\partial_0\phi_2 + V D_0\partial_1\phi_1 + V D_1\partial_0\phi_1 + \frac{1}{2}D_2 V\partial_0\phi_0 - \frac{1}{2}V^2\partial_0^2\phi_1.$$

The solvability condition for the zeroth harmonic in the above equation reads $D_2 V = 0$. Taken together with the previously obtained $D_1 V = 0$, this implies that the kink's velocity remains constant — at least up to times $\tau \sim \epsilon^{-3}$. This conclusion was not entirely obvious beforehand. What this result implies is that the radiation does not break the Lorentz invariance of the equation. The wobbling kink and its stationary radiation form a single entity that can be Lorentz-transformed from one coordinate frame to another — just like the "bare" kink alone.

The solvability condition for the first harmonic has the form of a nonlinear ODE:

$$i\frac{2\omega_0}{3}D_2 A + \zeta|A|^2 A - V^2 A = 0, \tag{9.15}$$

where the complex coefficient ζ is given by

$$\zeta = 12\int_0^\infty \text{sech}^2 X_0 \tanh^3 X_0\left[\frac{5}{2}\text{sech}^2 X_0 \tanh X_0 - 3X_0\text{sech}^2 X_0 + f_1(X_0)\right]dX_0, \tag{9.16}$$

with f_1 as in (9.11). The imaginary part of ζ admits an explicit expression,

$$\zeta_I = \frac{3\pi^2 k_0}{\sinh^2(\pi k_0/2)} = 0.046,$$

while the real part can be evaluated numerically: $\zeta_R = -0.85$.

Adding the equation (9.15) multiplied by ϵ^3 and the equation (9.14) multiplied by $-\frac{2}{3}i\omega_0\epsilon v$, we obtain [35]

$$i\frac{2\omega_0}{3}a_t + \zeta|a|^2 a + v^2 a + O(|a|^5) = 0, \tag{9.17}$$

where we have introduced the "unscaled" amplitude of the wobbling mode $a = \epsilon A$ and reverted to the original kink's velocity $v = \epsilon V$. We have also used that $A_t = \epsilon(D_1 - v\partial_1)A + \epsilon^2 D_2 A + O(\epsilon^3)$. The advantage of (9.17) over (9.15) is that the expression (9.17) remains valid for all times from $t = 0$ to $t \sim \epsilon^{-2}$ whereas the (9.15) governs the evolution only on long time intervals.

According to (9.17), the amplitude will undergo a monotonic decay:

$$|a(t)|^2 = \frac{|a(0)|^2}{1 + \omega_0 \zeta_I \, |a(0)|^2 t}. \tag{9.18}$$

This equation was originally derived by the Lindstedt method in [13] and using energy considerations in [13, 33].

The smallness of ζ_I can be deduced from (9.16) even without performing the exact integration. Indeed, the imaginary part of the integral (9.16) is of the form

$$\zeta_I = \int_0^\infty (F_0 \cos k\xi + G_0 \sin k\xi)\, \mathrm{d}\xi, \tag{9.19}$$

where $k > 1$ and the real functions $F_0(\xi)$ and $G_0(\xi)$ are even and odd, respectively. The functions $F_0(\xi)$, $G_0(\xi)$ and all their derivatives are bounded on $(0, \infty)$ and decay to zero as $\xi \to \infty$. A repeated integration by parts gives

$$\zeta_I(k) = \sum_{n=0}^{N} \frac{G_n(0)}{k^{n+1}} + O\left(\frac{1}{k^{N+1}}\right), \tag{9.20}$$

where

$$G_{n+1}(\xi) = -\frac{\mathrm{d}F_n}{\mathrm{d}\xi}, \quad F_{n+1}(\xi) = \frac{\mathrm{d}G_n}{\mathrm{d}\xi} \quad (n = 0, 1, 2, \ldots).$$

Because of the evenness of $F_0(\xi)$ and oddness of $G_0(\xi)$, all coefficients in the series (9.20) are zero. Therefore the integral (9.19) is smaller than any positive power of $1/k$ — that is, it is *exponentially* small as $k \to \infty$. As a result, even with a moderate value of k, $k = k_0 = \sqrt{8}$, we have ζ_I below 0.05.

The fact that the decay rate ζ_I is an exponentially decreasing function of the radiation wavenumber, has a simple physical explanation. The decay rate of the wobbling mode's energy is determined by the energy flux which, in turn, is proportional to the square of the radiation wave amplitude. On the other hand, the amplitude of radiation from any localised oscillatory mode (a localised external source, an impurity or internal mode) is an exponentially decreasing function of k. In particular, the amplitude of the second-harmonic radiation excited by the wobbling mode includes the factor J_∞ — see (9.13). This factor is evaluated to

$$J_\infty = \int_{-\infty}^{\infty} e^{ik_0\xi} \operatorname{sech}^2 \xi \, \mathrm{d}\xi = \frac{\pi k_0}{\sinh(\pi k_0/2)}.$$

Accordingly, if k_0 were allowed to grow rather than being set to $\sqrt{8}$, the energy flux would drop in proportion to $e^{-\pi k_0}$.

Thus the longevity of the wobbling mode is due to the wavelength of the second-harmonic radiation being several times shorter than the effective width of the kink (more specifically, πk_0 being about nine times greater than 1). As a result, the decay rate ζ_I ends up having a tiny factor of e^{-9}.

9.5 Radiation From a Distant Kink

The above treatment of the radiation is limited in two ways. Firstly, although higher terms in the expansion (9.3) were assumed to be smaller than the lower ones, that is, $\epsilon^{n+1}\phi_{n+1}/\epsilon^n\phi_n \to 0$ as $\epsilon \to 0$, the coefficient ϕ_1 becomes exponentially small while ϕ_2 remains of order 1 as $|X_0|$ grows. This means that our construction is only consistent for not too large values of $|X_0|$.

Second, the asymptotic expansion (9.3) with coefficients (9.5) and (9.7) describes a steadily oscillating kink but cannot account for any perturbations propagating in the system. Let, for simplicity, $V = 0$ and choose some initial condition for the amplitude A: $A = A_0$ at $T_2 = 0$. Then ϕ is equal to

$$\phi = -1 - \epsilon^2 \frac{2 - ik}{8} J_\infty A_0^2 e^{ik_0 X_0}$$

for *all* sufficiently large negative X_0 (and to the negative of this expression for all large positive X_0). This creates an impression that the initial perturbation has travelled a large distance instantaneously — while in actual fact the asymptotic solution that we have constructed describes a (slowly relaxing) stationary structure and cannot capture transients or perturbations.

To design a formalism for the propagation of nonstationary waves we perform a Lorentz transformation to the reference frame where $v = 0$. (As we have explained in the previous section, the radiation from the kink does not break the Lorentz invariance.) Consider large positive X_0 and expand the field as in

$$\phi = 1 + \epsilon^2\phi_2 + \epsilon^4\phi_4 + \cdots . \tag{9.21}$$

In a similar way, we let

$$\phi = -1 + \epsilon^2\phi_2 + \epsilon^4\phi_4 + \cdots \tag{9.22}$$

for large negative X_0. Substituting these, together with the derivative expansions (9.4), in equation (9.1), the order ϵ^2 gives

$$\phi_2 = \mathcal{J} B_+ e^{i(\omega_+ T_0 - k_+ X_0)} + \text{c.c.}, \quad X_0 > 0, \tag{9.23}$$

and

$$\phi_2 = -\mathcal{J} B_- e^{i(\omega_- T_0 - k_- X_0)} + \text{c.c.}, \quad X_0 < 0.$$

Here $\omega_\pm = \sqrt{k_\pm^2 + 4}$, and the amplitudes $B_\pm$ are functions of the stretched coordinates: $B_\pm = B_\pm(X_1, \ldots; T_1, \ldots)$. The coefficient $\mathcal{J}$ will be chosen at a later stage and the negative sign in front of B_- was also introduced for later convenience.

Consider a point $X_0 = \frac{1}{2} \ln \epsilon^{-1}$. Sending $\epsilon \to 0$ we have $X_0 \to \infty$, so that the "outer" expansion (9.21) with ϕ_2 as in (9.23) is valid. On the other hand, the ratio $(\epsilon^2 \phi_2)/(\epsilon \phi_1)$ with ϕ_1 and ϕ_2 as in (9.5) and (9.7), is $O(\epsilon^{1/2})$; hence the "inner" expansion (9.3) remains uniform at the chosen point.

The corresponding stretched coordinates $X_1 = (\epsilon/2) \ln \epsilon^{-1}$, $X_2 = (\epsilon^2/2) \ln \epsilon^{-1}$, ..., satisfy $X_1, X_2, \ldots \to 0$ as $\epsilon \to 0$. Consequently, the coefficient B_+ in (9.23) has zero spatial arguments: $B_+ = B_+(0, 0, \ldots; T_1, T_2, \ldots)$. In a similar way, the amplitude A^2 in (9.7) is $A^2(0, 0, \ldots; T_2, T_3, \ldots)$. Choosing $\mathcal{J} = \frac{1}{8}(2 - ik_0) J_\infty$ and equating (9.23) to (9.7), we obtain $\omega_+ = 2\omega_0$, $k_+ = k_0$, and

$$B_+(0, 0, \ldots; T_1, T_2, \ldots) = A^2(0, 0, \ldots; T_2, T_3, \ldots). \tag{9.24}$$

A similar matching at the point $X_0 = -\frac{1}{2} \ln \epsilon^{-1}$ leads to

$$B_-(0, 0, \ldots; T_1, T_2, \ldots) = A^2(0, 0, \ldots; T_2, T_3, \ldots). \tag{9.25}$$

The solvability condition for the nonhomogeneous equation arising at the order ϵ^3, gives a pair of linear transport equations

$$D_1 B_+ + c_0 \partial_1 B_+ = 0, \quad X_1 > 0, \tag{9.26}$$

$$D_1 B_- - c_0 \partial_1 B_- = 0, \quad X_1 < 0, \tag{9.27}$$

where $c_0 = k_0/(2\omega_0)$. Equation (9.26) should be solved under the boundary condition (9.24) while solutions of (9.27) should satisfy the condition (9.25).

Solutions of (9.26) propagate along the characteristic lines $X_1 = c_0 T_1 + \eta$ where η is a parameter $(-\infty < \eta < \infty)$. In a similar way, solutions of (9.27) travel along the characteristics $X_1 = -c_0 T_1 + \tilde{\eta}$. The velocity c_0 is the group velocity for wavepackets of second-harmonic radiation centred on the wavenumber k_0; as one can readily check, $c_0 = (d\omega/dk)|_{k=k_0}$, where $\omega = \sqrt{4 + k^2}$. This velocity is of course smaller than the speed of light: $c_0 < 1$.

If the amplitude B_+ satisfies the initial condition

$$B_+(X_1, 0) = \beta(X_1), \quad X_1 > 0$$

at the moment $T_1 = 0$, its subsequent evolution in the region $X_1 > c_0 T_1$ for $T_1 > 0$ is a mere translation:

$$B_+(X_1, T_1) = \beta(X_1 - c_0 T_1).$$

In a similar way, in the region $X_1 < -c_0 T_1$ with $T_1 > 0$ the amplitude B_- satisfies

$$B_-(X_1, T_1) = \beta(X_1 + c_0 T_1),$$

where $\beta(X_1)$ is the initial condition for this solution:

$$B_-(X_1, 0) = \beta(X_1), \quad X_1 < 0.$$

In the sector between the rays $X_1 = -c_0 T_1$ and $X_1 = c_0 T_1$, the solutions are defined by the boundary conditions instead of initial ones:

$$B_+(X_1, T_1) = A^2\big|_{X_1=T_1=0}, \qquad 0 < X_1 < c_0 T_1;$$
$$B_-(X_1, T_1) = A^2\big|_{X_1=T_1=0}, \qquad -c_0 T_1 < X_1 < 0.$$

If we choose the initial condition $\beta(X_1) = 0$ for all X_1, the field ϕ will have the form of a pair of fronts, or shock waves, propagating away from the origin. For $|X_1| > c_0 T_1$, the amplitudes will be zero, $B_\pm = 0$, whereas for $|X_1| < c_0 T_1$, these will assume nonzero constant values: $B_\pm(X_1, T_1) = A^2\big|_{X_1=T_1=0}$. In agreement with one's physical intuition, the wobbling kink only influences the adjacent domain $|X_1| < c_0 T_1$.

To describe the evolution of the radiation on a longer, T_2, scale, one needs to derive one more pair of amplitude equations for B_+ and B_-. The coefficient ϕ_4 in the expansions (9.21) and (9.22) can be determined if the following solvability conditions are satisfied — in the right and left half of the x-line, respectively:

$$i(2\omega_0 D_2 + k_0 \partial_2) B_\pm + \frac{1}{2}(D_1^2 - \partial_1^2) B_\pm = 0.$$

Eliminating $D_1 B_\pm$ using (9.26) and (9.27), the above equations are transformed to

$$i D_2 B_\pm \pm i c_0 \partial_2 B_\pm - \frac{\omega_{kk}}{2} \partial_1^2 B_\pm = 0. \tag{9.28}$$

Here

$$\omega_{kk} \equiv \left.\frac{d^2\omega}{dk^2}\right|_{k_0} = \frac{4\omega_0^2 - k_0^2}{8\omega_0^3}$$

is the dispersion of the group velocity of the radiation waves. Adding (9.28) and (9.26)–(9.27) multiplied by the appropriate powers of ϵ, we obtain a pair of Schrödinger equations in the laboratory coordinates:

$$i\partial_t B_\pm \pm i c_0 \partial_x B_\pm - \frac{\omega_{kk}}{2} \partial_x^2 B_\pm = 0. \tag{9.29}$$

Consider the top equation in (9.29) on the interval $\tilde{x} < x < \infty$, with the initial condition $B_+(x, 0) = 0$ and the boundary conditions

$$B_+(\tilde{x}, t) = A^2(\epsilon^2 t), \quad B_+(\infty, t) = 0.$$

Here $\tilde{x} = \frac{1}{2} \ln \epsilon^{-1}$. The solution evolving out of this combination of initial and boundary conditions will describe a step-like front propagating to the right with the velocity c_0 and dispersing on the slow time scale $t \sim \epsilon^{-2}$. In a similar way, the bottom equation in (9.29), considered on the interval $-\infty < x < -\tilde{x}$ with the initial condition $B_-(x, 0) = 0$ and boundary conditions $B_-(-\tilde{x}, t) = A^2(\epsilon^2 t)$, $B_-(-\infty, t) = 0$, will describe a slowly spreading shock wave moving to the left.

This completes the multiscale description of the freely radiating kink [35]. The uniformity of the asymptotic expansion at all scales is secured by introducing the long-range radiation variables which are related to the short-range radiations through boundary conditions but do not automatically coincide with those.

9.6 Damped Driven Wobbling Kink

The wobbling kink may serve as a stable source of radiation with a certain fixed frequency and wavenumber. However in order to sustain the wobbling, energy has to be fed into the system from outside.

A particularly efficient and uncomplicated way of pumping energy into a kink is by means of a resonant driving force. This type of driving does not have to focus on any particular location; the driving wave may fill, indiscriminately, the entire x-line. Only the object in possession of the resonant internal mode will respond to it. Furthermore, a small driving amplitude should be sufficient to sustain the oscillations as the free-wobbling decay rate is very low.

The two standard ways of driving an oscillator are by applying an external force in synchrony with its own natural oscillations or by varying, periodically, one of its parameters. The former is commonly called the *direct*, or *external*, forcing, whereas the latter goes by the name of *parametric* pumping. In the case of the harmonic oscillator, the strongest parametric resonance occurs if the driving frequency is close to twice its natural frequency.

There is an important difference between driving a structureless oscillator and pumping energy into the wobbling mode of the kink. The mode has an odd, antisymmetric, spatial structure so that the oscillation in the positive semi-axis is out of phase with the oscillation in the negative half line. It is not obvious whether the energy transfer from a spatially-uniform force to this antisymmetric mode is at all possible, and if yes — what type of driver (direct or parametric) and what resonant frequency would ensure the most efficient transfer. To find answers to these questions, we will examine driving forces of both types and with several frequencies.

In addition to the driving force, our analysis will take into account the dissipative losses. (The dissipation is an effect that is difficult to avoid in a physical system.) Accordingly, the resulting amplitude equation will have two types of damping terms: the linear damping accounting for losses due to friction, absorption, incomplete internal reflection of light or similar imperfections of the physical system — and cubic damping due to the emission of the second-harmonic radiation.

The damped-driven ϕ^4 equation was utilised to model the drift of domain walls in magnetically ordered crystals placed in oscillatory magnetic fields [16]; the Brownian motion of string-like objects on a periodically modulated bistable substrate [36]; ratchet dynamics of kinks in a lattice of point-like inhomogeneities [37] and rectification in Josephson junctions and optical lattices [38].

The first consistent treatment of the periodically forced ϕ^4 kink was due to Sukstanskii and Primak [16] who examined the joint action of external and parametric excitation, at a generic driving frequency. Using a combination of the Lindstedt method and the method of averaging, they have discovered the kink's drift with the velocity proportional to the product of the external and parametric driving amplitudes. It is important to note that this effect is not related to the wobbling of the kink; yet the drift velocity develops a peak when the driving frequency is near the frequency of the wobbling.

The individual effect of the external driving force was explored by Quintero, Sánchez and Mertens using an heuristic collective-coordinate Ansatz and disregarding radiation [39, 40]. As in [16], the driving frequency was generic, that is, not pegged to any particular value. The analysis of Quintero et al suggests that a resonant energy transfer from the driver to the kink takes place when the driving frequency is close to the frequency of the wobbling and, at a higher rate, when the driving frequency is near a half of that value. Although the resonant transfer is not quantified in the approach of [39, 40] and the reported collective coordinate behaviour is open to interpretation, the discovery of the subharmonic resonance (supported by simulations of the full PDE) identified an important direction for more detailed scrutiny. An interesting numerical observation of [39, 40] was a chaotic motion of the kink when the driving frequency is set to a half of the natural wobbling frequency.

In the subsequent publication [41], Quintero, Sánchez and Mertens extended their analysis to the kink driven parametrically — still at a generic frequency. As in their previous study, the authors observed a resonant energy transfer when the driving frequency is near the natural wobbling frequency of the kink. On the other hand, the parametric driving would not induce any translational motion of the kink.

Below, we examine the resonantly driven wobbling kink using our method of multiple scales. The exposition of this section follows [17].

9.6.1 Parametric Driving at the Natural Wobbling Frequency

The parametric driving of the harmonic oscillator is known to cause a particularly rapid (exponential) growth of the amplitude of its oscillations. In this section we examine the parametric driving of the kink close to its natural wobbling frequency.

The driven ϕ^4 equation has the form [41]

$$\frac{1}{2}\phi_{tt} - \frac{1}{2}\phi_{xx} - \phi + \phi^3 = -\gamma\phi_t + h\cos(\Omega t)\phi. \tag{9.30}$$

Here the driving frequency Ω is slightly detuned from $\omega_0 = \sqrt{3}$, the linear wobbling frequency of the undriven kink:

$$\Omega = \omega_0(1 + \rho).$$

It is convenient to express the small detuning ρ and the amplitude of the wobbling mode in terms of the same small parameter:

$$\rho = \epsilon^2 R,$$

where $R = O(1)$.

The small parameters h and γ in the equation (9.30) measure the driving strength and linear damping, respectively. It is convenient to choose the following scaling laws for these coefficients:

$$\gamma = \epsilon^2 \Gamma, \quad h = \epsilon^3 H,$$

where $\Gamma, H = O(1)$. The above choice ensures that the damping and driving terms appear in the amplitude equations at the same order as the cubic nonlinearity.

A pair of equations produced by the multiple-scale procedure includes an equation for the "unscaled" amplitude of the wobbling mode, $a = \epsilon A$ [17]:

$$\dot{a} = -\gamma a - i\omega_0 \left(\rho + \frac{v^2}{2} \right) a + \frac{i}{2}\omega_0 \zeta |a|^2 a - i\frac{\pi}{8}\omega_0 h + O(|a|^5).$$

Here the overdot indicates differentiation with respect to t and the complex coefficient ζ has been evaluated in Sect. 9.4. The second amplitude equation governs the velocity of the kink:

$$\dot{v} = -2\gamma v + O(|a|^5). \tag{9.31}$$

The damping term $\gamma \phi_t$ breaks the Lorentz invariance of the ϕ^4 model (9.30). The kink travelling at the velocity v is no longer equivalent to the kink at rest. Instead, according to (9.31), the travelling kink suffers deceleration — until v becomes of order $|a|^3$. Thus, after an initial transient of the length $t \sim \gamma^{-1}$, the dynamics is governed by a single equation for the complex wobbling amplitude:

$$\dot{a} + i\omega_0 \rho a - \frac{i}{2}\omega_0 \zeta |a|^2 a = -\gamma a - i\frac{\pi}{8}\omega_0 h. \tag{9.32}$$

Although the underlying PDE (9.30) is driven *parametrically*, the forcing term in (9.32) is characteristic for the *externally* driven Schrödinger equations (see [42, 43] and references therein). The reason is that the oscillator that we are trying to pump energy to, is the kink's internal mode rather than the kink itself. On the other hand, the leading part of the forcing term in the RHS of (9.30), $h \cos(\Omega t)\phi_0$, involves the kink (ϕ_0) rather than the internal mode (ϕ_1). As a result, the driver that was expected

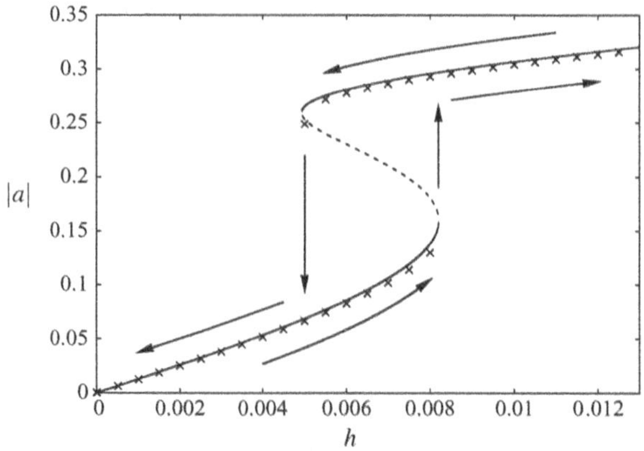

Fig. 9.3 The hysteresis loop of the ϕ^4 kink driven, parametrically, at the frequency close to its natural wobbling frequency. Continuous curves delineate stable and dashed curves unstable fixed points of the amplitude equation (9.32). Crosses represent numerical simulations of the full PDE (9.30) with the same γ and ρ. In the simulations, the driving strength h was increased from zero in small steps and then turned back to zero as indicated by the arrows. This figure was generated using $\gamma = 0.01$ and $\rho = -0.03$. Reproduced from [17], with permission from APS

to affect a parameter of the oscillatory mode, acts as an external periodic force with regard to this mode.

Equation (9.32) has no attractors other than fixed points. Assume the damping coefficient γ is fixed. When the detuning ρ satisfies $\rho > \rho_0$, where $\rho_0 = -1.14\gamma$, the dynamical system (9.32) has a single fixed point irrespectively of the value of h. The fixed point is stable and attracts all trajectories. If the detuning is chosen in the complementary domain $\rho < \rho_0$, the structure of the phase space depends on the value of h. Namely, there are two critical values h_+ and h_-, where $h_+ < h_-$, such that when h is smaller than h_+ or larger than h_-, the system's phase portrait features a single fixed point — and this point is attractive. The inner region $h_+ < h < h_-$ is characterised by two stable fixed points and exhibits hysteresis. (See Fig. 9.3). The values h_+ and h_- are expressible through ρ and γ [17].

The memory function associated with the bistability and hysteresis of the parametrically driven wobbling kink, endows this structure with potential applications in electronic circuits, devices based on ferromagnetism or ferroelectricity, and charge-density wave materials.

9.6.2 *Parametric Driving at Twice the Natural Wobbling Frequency*

The longest and widest Arnold tongue of the parametricaly driven harmonic oscillator corresponds to the subharmonic resonance where the driving frequency is close to twice the natural frequency of the oscillator. It is therefore interesting to examine this driving regime in the context of the wobbling kink. Would the 2 : 1 frequency ratio sustain a larger wobbling amplitude than the 1 : 1 regime considered in the previous subsection?

The driven equation in this case has the form

$$\frac{1}{2}\phi_{tt} - \frac{1}{2}\phi_{xx} - \phi + \phi^3 = -\gamma\phi_t + h\cos(2\Omega t)\phi, \tag{9.33}$$

where, as in the previous subsection, we set

$$\Omega = \omega_0(1+\rho).$$

The scaling laws for the weak detuning and small driving amplitude are chosen as before:

$$\rho = \epsilon^2 R, \quad \gamma = \epsilon^2 \Gamma.$$

This time, however, it is convenient to choose a different scaling law for the small driving amplitude:

$$h = \epsilon^2 H.$$

The resulting amplitude system is [17]

$$\dot{a} = -\gamma a - i\omega_0\left(\rho + \frac{v^2}{2}\right)a + \frac{i}{2}\omega_0\zeta|a|^2 a - \frac{i}{2}\omega_0\sigma h a^* + O(|a|^5),$$
$$\dot{v} = -2\gamma v + O(|a|^5).$$

Here the complex coefficient $\sigma = \sigma_R + i\sigma_I$ is given by an integral

$$\sigma = \int_{-\infty}^{\infty}\left[\frac{1}{2}\mathrm{sech}^2 X_0 \tanh^2 X_0 - 6\mathrm{sech}^2 X_0 \tanh^3 X_0 f_2(X_0)\right] \mathrm{d}X_0, \tag{9.34}$$

where f_2 consists of the second-harmonic radiation and a stationary standing wave—both induced by the driver:

$$f_2(X_0) = -\frac{1}{12}f_1(X_0) + \frac{1}{24}\tanh X_0(2\mathrm{sech}^2 X_0 - 3).$$

In the above expression, the function f_1 is given by the equation (9.11). The real and imaginary parts of the integral (9.34) are found to be

$$\sigma_R = 0.60, \quad \sigma_I = \frac{1}{12}\zeta_I = 0.0039.$$

After a transient of the length $t \sim \gamma^{-1}$, the dynamics is governed by a two-dimensional dynamical system

$$\dot{a} + i\omega_0 \rho a - \frac{i}{2}\omega_0 \zeta |a|^2 a = -\gamma a - \frac{1}{2} i\omega_0 \sigma h a^*. \tag{9.35}$$

Equation (9.35) is similar to (9.32) from the previous section; the only difference between the two expressions is the type of the driver. Unlike (9.32), the amplitude equation (9.35) features the parametric forcing in its standard Schrödinger form (see e.g. [44–47]).

The energy-transfer mechanism in the present case is different from the mechanism operating in the 1 : 1 forcing regime. In the partial differential equation (9.33), the function $\cos(2\Omega t)$ in the product $h\cos(2\Omega t)(\phi_0 + \epsilon\phi_1 + \cdots)$, acts as a parametric driver for the wobbling mode ϕ_1. In addition, the "external force" $h\cos(2\Omega t)\phi_0$ excites a standing wave with the frequency 2Ω and generates radiation at the same frequency. Both couple to the wobbling mode — parametrically, via the term $\epsilon^3\phi_0\phi_1\phi_2$ in the expansion of ϕ^3. As a result, in the case of equation (9.33) we have three concurrent mechanisms at work, and all three are of parametric nature. Combined, the three mechanisms produce the ha^*-term in the amplitude equation (9.35).

Turning to the analysis of the amplitude equation, we assume, first, that $\rho > \rho_0$, where $\rho_0 = -0.031\gamma$. There is a critical value of the driving amplitude $h_+(\gamma, \rho)$ such that the dynamical system (9.35) has two stable fixed points a_1 and $-a_1$ in the region $h > h_+$ and a single stable fixed point $a = 0$ in the complementary region $h < h_+$.

If the frequency detuning is taken to satisfy $\rho < \rho_0$, all trajectories flow to the origin for h smaller than h_- (where h_- is another critical value expressible through γ and ρ), and to one of the two nontrivial fixed points $\pm a_1$ for large h ($h > h_+$). In the intermediate region $h_- < h < h_+$ the system shows a tristability between $a = 0$ and the pair of points $a = \pm a_1$. See Fig. 9.4.

9.6.3 External Subharmonic Driving

Proceeding to the analysis of the direct driving force, we start with the case where the driving frequency is approximately a half of the natural wobbling frequency of the kink [39, 40]:

$$\frac{1}{2}\phi_{tt} - \frac{1}{2}\phi_{xx} - \phi + \phi^3 = -\gamma\phi_t + h\cos\left(\frac{\Omega}{2}t\right). \tag{9.36}$$

Here, as in the previous subsections, $\Omega = \omega_0(1 + \rho)$. We keep our usual scaling laws for the small detuning $\rho = \epsilon^2 R$ and damping coefficient $\gamma = \epsilon^2\Gamma$, but choose a fractional order of smallness for the driving amplitude: $h = \epsilon^{3/2} H$.

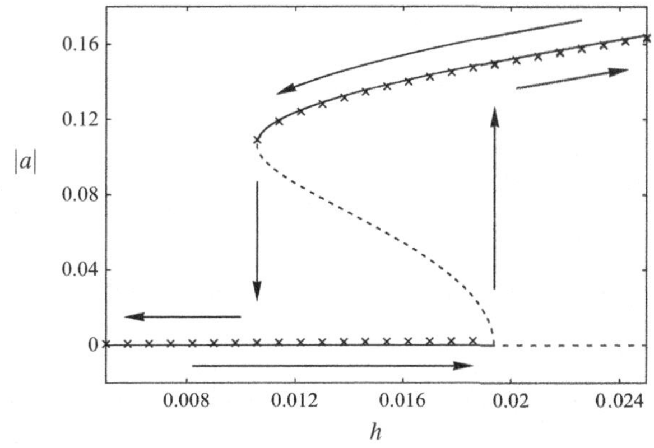

Fig. 9.4 The hysteresis loop of the kink driven, parametrically, at the frequency close to twice its natural wobbling frequency. Continuous and dashed curves delineate stable and unstable fixed points of the amplitude equation (9.35) with $\gamma = 0.005$ and $\rho = -0.005$. Crosses represent results of the numerical simulations of the equation (9.33) with the same γ and ρ. The driving strength h was increased from 0.005 to 0.025 in small steps and then turned back to 0.005 as indicated by the arrows. Reproduced from [42], with permission from APS

One amplitude equation is standard, $\dot{v} = -2\gamma v + O(|a|^5)$. Once the velocity has become of order $|a|^3$, it drops out of the second equation which then acquires the form [17]

$$\dot{a} + i\omega_0 \left(\rho + \frac{\lambda}{2} h^2 \right) a - \frac{i}{2} \omega_0 \zeta |a|^2 a = -\gamma a + i \frac{60}{169} \pi \omega_0 h^2. \qquad (9.37)$$

Note that the term proportional to λ is of higher order than all other terms in this equation. We had to include this forth-order correction to improve the agreement between the fixed points of the dynamical system (9.37) and results of simulation of the full partial differential equation (9.36). The fact that a higher-order term makes a significant contribution is due to its large coefficient: $\lambda = -7.47 - 1.68\,i$.

Equation (9.37) coincides with the equation (9.32) that we considered earlier, with ρ replaced with $\rho' = \rho + \lambda h^2/2$, h replaced with $h' = \frac{480}{169} h^2$, and $a' = -a$. Hence the dynamics of the 1 : 2 externally driven wobbllng kink reproduce those of the kink driven by the 1 : 1 parametric force.

As in that earlier system, the absence or presence of hysteresis in the dynamics depends on whether ρ' is above or below the critical value $\rho_0 = -1.14\gamma$. If the difference $\rho' - \rho_0$ is positive, all trajectories are attracted to a single fixed point. If, on the other hand, $\rho' - \rho_0 < 0$, we have two stable fixed points for each h' in a finite interval (h_+, h_-), where h_+ and h_- are expressible through ρ' and γ. This bistability leads to the hysteretic phenomena similar to the one depicted in Fig. 9.3. Outside the interval (h_+, h_-), all trajectories flow to a single fixed point. See Fig. 9.5.

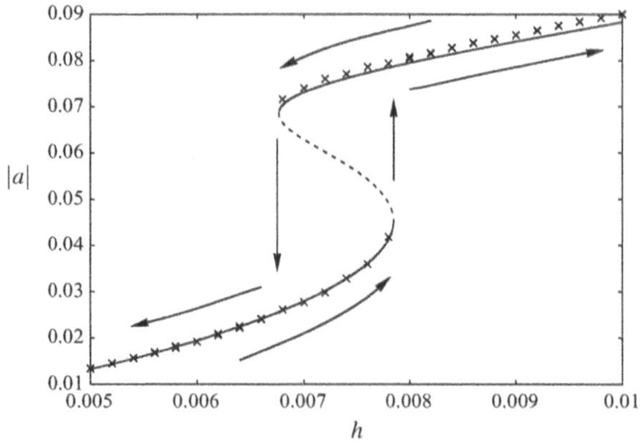

Fig. 9.5 The hysteresis loop of the kink driven by an external subharmonic force. The continuous and dashed lines depict the stable and unstable fixed points of the dynamical system (9.37). Crosses result from the numerical simulations of the partial differential equation (9.36), with the same γ and ρ. (In this plot, $\gamma = 10^{-3}$ and $\rho = -2 \times 10^{-3}$.) The driving strength h is increased in small increments and then turned down. The dependence of $|a|$ on h is similar to the one in Fig. 9.3; yet the same driving strength sustains a much smaller wobbling amplitude than in Fig. 9.3. Reproduced from [42], with permission from APS

The energy transfer mechanism associated with the 1 : 2 external pumping deserves to be commented. It was suggested [39, 40] that the mechanism consists in the coupling of the wobbling to the translation mode. Our asymptotic expansion furnishes [17] a different explanation though.

Expanding ϕ in powers of $\epsilon^{1/2}$,

$$\phi = \phi_0 + \epsilon\phi_1 + \epsilon^{3/2}\phi_{3/2} + \epsilon^2\phi_2 + \epsilon^{5/2}\phi_{5/2} + \cdots,$$

we observe that the external driver excites an even-parity standing wave $\phi_{3/2}$ at its frequency $\Omega/2$. The standing wave undergoes frequency doubling and parity transmutation via the term $\epsilon^3\phi_0\phi_{3/2}^2$ in the expansion of ϕ^3. It is this latter term that serves as an external driver for the wobbling mode. It has the resonant frequency Ω and its parity coincides with the parity of the mode.

This energy-pumping mechanism is a two-stage process and the resulting effective driving strength is proportional to h^2 rather than h. As a result, the external subharmonic driving sustains a much smaller wobbling amplitude than the harmonic parametric forcing. This conclusion is obvious if one compares the vertical scales in Figs. 9.5 and 9.3.

9.6.4 External Harmonic Driving

Driving the kink by an external periodic force with the wobbling frequency leads to the most interesting phenomenology. In this case the ϕ^4 equation has the form [39, 40]

$$\frac{1}{2}\phi_{tt} - \frac{1}{2}\phi_{xx} - \phi + \phi^3 = -\gamma\phi_t + h\cos(\Omega t), \tag{9.38}$$

where $\Omega = \omega_0(1+\rho)$. As in the previous three cases, we let $\rho = \epsilon^2 R$ and $\gamma = \epsilon^2\Gamma$. Choosing a linear scaling for the driving amplitude, $h = \epsilon H$, and assuming that the velocity $v = \epsilon^2 V$ (rather than ϵV as before), we obtain the following system of two amplitude equations [17]:

$$\dot{a} + i\omega_0\left(\rho + \frac{\nu}{2}h^2\right)a - \frac{i}{2}\omega_0\zeta|a|^2 a = -\gamma a + \frac{3\pi}{4}vh - \frac{i}{2}\omega_0\mu h^2 a^* + O(|a|^5), \tag{9.39}$$

$$\dot{v} = -2\gamma v + \frac{3}{2}h\left[\frac{\pi}{2}\omega_0\left(\omega_0\rho + i\gamma\right) - \eta|a|^2 - \chi h^2\right]a + \text{c.c.} + O(|a|^5). \tag{9.40}$$

Here ν, μ, η, and χ are numerical coefficients:

$$\nu = 4.16 - 0.33i, \quad \mu = 1.02 + 0.16i,$$
$$\eta = -2.00 - 0.38i, \quad \chi = -12.21 - 0.57i.$$

According to the equation (9.40), the velocity does not drop out of the system; instead, it plays an important role in the dynamics. Unlike all types of driving that we considered so far, the harmonic external driving can sustain the translational motion of the kink. Figure 9.6 shows an example of the kink accelerated by the 1:1 direct driving force which simultaneously excites the wobbling [17].

When h is small, the fixed point at $a = v = 0$ is the only attractor in the system. As h is increased while keeping the parameters γ and ρ constant, a pair of fixed points (a, v) and $(-a, -v)$, with nonzero a and v, bifurcates from the trivial fixed point (see Fig. 9.7).

As the driving strength approaches some critical value h_c, the $|a|$- and v-component of these points grow to infinity. No stable fixed points exist in the parameter region beyond h_c. In that region, all trajectories escape to infinity: $|a(t)| \to \infty$, $v(t) \to \infty$ as $t \to \infty$. The critical value h_c can be determined assuming that the growth is self-similar, that is, that v grows as a power of $|a|$. This assumption leads to a simple exponential asymptote $e^{\mathfrak{r}t}$ for $|a(t)|$, with the growth rate [17]

$$\mathfrak{r} = -\frac{2}{3}\gamma - \frac{3\pi i}{4\omega_0}\frac{\eta\zeta^* - \zeta\eta^*}{|\zeta|^2}h^2.$$

Setting $\mathfrak{r} = 0$ gives

$$h_c = 0.65\gamma^{1/2}. \tag{9.41}$$

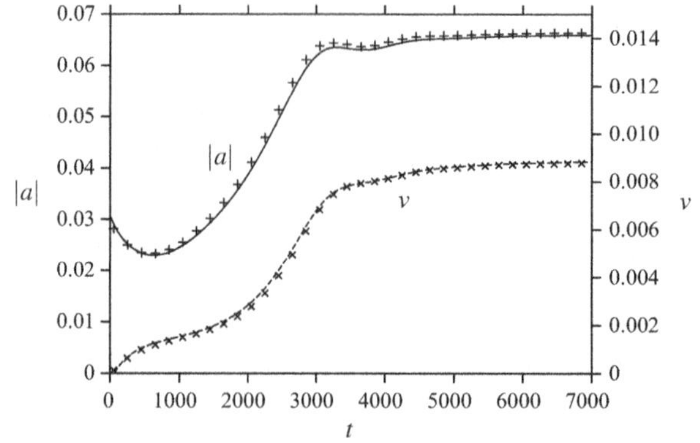

Fig. 9.6 The wobbling kink set in motion by the harmonic external force. The curves were obtained by solving the amplitude equations (9.39)–(9.40). The crosses represent numerical simulations of the PDE (9.38) with the corresponding initial conditions. In this plot, $h = 0.012$, $\gamma = 0.001$ and $\rho = 0$. Reproduced from [42], with permission from APS

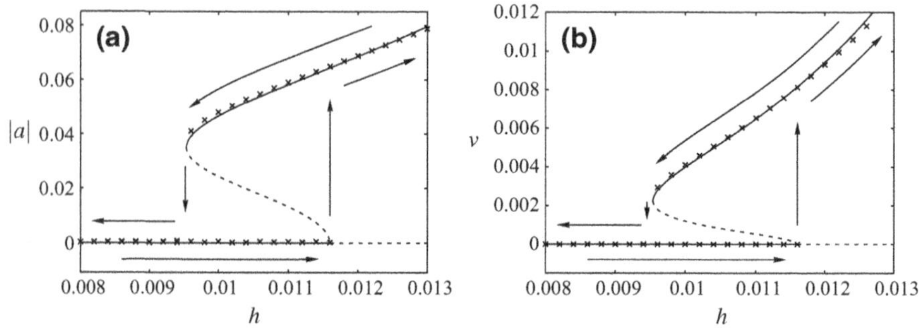

Fig. 9.7 The hysteresis loop of the wobbling kink driven by an external harmonic force. The continuous and dashed curves delineate the stable and unstable fixed points of the amplitude equations (9.39)–(9.40), with $\gamma = 10^{-3}$ and $\rho = -10^{-4}$. The crosses indicate results of the numerical simulations of the PDE (9.38) with the same γ and ρ. The driving amplitude is increased from $h = 8 \times 10^{-3}$ to 13×10^{-3} in small steps and then turned back to 8×10^{-3}. Reproduced from [42], with permission from APS

In the vicinity of the critical value (9.41) our smallness assumptions about $|a|$ and v are no longer met and the amplitude system (9.39)–(9.40) is no longer valid. The simulations of the full PDE (9.38) with h just below and just above h_c, reveal that the kink does settle to the wobbling with finite amplitude here. This is accompanied by its motion with constant velocity. However, the observed value of the wobbling amplitude is $O(\gamma^{1/3})$ rather than $O(\gamma^{1/2})$ as we assumed in the derivation of (9.39)–(9.40), and the measured value of the kink velocity is $O(\gamma^{1/3})$ rather than $O(\gamma)$. This change of scaling accounts for the breakdown of our asymptotic expansion in the vicinity of h_c.

Since the spatially uniform (even) driving force has an opposite parity to that of the (odd) wobbling mode, the energy transfer mechanism has to be indirect here. In fact there are two mechanisms at work, both involving a standing wave with the frequency Ω excited by the external driver. In the first mechanism, the square of this wave, its zeroth and second harmonics as well as the radiation excited by this wave, couple, parametrically, to the wobbling mode. The second mechanism exploits the fact that when the kink moves relative to the standing wave, the wobbling mode acquires an even-parity component. It is this part of the mode that couples — directly — to the standing wave.

9.7 Concluding Remarks

The first objective of this chapter was to review the multiscale singular perturbation expansion for the wobbling kink of the ϕ^4 model. The advantage of this approach over the Lindstedt method, is that it takes into account the existence of a hierarchy of space and time scales in the system. In particular, the multiscale expansion provides a consistent treatment of the long-range radiation from the oscillating kink.

The central outcome of the asymptotic analysis is the equation (9.17) for the amplitude of the wobbling mode. The nonlinear frequency shift and the lifetime of the wobbling mode are straightforward from this amplitude equation. We have identified the main factor ensuring its longevity. This is a small amplitude of radiation due to the significant difference between the wavelength of the radiation and the characteristic width of the wobbling kink.

The second part of this brief review (Sect. 9.6) concerned ways of sustaining the wobbling of the kink indefinitely. We discussed four resonant frequency regimes, namely the $1:1$ and $2:1$ parametric driving, and $1:2$ and $1:1$ external forcing.

It is instructive to compare the amplitude of the stationary wobbling mode sustained by these four types of resonance. For the given driving strength h, the harmonic $(1:1)$ parametric driver ensures the strongest possible response. In this case the amplitude of the stationary kink oscillations is of the order of $h^{1/3}$. The "standard" parametric resonance, where the driving frequency is chosen to be near twice the natural wobbling frequency of the kink, is second strongest. In this case the wobbling amplitude is of the order of $h^{1/2}$. The direct driving at half the wobbling frequency produces oscillations with the amplitude $a \sim h^{2/3}$. Finally, the harmonic $(1:1)$ direct resonance is the weakest of the four responses considered. In this case, the amplitude of the forced oscillations of the kink is $a \sim h$.

Another resonance characteristic worth comparing, is the width of the "Arnold tongue" — the domain on the (Ω, h) parameter plane where the driver sustains stable stationary wobbling of the kink. The $1:1$ parametric driver produces the widest tongue; in this case the resonant region is bounded by the curve $h \sim \rho^{3/2}$. The "standard" $(2:1)$ parametric resonance is the second widest, bounded by the curve $h \sim \rho$. The Arnold tongue for the external driving force with the frequency $\Omega/2$, is bounded by $h \sim \rho^{3/4}$. Finally, the harmonic direct resonance is the narrowest of the four, with the boundary curve $h \sim \rho^{1/2}$.

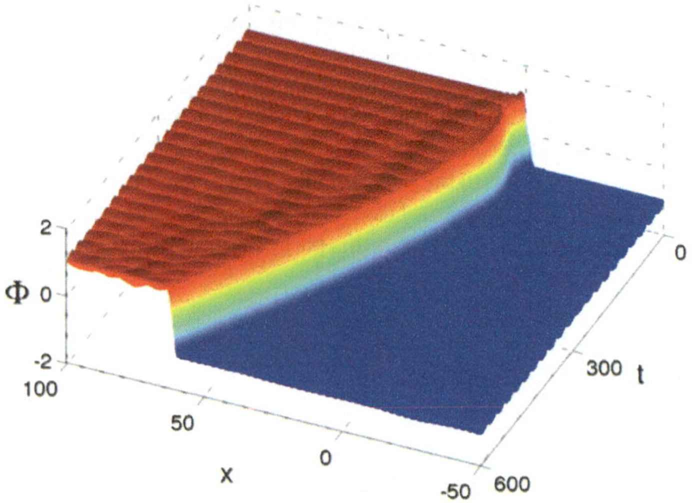

Fig. 9.8 A kink accelerates from rest as its wobbling mode couples to the standing wave excited by the resonant harmonic force. The figure is produced by numerical simulation of equation (9.38) with $\gamma = 10^{-3}$, $\rho = 10^{-2}$, and $h = 4 \times 10^{-2}$. The initial conditions are $\phi_t = 0$ and $\phi = \tanh x + 2a \tanh x \operatorname{sech} x$ with $a = 0.3$

Our comparison would be incomplete without noting that neither the harmonic parametric driver nor the $\Omega/2$ direct driving force have to overcome any thresholds in order to sustain stationary wobbling with a nonzero amplitude. In contrast, the subharmonic (2 : 1) parametric and harmonic (1 : 1) direct resonances occur only if the driving amplitude exceeds a certain threshold value.

Thus, the harmonic external driving emerges as the least efficient way of sustaining the steady wobbling of the kink. Out of the four driving techniques considered in Sect. 9.6, it produces the weakest response, requires the finest tuning of the driving frequency while the corresponding driving strength has to overcome a threshold set by the dissipation. Although these factors are clearly disadvantageous, the harmonic external driving has some aspects that are worth further investigation. In particular, it gives rise to an interesting "rack and pinion" mechanism that converts the energy of external oscillation to the translational motion of the kink (Fig. 9.8). This mechanism may prove useful for the control of solitons with internal modes in other systems.

9.8 Epilogue

This brief review is a tribute to Boris Getmanov, a musician and nonlinear scientist. A whole generation of former school kids still recalls gyrating to his band's boogie grooves while those with a taste for integrable systems remember Getmanov's discovery of the complex sine-Gordon [48–50]. (Incidentally, that discovery was only possible due to his numerical experimentation with the ϕ^4 model.) While Bob's piano is no longer heard at Dubna parties, the ϕ^4 and sine-Gordon equations are still around, alluring their new insomniacs and artists.

Acknowledgements Most of the results reviewed in this chapter were obtained jointly with Oliver Oxtoby [17, 35]. I am grateful to Oliver for his collaboration on the wobbling kink project. Special thanks go to Nora Alexeeva for generating Figs. 9.1, 9.2 and 9.8 for this piece.

References

1. B.S. Getmanov, JETP Lett. **24**, 291 (1976)
2. D.K. Campbell, J.F. Schonfeld, C.A. Wingate, Phys. D **9**, 1 (1983)
3. M. Peyrard, D.K. Campbell, Phys. D **9**, 33 (1983)
4. D.K. Campbell, M. Peyrard, Phys. D **18**, 47 (1986)
5. D.K. Campbell, M. Peyrard, P. Sodano, Phys. D **19**, 165 (1986)
6. T.I. Belova, A.E. Kudryavtsev, Phys. D **32**, 18 (1988)
7. P. Anninos, S. Oliveira, R.A. Matzner, Phys. Rev. D **44**, 1147 (1991)
8. R.H. Goodman, R. Haberman, Phys. Rev. Lett. **98**, 104103 (2007)
9. I. Takyi, H. Weigel, Phys. Rev. D **94**, 085008 (2016)
10. T.I. Belova, Sov. Phys. JETP **89**, 587 (1996)
11. T.I. Belova, A.E. Kudryavtsev, Phys.-Uspekhi **40**, 359 (1997)
12. Z. Fei, Yu.S. Kivshar, L. Vázquez, Phys. Rev. A **46**, 5214 (1992)
13. N.S. Manton, H. Merabet, Nonlinearity **10**, 3 (1997)
14. T. Romańczukiewicz, Acta Phys. Pol. B **36**, 3877 (2005)
15. T. Romańczukiewicz, J. Phys. A Math. Theor. **39**, 3479 (2006)
16. A.L. Sukstanskii, K.I. Primak, Phys. Rev. Lett. **75**, 3029 (1995)
17. O.F. Oxtoby, I.V. Barashenkov, Phys. Rev. E **80**, 026609 (2009)
18. Z. Fei, V.V. Konotop, M. Peyrard, L. Vázquez, Phys. Rev. E **48**, 548 (1993)
19. B.A. Malomed, J. Phys. A Math. Gen. **25**, 755 (1992)
20. N.A. Voronov, I.Yu. Kobzarev, N.B. Konyukhova, JETP Lett. **22**, 290 (1975)
21. I.L. Bogolyubskii, V.G. Makhankov, JETP Lett. **24**, 12 (1976)
22. N.A. Voronov, I.Yu. Kobzarev, JETP Lett. **24**, 532 (1976)
23. I.L. Bogolyubskii, V.G. Makhankov, JETP Lett. **25**, 107 (1977)
24. T.I. Belova, N.A. Voronov, I.Yu. Kobzarev, N.B. Konyukhova, Sov. Phys. JETP **46**, 846 (1977)
25. M.J. Rice, A.R. Bishop, J.A. Krumhansl, S.E. Trullinger, Phys. Rev. Lett. **36**, 432 (1976)
26. M.J. Rice, Phys. Lett. A **71**, 152 (1979)
27. M.J. Rice, J. Timonen, Phys. Lett. A **73**, 368 (1979)
28. M.J. Rice, E.J. Mele, Solid State Commun. **35**, 487 (1980)
29. M.J. Rice, Phys. Rev. B **28**, 3587 (1983)
30. H. Segur, J. Math. Phys. **24**, 1439 (1983)
31. M. Ślusarczyk, Acta Phys. Pol. B **31**, 617 (2000)
32. R. Pełka, Acta Phys. Pol. B **28**, 1981 (1997)
33. T. Romańczukiewicz, Acta Phys. Pol. B **35**, 523 (2004)
34. O.M. Kiselev, Russ. J. Math. Phys. **5**, 29 (1994)
35. I.V. Barashenkov, O.F. Oxtoby, Phys. Rev. E **80**, 026608 (2009)
36. M. Borromeo, F. Marchesoni, Chaos **15**, 026110 (2005)
37. L. Morales-Molina, F.G. Mertens, A. Sánchez, Phys. Rev. E **72**, 016612 (2005)
38. L. Morales-Molina, N.R. Quintero, A. Sánchez, F.G. Mertens, Chaos **16**, 013117 (2006)
39. N.R. Quintero, A. Sánchez, F.G. Mertens, Phys. Rev. Lett. **84**, 871 (2000)
40. N.R. Quintero, A. Sánchez, F.G. Mertens, Phys. Rev. E **62**, 5695 (2000)

41. N.R. Quintero, A. Sánchez, F.G. Mertens, Phys. Rev. E **64**, 046601 (2001)
42. I.V. Barashenkov, E.V. Zemlyanaya, Phys. D **132**, 363 (1999)
43. I.V. Barashenkov, E.V. Zemlyanaya, J. Phys. A Math. Theor. **44**, 465211 (2011)
44. I.V. Barashenkov, E.V. Zemlyanaya, M. Bär, Phys. Rev. E **64**, 016603 (2001)
45. I.V. Barashenkov, S.R. Woodford, Phys. Rev. E **71**, 026613 (2005)
46. I.V. Barashenkov, S.R. Woodford, E.V. Zemlyanaya, Phys. Rev. E **75**, 026604 (2007)
47. I.V. Barashenkov, E.V. Zemlyanaya, Phys. Rev. E **83**, 056610 (2011)
48. B.S. Getmanov, JETP Lett. **25**, 119 (1977)
49. B.S. Getmanov, Theor. Math. Phys. **38**, 124 (1977)
50. B.S. Getmanov, Theor. Math. Phys. **48**, 572 (1981)

Chapter 10
ϕ^4 Solitary Waves in a Parabolic Potential: Existence, Stability, and Collisional Dynamics

Ryan M. Ross, Panayotis G. Kevrekidis, David K. Campbell, Robert Decker and Aslihan Demirkaya

Abstract We explore a ϕ^4 model with an added external parabolic potential term. This term dramatically alters the spectral properties of the system. We identify single and multiple kink solutions and examine their stability features; importantly, all of these stationary structures turn out to be unstable. We complement these with a dynamical study of the evolution of a single kink in the trap, as well as of the scattering of kink and antikink solutions of the model. We observe that some of the key characteristics of kink-antikink collisions, such as the critical velocity and the multi-bounce windows, are sensitively dependent on the trap strength parameter, as well as the initial displacement of the kink and antikink.

R. M. Ross · P. G. Kevrekidis
Department of Mathematics and Statistics, University of Massachusetts, Amherst, MA 01003-4515, USA
e-mail: ryanross1023@gmail.com

P. G. Kevrekidis
e-mail: kevrekid@math.umass.edu

D. K. Campbell
Department of Physics, Boston University, 590 Commonwealth Ave, Boston, MA, USA
e-mail: dkcampbe@bu.edu

R. Decker · A. Demirkaya (✉)
Mathematics Department, University of Hartford, 200 Bloomfield Ave, West Hartford, CT 06117, USA
e-mail: demirkaya@hartford.edu

R. Decker
e-mail: rdecker@hartford.edu

P. G. Kevrekidis and J. Cuevas-Maraver (eds.), *A Dynamical Perspective on the ϕ^4 Model*, Nonlinear Systems and Complexity 26,
https://doi.org/10.1007/978-3-030-11839-6_10

10.1 Introduction

Models of the sG [1] and more generally of the nonlinear KG type, such as ϕ^4 [2] have been studied intensely via a combination of analytical and numerical techniques for well over three decades now. Part of the reason for this widespread appeal concerns the diverse set of physical applications for which such models have been argued to be of relevance. These start from the simplest (coupled torsion pendula, motion of dislocations etc. in the case of the discrete sG [3]) and extend far and wide. Case examples include, but are not limited to structural phase transitions [4, 5], domain walls arising in cosmological models [6, 7], simple polymeric models, as well as models of uniaxial ferroelectrics; see, e.g., [8, Chap. 9].

However, there is another source of the fascination, especially from an applied mathematics perspective, in non-integrable models such as ϕ^4. This has to do with their remarkable properties related to the collisional dynamics of kinks and antikinks. There, resonant phenomena between translational, internal and extended modes have been identified and a phenomenological "resonance energy exchange mechanism" [9] has been used to characterize features such as the "multi-bounce windows", the fractal emergence of such windows, and the universal nature of these phenomena not only in the ϕ^4 model, but also in other models such as the double, and the parametric sG equations [7, 9–11]. While recent years have seen both a more rigorous mathematical analysis [12] and an experimental realization of a system reminiscent of the effective kink-antikink interactions [13], there have also been further studies causing a renewed interest in the phenomenological approach and towards posing new questions. For instance, particular variations of the ϕ^6 model have been shown numerically to possess multi-bounce windows without having internal modes [14]. It has also been found that by parametrically deforming the ϕ^4 model, one can introduce more internal modes and suppress the two-bounce windows [15]. Lastly, the recent work of [16, 17] has performed a careful bookkeeping of the collective coordinate approaches intended to provide an effective description of the collision events. In so doing, a number of nontrivial problems regarding both computing with such reduced models and the conclusions drawn from them have arisen. Thus, despite its time-honored tradition, the study of kink-antikink collisions in non-integrable models remains a surprisingly active field of investigation to this day [18–20].

In the present work, we examine a different variant of the ϕ^4 problem, considering the scenario where an external potential has been imposed on the field theory. In particular, motivated by the extensive work on atomic Bose-Einstein condensates (BEC), we consider a parabolic trap, which in the latter context emulates the role of a generic magnetic confinement [21–23]. This possibility turns out to alter dramatically the properties of the kink waveforms. In particular, even the single kink turns out to be dynamically unstable in the emerging energy landscape. We examine the case of the single kink and its existence, stability and dynamical properties, as well as that of the kink-antikink state. In the latter, we examine the kink-antikink collisions and attempt to capture them by means of low dimensional collective coordinate models.

We highlight both the favorable, as well as the unfavorable, traits of these models with respect to their ability to capture the full PDE dynamics and raise some relevant possibilities towards future studies.

10.2 The Model, Computational Approach and Ground State

Our starting point for the considerations that follow is the classical ϕ^4 equation with an added x-dependent term:

$$u_{tt} = u_{xx} - \frac{\partial V}{\partial u} \tag{10.1}$$

where the potential function is given by

$$V(u, x) = \frac{1}{2}(u^2 - 1)^2 - \frac{1}{2} + \frac{1}{4}\Omega^2 x^2 u^2. \tag{10.2}$$

Ω is the trap strength constant and typically (in settings related to our motivating example of atomic BECs [21–23]) assumes values in (0, 1). However, we mostly restrict our attention to the dynamics for $\Omega \ll 1$, using $\Omega = 0.15$ for most of the data runs.[1] The case $\Omega = 0$ has been well-studied in the past and in its analysis, the perturbation theory and collective coordinate methods have been used [2, 7, 9, 16, 24]. We extend these methods to study our ϕ^4 variant bearing a spatially dependent potential in different parameter regimes.

To explore the properties of (10.1) we employ the following numerical scheme. First, we use a fourth-order central difference in space (see e.g. [7, 25] for this and other similar difference schemes). The resulting system of ODEs is then evolved in time using fourth-order Runge–Kutta methods, ensuring the accuracy of the conservation of the energy. The average value of the energy is of O(1), while the deviations from the mean are no more than $O(10^{-4})$. We apply free boundary conditions, although as we shall see below this is generally inconsequential because the field effectively vanishes well before the edge of our computational domain. For all our data runs, a spatial step size of $\Delta x = 0.02$ was used. Our computations typically use $\Omega = 0.15$ and a spatial grid running in the interval $[-30, 30]$. For smaller values of Ω, we use a larger interval as necessary to ensure the decay of the wavefunction within the computational domain.

[1] Although the BEC example is simply of motivating nature, it turns out that in that setting Ω plays the role of longitudinal to transverse trapping, hence to achieve effective one-dimensionality of the system the condition $\Omega \ll 1$ needs to be enforced. This is what motivates our selection here too.

The stationary ground state of the system $u_\Omega(x)$ solves the ODE

$$\frac{d^2u_\Omega}{dx^2} + 2(u_\Omega - u_\Omega^3) - \frac{1}{2}\Omega^2 x^2 u_\Omega = 0 \tag{10.3}$$

u_Ω is not known explicitly but in the case $\Omega \ll 1$, it is well-estimated by the Thomas–Fermi (TF) approximation which neglects the second derivative term (again motivated by similar considerations in atomic condensates [21, 22]) defined by

$$u_{\mathrm{TF}}(x) = \max\left\{0, \sqrt{1 - \frac{1}{4}\Omega^2 x^2}\right\}. \tag{10.4}$$

Equation (10.4) solves (10.3) when the u_{xx} term is neglected. We use this approximation as our initial guess for the nonlinear iterations converging to the true numerical ground state of the system. The comparison of the TF approximation with the true numerical result is given in Fig. 10.1. At the level of the TF approximation, we note that the ground state is zero whenever $|x| > x_s$, where $u_{\mathrm{TF}}(x_s) = 0$, which gives

$$x_s = \frac{2}{\Omega} \tag{10.5}$$

As we can observe in the figure, the TF approximation is very good except for the vicinity of x_s, where its disagreement with the true ground state has been quantified in a series of rigorous studies [26, 27].

To study the stability of the ground state, we consider the linearization around the solution using the ansatz:

$$u(x,t) = u_\Omega(x) + \epsilon e^{\lambda t}\chi(x). \tag{10.6}$$

Here u_Ω is the numerically obtained ground state, and (λ, χ) represent the corresponding eigenvalues and eigenvectors of the linearization respectively. The equation that they satisfy at $O(\epsilon)$ is of the form:

$$\lambda^2\chi = \chi'' - V''(u_\Omega)\chi \tag{10.7}$$

For the ground state we find, as is shown in the right panel of Fig. 10.1 that all the eigenvalues lie on the imaginary axis, hence the relevant state is spectrally stable; for instability to arise, at least one of the relevant eigenstates should correspond to growth, i.e., to $\mathrm{Re}(\lambda) > 0$.

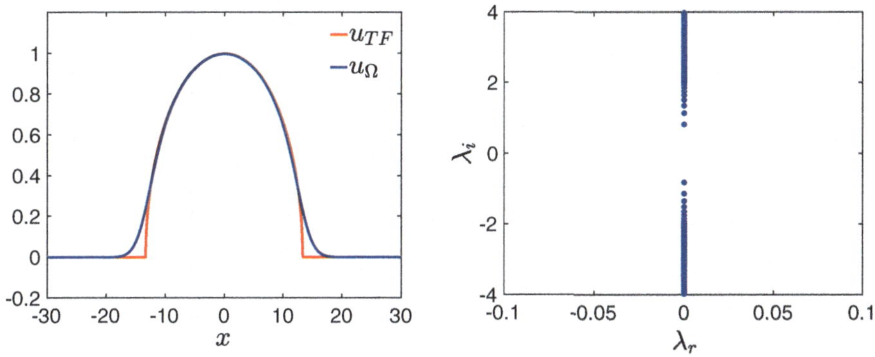

Fig. 10.1 The left panel shows the Thomas–Fermi approximation $u_{\mathrm{TF}}(x)$ (in green) from (10.4) and the numerical ground state u_Ω obtained through the Newton–Raphson iterations (in blue) for $\Omega = 0.15$. The right panel shows the spectral plane (λ_r, λ_i) of linearization eigenvalues $\lambda = \lambda_r + i\lambda_i$ corresponding to the ground state; the eigenvalues are all purely imaginary, indicating the spectral stability of the ground state

10.3 Single Kink Solutions

We now turn to excited states in the system in the form of kinks and antikinks, as well as bound states thereof. We start with the single static kink whose solution is not known explicitly but can be numerically approximated by

$$u_{0,K}(x) \approx u_\Omega(x)\tanh(x), \qquad u_{0,\bar{K}}(x) \approx -u_\Omega(x)\tanh(x), \tag{10.8}$$

once again in the TF approximation. For various values of Ω, the numerically calculated solutions are shown in Fig. 10.2. In the figure, the spectrum of linearization around the single kink is also shown (right panel). We see that in all the cases of nonzero Ω, there exists a pair of nonzero real eigenvalues. More concretely, the eigenvalue pair that used to be at $\lambda = 0$ due to translational invariance now exists as real and leads to the instability of the kink, which is progressively stronger, the larger the value of Ω.

10.3.1 Numerical Results (PDE)

We now examine the dynamical evolution of such unstable kinks. As a means of obtaining moving kink solutions u_K (or antikink solutions $u_{\bar{K}}$) of (10.1), we can employ the Lorentz transformation, $x \to \gamma(x - x_0 - vt)$ where $\gamma = 1/\sqrt{1-v^2}$, where x_0 is the initial position and v is the velocity of the moving kink. The initial conditions of the PDE are

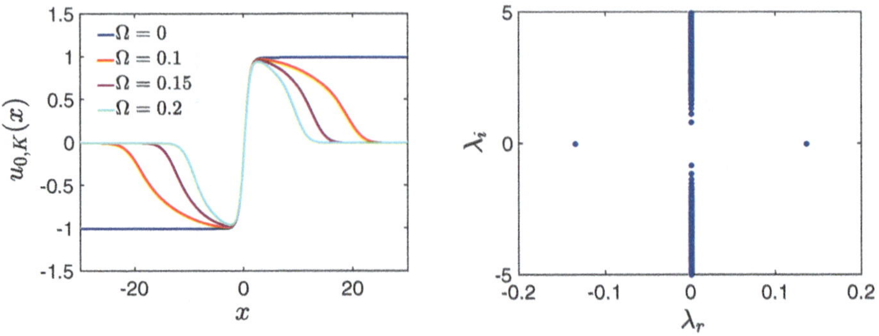

Fig. 10.2 The left panel shows the single kink solution as Ω varies. The right panel shows the spectral plane when $\Omega = 0.15$. The nonzero real eigenvalues are $\approx \pm 0.13$

$$u(x, 0) = u_\Omega(x) \tanh(\gamma(x - x_0));$$
$$u_t(x, 0) = -v\gamma u_\Omega(x) \operatorname{sech}^2(\gamma(x - x_0));$$

Of course the model in the presence of the potential is no longer Lorentz invariant (as it is in the case of $\Omega = 0$). Nevertheless, our results have shown that from a numerical perspective producing a Lorentz-boosted kink and multiplying it by the (stationary) TF background is an efficient method to produce "moving" waveforms starting from standing ones.

The presence of the the parabolic x-dependent term causes the velocity of the kink not to remain constant (as it would under stationary initialization in the untrapped, translationally invariant problem). When the initial velocity (v_{in}) is zero, the numerical results show that the motion of the kink solution depends sensitively on its starting position. When the kink starts to the left of the origin ($x_0 < 0$), the kink slides to the left, accelerating as it does so. If the kink starts to the right of the origin ($x_0 > 0$), then it slides to the right. In both cases, the kink eventually ends up being expelled from the system. Figure 10.3 shows examples of initializing the kink both to the left and to the right of the fixed point at $x = 0$ with no speed and illustrates how the kink slides along the side where it starts. This corroborates in a definitive way the results of the stability analysis in that the kink now (in the presence of the parabolic trap) encounters a saddle point at $x = 0$ (the center of the trap). Hence, on each side of this unstable equilibrium, it will slide along the unstable manifold. In fact, it is interesting to note that in panel (c) of the figure the kink is initialized *at* $x = 0$ (up to numerical roundoff error). We can see that in that case, it stays at the origin for a long time, until eventually the projection of the roundoff error in the unstable eigendirection grows to O(1) and kicks the kink away from the unstable equilibrium (in the case of this particular example its center moves to $x > 0$, but it can just as well move to the $x < 0$, depending on the perturbation).

With nonzero initial velocity, the behavior of the kink depends on the magnitude of the velocity. For a kink starting away from the origin, the saddle point at the $x_0 = 0$

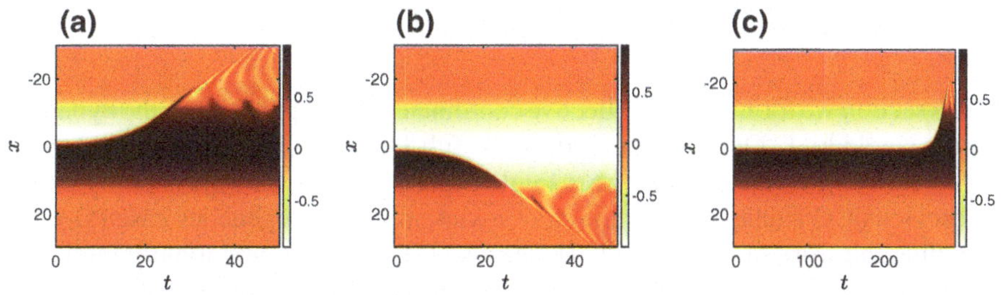

Fig. 10.3 **a** Initialized with center position $x_0 = -1$, the kink slides away from zero – to the left, because x_0 is negative – and gradually accelerates. **b** With $x_0 = 1$, the kink slides away from zero to the right, because x_0 is positive and gradually accelerates. **c** With $x_0 = 0$, the kink preserves its steady form for a while but eventually as its instability manifests, it slides (in this particular case) to the right

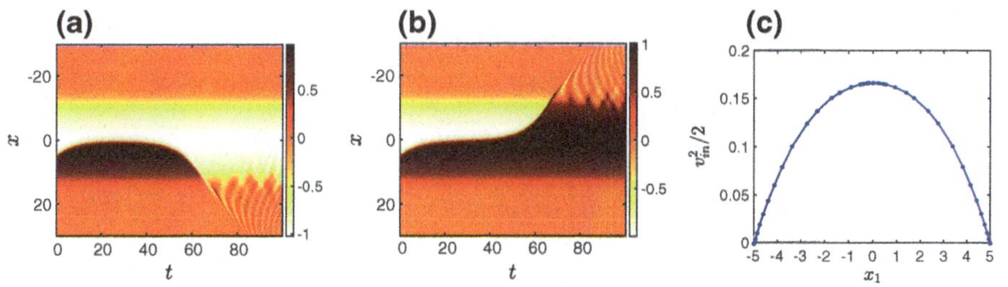

Fig. 10.4 A moving kink with $x_0 = 5$ initially with **a** $v_{in} = 0.577$, lacks the kinetic energy to overcome the potential energy barrier at $x = 0$. On the other hand, in **b** with $v_{in} = 0.578$, its kinetic energy suffices for it to break through to the other side. **c** shows the effective potential of the moving kink with $x_0 = 5$, approximated as explained in the text

represents a finite (potential) energy barrier which the kink must overcome to break through to the other side. Given sufficient kinetic energy, a kink moving towards the origin will overcome this potential energy barrier. On the other hand, if the velocity (and hence the kinetic energy) of the kink is "subcritical", i.e., below the value needed to overcome the potential barrier then the kink will bounce off the barrier and return to be expelled on the side that it started. Figure 10.4a, b demonstrate these two scenarios in a convincing way numerically. In particular, between $v_{in} = 0.577$ and $v_{in} = 0.578$ of the left and right panel, respectively, there is clearly a critical velocity as the former results in a bounce-back of the kink from $x = 0$, while the latter leads to its transmission through the barrier and on to the other side. For any $v_{in} < 0.577$, the kink stops at $x = x_1$, called a turning point and then moves in the opposite direction. The Fig. 10.4c shows the relation between the effective potential approximated by the associated kinetic energy (per unit mass) $\frac{1}{2}v_{in}^2$ and the turning point x_1. Of course, in a way consistent with the above picture, if the kink is initially moving away from the origin, then it will continue to move away and gradually accelerate over time.

10.3.2 Collective Coordinate Approach (ODE)

In this section we formulate a "collective coordinate" (CC) approach for the dynamics for the single kink. Our aim is to reduce the full PDE with infinitely many degrees of freedom to an ODE model with a single degree of freedom. This is partly driven by previous CC studies of the motion of a kink in the homogeneous case ($\Omega = 0$), including the work of Sugiyama [24] (see also [7]). It is also importantly motivated by the nature of the motion in the vicinity of the saddle point discussed above, and by the desire to obtain a simplified description of this motion.

To begin, we note that the full PDE conserves the energy (Hamiltonian)

$$\mathcal{H} = \frac{1}{2}\int_{-\infty}^{\infty} u_t^2 + u_x^2 + (u^2-1)^2 + \frac{1}{2}\Omega^2 x^2 u^2 \,\mathrm{d}x = \mathcal{T} + \mathcal{V}, \tag{10.9}$$

where the kinetic and potential energies of the kink, respectively, are

$$\begin{aligned}\mathcal{T}(t) &= \frac{1}{2}\int_{-\infty}^{\infty} u_t^2 \,\mathrm{d}x\\ \mathcal{V}(t) &= \frac{1}{2}\int_{-\infty}^{\infty} u_x^2 + (u^2-1)^2 + \frac{1}{2}\Omega^2 x^2 u^2 \,\mathrm{d}x.\end{aligned}$$

The corresponding Lagrangian is

$$\begin{aligned}\mathcal{L}(t) &= \mathcal{T}(t) - \mathcal{V}(t)\\ &= \frac{1}{2}\int_{-\infty}^{\infty} u_t^2 - u_x^2 - (u^2-1)^2 - \frac{1}{2}\Omega^2 x^2 u^2 \,\mathrm{d}x.\end{aligned} \tag{10.10}$$

Assuming that we operate in the TF limit of $\Omega \ll 1$, we seek to characterize the kink motion by utilizing the ansatz

$$u_1(x,t) = u_\Omega(x)\tanh(x - X(t)),$$

where $X(t)$ is the time-dependent displacement of the kink from the origin and u_Ω is the ground state of the system in the presence of the trap. Then, the Lagrangian $\mathcal{L}$ becomes as follows:

$$\mathcal{L}(X,\dot{X}) = \int_R L(u_1(x, X(t)))\,\mathrm{d}x = a_0(X)\dot{X}^2 - a_1(X),$$

where

$$\begin{aligned}a_0(X) &= \frac{1}{2}\int_R u_\Omega^2(x)\,\mathrm{sech}^4(x-X)\,\mathrm{d}x\\ a_1(X) &= \frac{1}{2}[I_1(X) + I_2(X) + I_3(X) + I_4(X) + I_5(X)]\end{aligned}$$

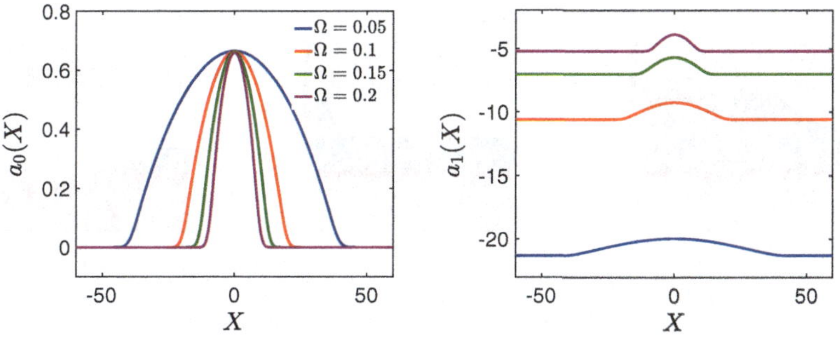

Fig. 10.5 The left panel shows $a_0(X)$ and the right panel shows $a_1(X)$ as Ω varies: 0.05 (blue), 0.1 (red), 0.15 (green), 0.2 (purple)

$$I_1(X) = \int_R [u'_\Omega(x)]^2 \tanh^2(x - X)\,\mathrm{d}x$$
$$I_2(X) = \int_R 2u'_\Omega(x)u_\Omega(x)\tanh(x - X)\,\mathrm{sech}^2(x - X)\,\mathrm{d}x$$
$$I_3(X) = \int_R u_\Omega^2(x)\,\mathrm{sech}^4(x - X)\,\mathrm{d}x$$
$$I_4(X) = \int_R (u_\Omega^2(x)\tanh^2(x - X) - 1)^2 - 1\,\mathrm{d}x$$
$$I_5(X) = \int_R \frac{1}{2}\Omega^2 x^2 u_\Omega^2(x)\tanh^2(x - X)\mathrm{d}x.$$

Since we have an accurate numerical solution for u_Ω, we compute the coefficients a_0 and a_1 and solve the Euler–Lagrange equation numerically. Figure 10.5 shows the plots of a_0 and a_1 as a function of X for various values of Ω. By applying the Euler–Lagrange equation, we obtain the dynamical evolution:

$$\begin{aligned}\dot{X} &= Y\\ \dot{Y} &= -\frac{1}{2}\frac{a_0'(X)}{a_0(X)}Y^2 - \frac{1}{2}\frac{a_1'(X)}{a_0(X)}.\end{aligned} \tag{10.11}$$

We solve these equations numerically by using the initial conditions $X(0) = x_0$ and $X'(0) = v_{\mathrm{in}}$ where x_0 is the initial position of the kink and v_{in} is the initial velocity of the kink. In the numerical computations, the Dormand-Prince integrator is used, which is implemented in Matlab by `ode45` with built-in error control. Note that here, overdots denote time derivatives.

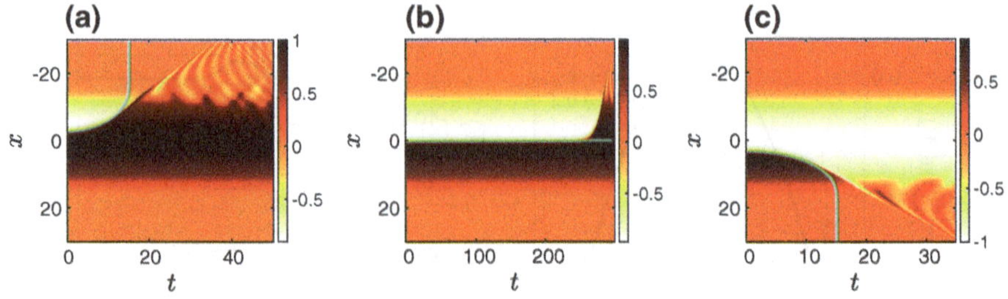

Fig. 10.6 Overlays of the ODE solution $X(t)$ (in solid line) on top of PDE results for initially stationary ($v_{in} = 0$) kinks with **a** $x_0 = -3$ **b** $x_0 = 0$ **c** $x_0 = 3$

10.3.2.1 Connection to Numerical Results

We now compare our full numerical results (PDE) with the findings of the CC ODE Method. The numerical CC computations show that when the initial velocity is zero ($v_{in} = 0$), with nonzero displacement ($x_0 < 0$ or $x_0 > 0$), the ODE results agree with PDE results quite well up to the time/location when/where the kink is expelled from the system; from there on the kink cannot be clearly discerned anyway, and hence it is not particularly meaningful to seek to track it in the PDE or to match it to the ODE results. Figure 10.6a, c show that when $x_0 = -3$ and $x_0 = 3$, with zero initial velocity, the results mostly agree until $t \approx 11$ when the kink is expelled from the TF background.

In the case of zero displacement ($x_0 = 0$) and zero velocity ($v_{in} = 0$) of Fig. 10.6b, both the ODE and the PDE show the kink as residing on the saddle point for a long time interval. In the ODE case, the absence of additional degrees of freedom and the conservation of the energy does not allow the kink to depart from the unstable equilibrium over the time horizons considered. In the infinite dimensional system, the instability is manifested at $t \approx 250$.

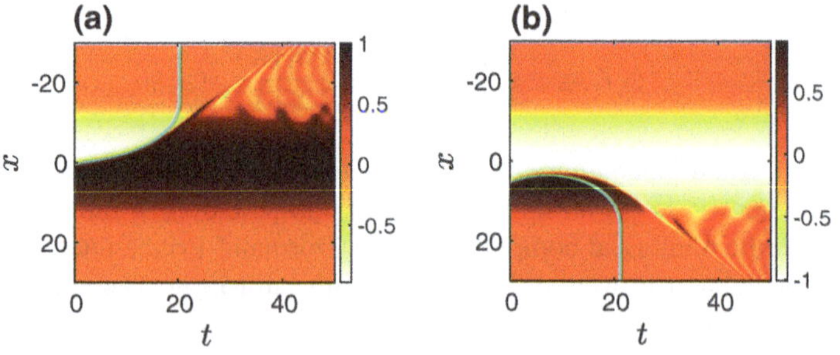

Fig. 10.7 **a** A moving kink with $x_0 = 0$, $v_{in} = 0.2$, is shown again in comparison between ODE and PDE. **b** $x_0 = 5$, $v_{in} = 0.5$, A kink moving towards the origin with almost enough energy to break through the barrier at 0. Both the ODE and PDE show it being reflected from the barrier at $x = 0$

When the initial velocity is nonzero, we find similar results. The ODE results agree with the PDE results until the ODE kink is expelled from the system. Figure 10.7a exhibits the ODE and PDE agreement until $t \approx 15$ when $x_0 = 0$ and $v_{\text{in}} = 0.2$. Figure 10.7b exhibits this type of agreement until $t \approx 23$ when $x_0 = 5$ and $v_{\text{in}} = 0.5$. In summary, we can infer that while the kink remains within the region where the TF background is accurate, the CC method appears to yield an excellent qualitative and a very good quantitative characterization of the resulting motion.

10.4 Kink-Antikink Solutions

Armed with our understanding of the single kink energy landscape, we now turn to case of a kink-antikink pair. Such a state can in fact be *stationary* in our model, although it is never genuinely stationary in the case of $\Omega = 0$. The reason is fairly similar to the corresponding existence of a stationary state in a NLS model with a trap analyzed in detail, e.g., in [28]. However, there is a *crucial* difference. In the latter case, the trap has a restoring contribution, while the dark solitons (the kinks) of the NLS repel each other. Here, the situation is reversed: namely, as we saw in the previous section, the trap tends to expel the kink, while (as is known from the study of kink-antikink interactions in the homogeneous $\Omega = 0$ case) the kink and antikink attract each other [7, 9], i.e., the role of both contributions is reversed. As a result, the equilibrium point instead of being a center as it is in the case of NLS, here, again, it is a saddle. Indeed, if the kink pair is separated by a small distance, then the interaction prevails and forces the kinks to approach each other. On the other hand, a large separation leads to a dominant effect due to the trap, and the kink-antikink pair are pushed apart.

The static kink-antikink solution of (10.1) can be numerically approximated by

$$u_{K,\bar{K}}(x) \approx u_\Omega(x)(\tanh(x + x_0) - \tanh(x - x_0) - 1). \tag{10.12}$$

where x_0 is unique and dependent on Ω. We define that particular x_0 as x_{cr}, since it operates as a critical value for the kink and antikink pair separating between expulsion from and attraction to the center. For various values of Ω, the numerically calculated stationary kink-antikink solutions are shown in Fig. 10.8. As usual, the right panel of the figure identifies the spectrum of the linearized operator about the static (in this case kink-antikink) solution. Here, we get two pairs of real eigenvalues, indicating instability. The pattern of the spectral problem is thus transparent: for one kink, there is a single unstable mode, for two kinks, two such modes and so on. Another way to state this is that all the negative energy eigenstates of [28] pertaining to the normal modes of vibration of the kinks in the real-field version of the model lead to unstable eigendirections.

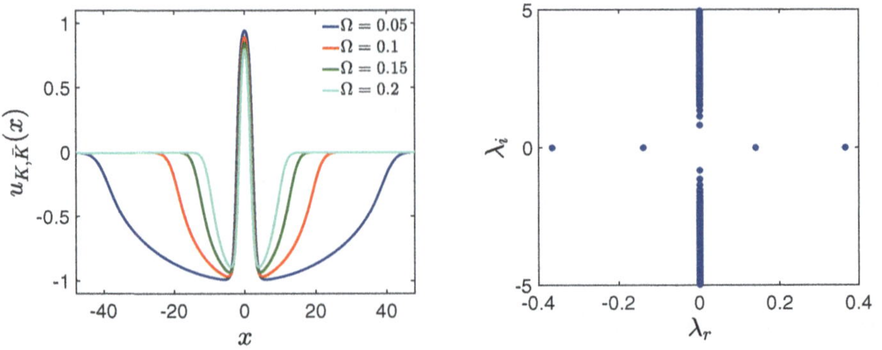

Fig. 10.8 Kink-antikink stationary solutions for different values of Ω. For $\Omega = 0.05$, the equilibrium separation is $x_0 = 2.173$, for $\Omega = 0.1$, it is $x_0 = 1.8669$, for $\Omega = 0.15$, it is $x_0 = 1.6907$, while for $\Omega = 0.2$, it is $x_0 = 1.5661$. The spectral plane reveals two real eigenvalue pairs, again signaling the instability when $\Omega = 0.15$ (the case shown), as well as for all other values of $\Omega \neq 0$

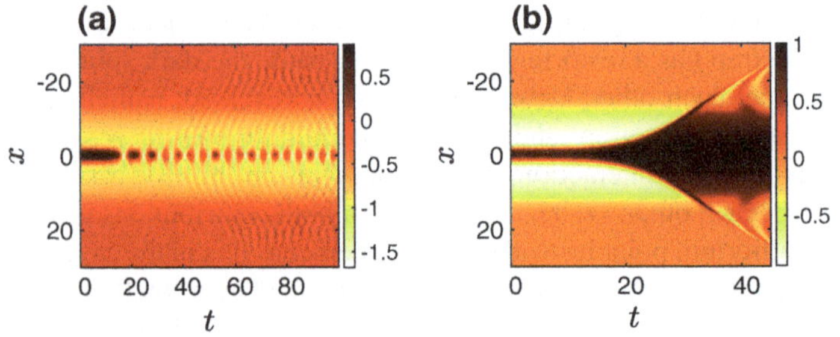

Fig. 10.9 **a** ($x_0 = 1.69$) a kink-antikink system with zero initial velocity for $x_0 < x_{\rm cr}$ forms a bound state. **b** ($x_0 = 1.7$), a zero-velocity configuration with $x_0 > x_{\rm cr}$ exhibits expulsion

10.4.1 Numerical Results (PDE)

We can obtain moving kink-antikink solutions by picking the initial conditions

$$u(x,0) = u_\Omega(x)(\tanh(\gamma(x+x_0)) - \tanh(\gamma(x-x_0)) - 1);$$
$$u_t(x,0) = -v\gamma u_\Omega(x)(\mathrm{sech}^2(\gamma(x+x_0)) + \mathrm{sech}^2(\gamma(x-x_0)));$$

where $\gamma = 1/\sqrt{1-v^2}$, $\pm x_0$ are the initial positions and $\pm v$ are the velocities of the moving kink and antikink. Once again, we utilize the (no longer representing an invariance) Lorentz boost and find it to be an efficient method of producing moving kinks-antikinks.

Figure 10.9 shows that when the velocity of the kink and antikink is initially zero, then for any $x_0 < x_{\rm cr} \approx 1.6907$ (for $\Omega = 0.15$), the kink-antikink system forms a bound state. For any $x_0 > x_{\rm cr}$, the system exhibits expulsion. This confirms our spectral understanding of the relevant equilibrium representing a saddle point. Now

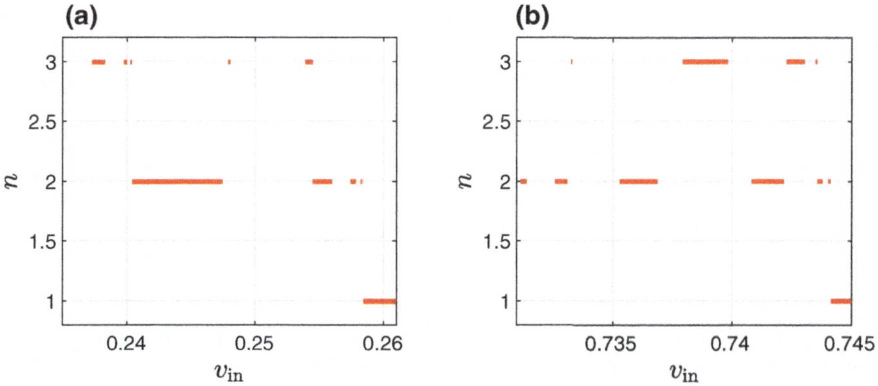

Fig. 10.10 Plots of number of bounces versus initial velocity with **a** $x_0 = 1.4$ for $v_{in} \in (0.235, 0.261)$ and **b** $x_0 = 7$ for $v_{in} \in (0.731, 0.745)$. As in the homogeneous case, we observe ranges for which multiple bounces occur

the unstable manifold consists of two eigendirections, namely the in-phase and out-of-phase motion of the pair, although here only the out-of-phase mode is excited.

When $\Omega = 0$, the kink-antikink system is relatively insensitive to changes in the initial position x_0 (provided that the kinks are not too close). This, however, changes in the presence of the parabolic trapping term. For x_0 sufficiently small, the kink and antikink approach each other and collapse into a bound state for small input velocities, even with zero initial velocity. For $x_0 < x_{cr}$ and initial velocities sufficiently large we observe the typical behaviors that are familiar from the homogeneous case [7, 9]: reflection, n-bounce windows (i.e., windows where the kinks escape each other's attraction upon n collisions), and "bion/breather" [9, 29] formation (i.e., emergence of a bound state that oscillates in time and gradually decays). In fact, the types of resonance windows observed in the homogeneous case [7, 9, 12] persist in the inhomogeneous case and present a similar type of pattern. A visualization of this can be seen in Fig. 10.10. In this figure and in subsequent figures, we opt not to plot velocities for which four or more bounces occur, as the windows become exceedingly narrow for this many bounces and it is not practical to display them together with one- and two-bounce windows. Just as when $\Omega = 0$, the regions in between resonance windows correspond to "bion" formation (numerically these are treated as events with many bounces, so they are omitted from the figures).

For $x_0 > x_{cr}$, the presence of the saddle point of the energy once again imposes a barrier that the kinks need to overcome in order to interact. Thus, in this case, there is a critical velocity associated with x_{cr}, which we denote by v_{cr}, such that whenever $x > x_{cr}$ and $v < v_{cr}$ the kinks do not have enough kinetic energy to reach the origin, so they stop (at a turning point) before they reach it. Then they recede towards spatial infinity without having collided ever.

If both $x > x_{cr}$ and $v > v_{cr}$, then the kinks will have enough kinetic energy to overcome the barrier, reach the origin and collide, and then we observe the typical phenomena associated with the ϕ^4 model, including most notably the existence of multi-bounce windows. These can be visualized in Figs. 10.10 and 10.16 with both small and large separation x_0, respectively. The velocity interval data that was used to create these figures is contained in Tables 10.1, 10.2, 10.3 and 10.4 in Appendix 1.

10.4.2 Collective Coordinate Approach (ODE)

To complement our PDE simulations, we formulate a collective coordinate model for a kink-antikink configuration. As before, our aim is to reduce the full PDE with infinitely many degrees of freedom to a simple model with two degrees of freedom in a way similar to what has been previously done in the homogeneous ϕ^4 case [7, 9, 12, 13]. To account for the internal modes of the kink and antikink, we introduce a new unknown $A(t)$, representing the amplitude of the internal mode. We use only one function for both modes since the collision picture is symmetric between kink and antikink, so we expect the amplitudes of the kink mode and antikink mode to coincide if the initial configuration is symmetric. Our corresponding ansatz in this case will be:

$$u(x,t) = u_\Omega(x)(u_0(x+X) - u_0(x-X) - 1) + A(t)(\chi_1(x+X) - \chi_1(x-X)) \tag{10.13}$$

where $u_0(x) = \tanh(x)$, and χ_1 is an eigenfunction corresponding to the smallest positive eigenfrequency that satisfies the linearization equation for a single kink.

The particular eigenfunction we chose is given in Fig. 10.11 for three different Ω values.

Surprisingly, these modes are related chiefly to the background state u_Ω rather than to the internal vibrational mode of the kink. More specifically, they are localized near x_s. In the homogenous model, the only nonzero discrete eigenfrequency corresponds to the internal vibrational mode, and it is this mode that is used for the collective

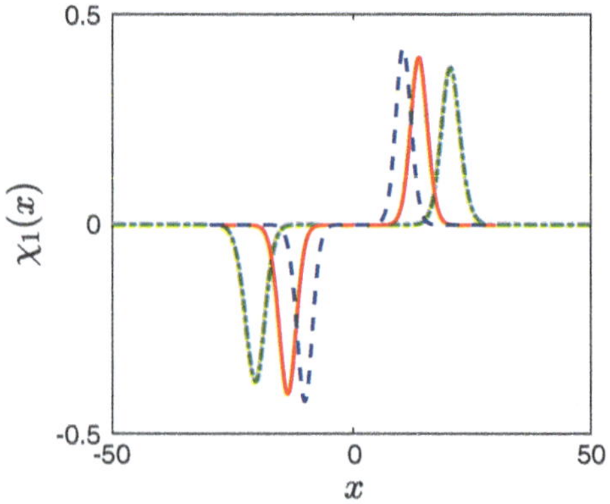

Fig. 10.11 Plot of the eigenfunction used in the collective coordinates method. Green dashed-dots correspond to $\Omega = 0.1$, red solid corresponds to $\Omega = 0.15$ and the dashed blue corresponds to $\Omega = 0.2$

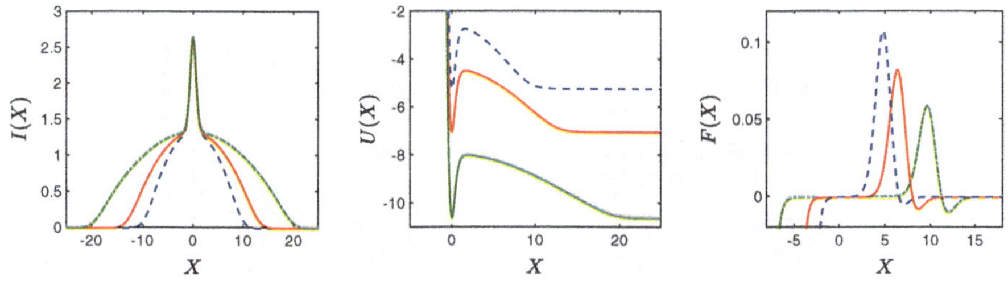

Fig. 10.12 Plots of the functions $I(X)$ (left), $U(X)$ (middle) and $F(X)$ (right) as a function of the position X of the kink's center, evaluated through the method of collective coordinates, as discussed in the text. Green dashed-dots correspond to $\Omega = 0.1$, red solid corresponds to $\Omega = 0.15$ and the dashed blue corresponds to $\Omega = 0.2$

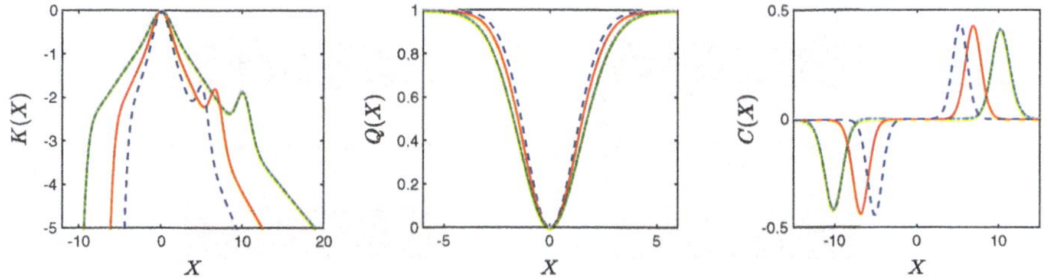

Fig. 10.13 Plots of the functions $K(X)$ (left), $Q(X)$ (middle) and $C(X)$ (right) as a function of the position X of the kink's center, evaluated through the method of collective coordinates, as discussed in the text. Green dashed-dots correspond to $\Omega = 0.1$, red solid corresponds to $\Omega = 0.15$ and the dashed blue corresponds to $\Omega = 0.2$

coordinate ansatz. For our model, the mode with the smallest eigenfrequency results in a better match of collective coordinate and PDE results than when the (analogue of the homogeneous case kink) internal mode is used. This may be due to the former mode incorporating more adequately the role of the potential-induced background, although, admittedly, this issue warrants further theoretical investigation.

As proposed in [24], we work on a reduced effective Lagrangian that captures the fundamental features. The derivations of the formulas are in Appendix 2.

$$L(X, \dot{X}, A, \dot{A}) = I(X)\dot{X}^2 - U(X) + 2F(X)A + K(X)A^2 + Q(X)\dot{A}^2 + 2C(X)\dot{A}\dot{X}. \tag{10.14}$$

Figures 10.12 and 10.13 show the collective coordinate coefficients as functions of X for various values of Ω.

The Euler–Lagrange equations in this case with respect to X and A give

$$\begin{aligned} C(X)\ddot{X} + Q(X)\ddot{A} &= -Q'(X)\dot{X}\dot{A} - C'(X)\dot{X}^2 + F(X) + K(X)A \\ 2I(X)\ddot{X} + 2C(X)\ddot{A} &= -I'(X)\dot{X}^2 - U'(X) + 2F'(X)A \\ &\quad + K'(X)A^2 + Q'(X)\dot{A}^2 \end{aligned} \tag{10.15}$$

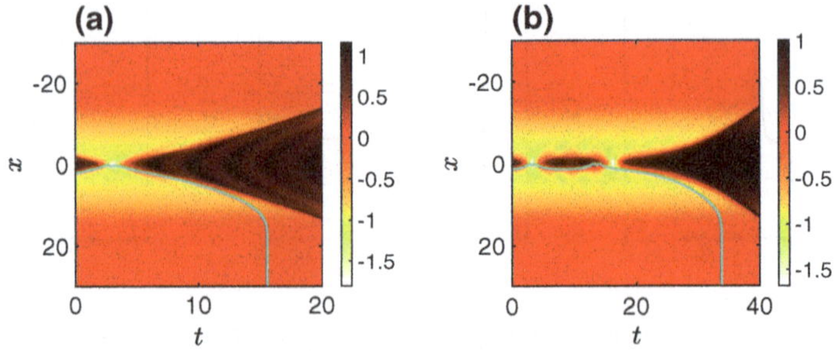

Fig. 10.14 **a** $x_0 = 2$, $v_{in} = 0.7$ **b** $x_0 = 1.4$, $v_{in} = 0.245$ a reflection and two-bounce, for which choosing the same initial speed for the ODE as the PDE yields accurate results

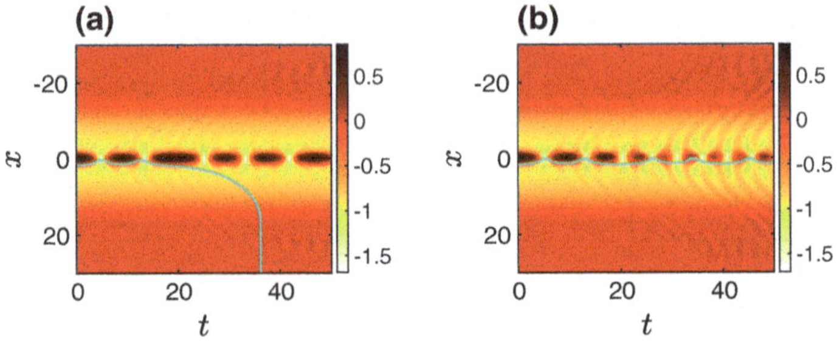

Fig. 10.15 **a** $x_0 = 1.607$, $v_{in} = 0.13$ **b** $x_0 = 1.607$, $v_{in} = 0.11$ a bion, whose formation is not captured very well when the ODE and PDE speeds are the same. The rightmost ODE solution clearly is in better (but still not perfect) agreement with the PDE solution

Here, overdots denote time derivatives. We solve these equations numerically by using the initial conditions $A(0) = 0$, $\dot{A}(0) = 0$, $X(0) = x_0$ and $\dot{X}(0) = v_{in}$ where x_0 is the initial half-distance between the kink and the antikink, and v_{in} is the initially prescribed speed of the kink (and antikink). Figure 10.14 shows the ODE model (10.15) solution $X(t)$ on top of the PDE model (10.2) contour plot. It shows reasonably good agreement between ODE and PDE solutions for one and two-bounce windows. Figure 10.15 presents infinite-bounce windows. Using the same $x_0 = 1.607$ value and different v_{in} values of 0.11 and 0.13, we see that the agreement is better for $v_{in} = 0.11$. Note that (10.15) is the reduced system where the higher order terms of $A(t)$ and $\dot{X}(t)$ are ignored. We make this assumption based on the previous work [24], that higher terms of A and $\dot{X}$ can be dropped because they are small. The expressions involving integrals which are X-dependent in (10.15) are computed by numerical integration.

For Ω values not much larger than the value considered so far (of 0.15), the phenomenology is virtually identical to that of the previous section. All of the interactions we have come across so far, namely bions, multi-bounce interactions, and expulsion, are observed for kink-antikink systems as the initial velocity is varied. Figure 10.16 shows the presence of two- and three-bounce windows and a threshold reflection

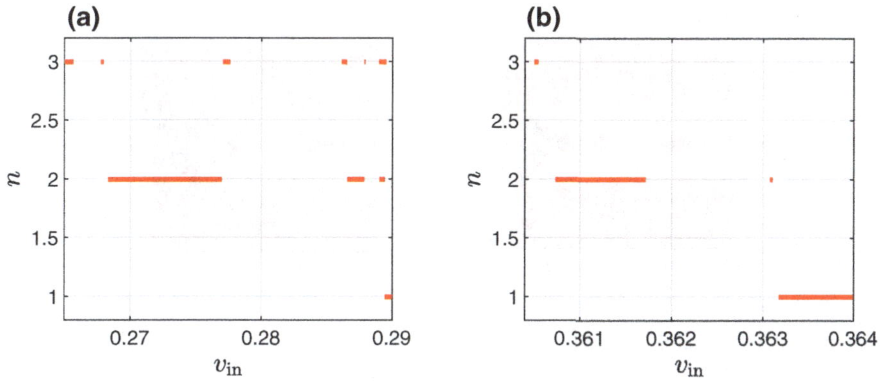

Fig. 10.16 **a** Plots of number of bounces versus initial velocity for fixed $x_0 = 2$. **a** $\Omega = 0.2$ for $v_{\text{in}} \in (0.265, 0.29)$, **b** $\Omega = 0.3$ for $v_{\text{in}} \in (0.359, 0.364)$. As Ω increases, existing multi-bounce windows become significantly narrower (note the change in the velocity increment in the rightmost figure), and many disappear completely

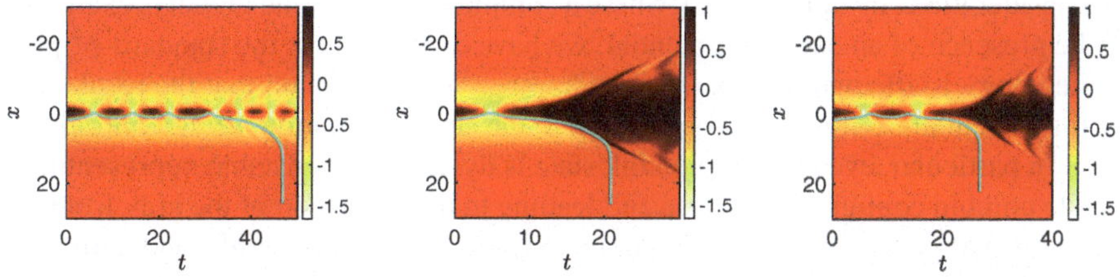

Fig. 10.17 Collective coordinate results for $\Omega = 0.2$ and $\Omega = 0.3$. Here the collective coordinate model performs about as well as when $\Omega = 0.15$ in our earlier data runs. $x_0 = 2$, $v_{\text{in}} = 0.264$, $\Omega = 0.2$ (left), $x_0 = 2$, $v_{\text{in}} = 0.364$, $\Omega = 0.2$ (middle), $x_0 = 2$, $v_{\text{in}} = 0.361$, $\Omega = 0.3$ (right)

velocity for $\Omega = 0.2$ and $\Omega = 0.3$. However, the widths of existing bounce windows shrink drastically as the trap strength is increased. Increasing Ω from 0.2 to 0.3 as in the same figure shrinks the resonance windows by approximately a factor of $\frac{1}{10}$. Our collective coordinate model also exhibits these features – it is capable of capturing one- and two-bounce interactions, and exhibits the same sensitive dependence on the initial speed we observe when Ω is small for configurations with more bounces. Examples can be seen in Fig. 10.17.

Thus far we have confined ourselves to the case $\Omega \ll 1$. However, the behaviors for larger Ω are interesting in their own right and can differ significantly from those observed for weak trapping. Intuitively, we expect that a large value of Ω will constrain the dynamics severely (as it only allows the kinks to move within a restricted TF region). Our numerical simulations suggest that two-bounce and three-bounce still exist there, yet they are extremely narrow and shrink rapidly as Ω increases. Figure 10.18 shows some prototypical examples of multi-bounce cases for $\Omega = 0.4$.

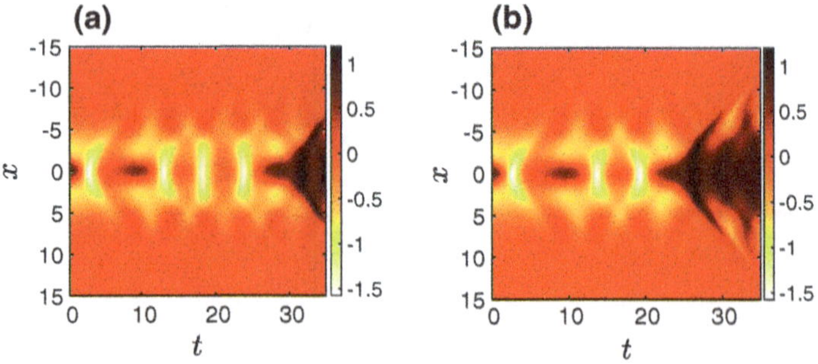

Fig. 10.18 Kink-antikink system for $\Omega = 0.4$, $x_0 = 1.4$, **a** $v_{in} = 0.446$ **b** $v_{in} = 0.4463$

10.5 Conclusions and Future Challenges

In this work we have taken the well-known ϕ^4 model in a novel direction: motivated by considerations in atomic BECs where a complex variant of the model is examined in the presence of an external potential, we have explored how the standard ϕ^4 phenomenology is affected by a parabolic trapping. We have found that the results can be fundamentally different in comparison to the standard homogeneous (untrapped) case. In particular, even the single kink state is dynamically unstable, representing a saddle point in configuration space and leading to the expulsion of the kink from the trap center of symmetry. Moreover, in the presence of multiple kinks and antikinks, the number of coherent structures determines the number of unstable eigendirections (and the dimension of the corresponding unstable manifold). It is interesting to also observe the reverse role of the potential and of the kink-antikink interaction in comparison with the atomic BEC case. These features generally favor the expulsion of the kinks from the system unless the kinks are sufficiently close to each other that their interaction dominates. In the setting considered herein, n-bounce windows may still be present, although these windows were found to shrink radically as the potential parameter Ω is increased. Lastly, we have developed a collective coordinate low-dimensional ODE approach, under relevant approximations (such as the truncation of the higher power terms) which yields, especially for low numbers of bounces, reasonably consonant results between the PDE field theory and the ODEs. On the other hand, when presenting a more "complete" variant of the equations, issues similar to those encountered by [16, 17] in the homogeneous variant of the model emerge. Indeed, the prospect of resolving these collective coordinate issues and improving the ODE models in this inhomogeneous ϕ^4 variant (and even for the homogeneous case within the model) is an important avenue for possible future work.

From a mathematical standpoint, understanding better the stability properties of the model could be an interesting direction for further work. Also, extending inhomogeneous model considerations to other models of recent interest such as the ϕ^6 [14, 16, 18, 19, 30], and ϕ^8 models [20, 31, 32] would be valuable in understanding the genericity (or not) of the conclusions drawn in this study. Some of these directions are currently under consideration and will be reported in future publications.

Acknowledgements This material is based upon work supported by the National Science Foundation under Grant No. DMS-1602994 (PGK).

Appendix 1: Tables for *n*-Bounce Windows

We list the velocity intervals $[v_1, v_2]$ that result in multi-bounce windows when the initial velocity v_{in} is picked within these intervals. In Tables 10.1 and 10.2, we list those intervals for a small separation ($x_0 = 1.4$) and a big separation ($x_0 = 7$) with $\Omega = 0.15$. In Tables 10.3 and 10.4, we list those intervals for bigger values of Ω (0.2 and 0.3 respectively) with fixed $x_0 = 2$.

Appendix 2: Derivation of the Coefficients in (10.14)

We define $\phi_{\pm} = \pm\phi_0(x \pm X(t))$ and $\chi_{\pm} = \pm\chi_1(x \pm X(t))$, $\phi'_{\pm} = \pm\phi'_0(x \pm X(t))$ and $\chi'_{\pm} = \pm\chi'_1(x \pm X(t))$. Then (10.13) becomes

$$u(x,t) = u_{\Omega}(\phi_+ + \phi_- - 1) + A(\chi_+ + \chi_-). \tag{10.16}$$

Table 10.1 n-bounce windows for $x_0 = 1.4$. One-bounce windows occur for $v_{in} > 0.25845$

n	v_1	v_2	Δv_n
3	0.23729	0.23827	0.00098
3	0.23973	0.23996	0.00023
3	0.24023	0.24035	0.00012
2	0.24038	0.24744	0.00706
3	0.24786	0.24801	0.00015
3	0.25396	0.2545	0.00054
2	0.25453	0.25601	0.00148
2	0.25754	0.25787	0.00033
2	0.25824	0.25831	0.00007
2	0.2584	0.25841	0.00001

Table 10.2 n-bounce windows for $x_0 = 7$. One-bounce windows occur for $v_{\text{in}} > 0.74414$

n	v_1	v_2	Δv_n
2	0.73116	0.7314	0.00024
2	0.73260	0.7331	0.00050
3	0.73327	0.73331	0.00004
2	0.73527	0.73686	0.00159
3	0.73791	0.73981	0.0019
2	0.74083	0.74214	0.00131
3	0.74226	0.74228	0.00002
3	0.74346	0.74353	0.00007
2	0.74355	0.74376	0.00021
2	0.74401	0.74411	0.0001

Table 10.3 n-bounce windows for $x_0 = 2, \Omega = 0.2$. One-bounce windows occur for $v_{\text{in}} > 0.28942$

n	v_1	v_2	Δv_n
3	0.26501	0.26565	0.00064
3	0.26775	0.26796	0.00021
2	0.26833	0.27694	0.00861
3	0.27703	0.27756	0.00053
3	0.28611	0.28649	0.00038
2	0.28654	0.28783	0.00129
3	0.28784	0.2879	0.00006
3	0.28896	0.28897	0.00001
2	0.28902	0.28941	0.00039

Table 10.4 n-bounce windows for $x_0 = 2, \Omega = 0.3$. One-bounce windows occur for $v_{\text{in}} > 0.36317$

n	v_1	v_2	Δv_n
3	0.3605	0.36054	0.00004
2	0.36073	0.36171	0.00098
2	0.36307	0.3631	0.00003

Substituting (10.16) into (10.10) gives

$$
\begin{aligned}
L = & \int \left\{ \frac{1}{2} \left[u_\Omega \left(\phi'_+ - \phi'_- \right) \dot{X} + \dot{A}(\chi_+ + \chi_-) + A(\chi'_+ - \chi'_-)\dot{X} \right]^2 \right\} \mathrm{d}x \\
& - \int \left\{ \frac{1}{2} \left[u_\Omega \left(\phi'_+ + \phi'_- \right) + u'_\Omega(\phi_+ + \phi_- - 1) + A(\chi'_+ + \chi'_-) \right]^2 \right\} \mathrm{d}x \qquad (10.17) \\
& - \int V(u) \mathrm{d}x.
\end{aligned}
$$

To handle the $V(u)$ terms we first write $u = u_a + u_b$ where $u_a = u_\Omega(\phi_+ + \phi_- - 1)$ and $u_b = A(\chi_+ + \chi_-)$. Then using the Taylor series expansion, we get

$$\begin{aligned} V(u) &= V(u_a + u_b) \\ &= V(u_a) + V'(u_a)u_b + \frac{V''(u_a)}{2!}u_b^2 + \frac{V'''(u_a)}{3!}u_b^3 + \frac{V^{(4)}(u_a)}{4!}u_b^4 \end{aligned} \tag{10.18}$$

The corresponding reduced Lagrangian (ignoring higher order terms) that is used in our simulations is given by (10.14). Applying Euler–Lagrange equations, we obtain (10.15).

We will list the formulae of the coefficients below. Note that they are all functions of $X(t)$. Since χ_1 is not known explicitly, the coefficients presented are in the integral form. These coefficients are calculated numerically.

$$\begin{aligned} U(X) = &\int \frac{1}{2}\left(u_\Omega\left(\phi'_+ + \phi'_-\right) + u'_\Omega(\phi_+ + \phi_- - 1)\right)^2 \mathrm{d}x \\ &+ \int V(u_\Omega(\phi_+ + \phi_- - 1))\,\mathrm{d}x \end{aligned}$$

$$\begin{aligned} F(X) = &-\frac{1}{2}\int (u_\Omega\left(\phi'_+ + \phi'_-\right) + u'_\Omega(\phi_+ + \phi_- - 1))(\chi'_+ + \chi'_-)\,\mathrm{d}x \\ &-\frac{1}{2}\int V'(u_\Omega(\phi_+ + \phi_- - 1))(\chi_+ + \chi_-)\,\mathrm{d}x \end{aligned}$$

$$K(X) = -\frac{1}{2}\int (\chi'_+ + \chi'_-)^2\,\mathrm{d}x - \frac{1}{2}\int V''(u_\Omega(\phi_+ + \phi_- - 1))(\chi_+ + \chi_-)^2\,\mathrm{d}x$$

$$I(X) = \frac{1}{2}\int u_\Omega^2\left(\phi'_+ - \phi'_-\right)^2\,\mathrm{d}x, \qquad Q(X) = \frac{1}{2}\int (\chi_+ + \chi_-)^2\,\mathrm{d}x,$$

$$C(X) = \frac{1}{2}\int u_\Omega\left(\phi'_+ - \phi'_-\right)(\chi_+ + \chi_-)\,\mathrm{d}x$$

References

1. J. Cuevas-Maraver, P.G. Kevrekidis, F. Williams (eds.), *The sine-Gordon Model and Its Applications* (Springer, Heidelberg, 2014)
2. T.I. Belova, A.E. Kudryavtsev, Phys. Usp. **40**, 359 (1997)
3. O.M. Braun, Yu.S. Kivshar, *The Frenkel-Kontorova Model, Concepts, Methods and Applications* (Springer, Berlin, 2004)
4. S.N. Behera, A. Khare, Pramana **15**, 245 (1980)
5. Y.M. Gufan, *Structural Phase Transitions* (Nauka, Moscow, 1982). [In Russian]
6. A. Vilenkin, E.P.S. Shellard, *Cosmic Strings and Other Topological Defects* (Cambridge University Press, Cambridge, 2000)
7. P. Anninos, S. Oliveira, R.A. Matzner, Phys. Rev. D **44**, 1147 (1991)

8. T. Vachaspati, *Kinks and Domain Walls: An Introduction to Classical and Quantum Solitons* (Cambridge University Press, Cambridge, 2006)
9. D.K. Campbell, J.F. Schonfeld, C.A. Wingate, Phys. D **9**, 1 (1983)
10. D.K. Campbell, M. Peyrard, Phys. D **18**, 47 (1986)
11. D.K. Campbell, M. Peyrard, Phys. D **19**, 165 (1986)
12. R.H. Goodman, R. Haberman, SIAM J. Appl. Dyn. Syst. **4**, 1195 (2005)
13. R.H. Goodman, A. Rahman, M.J. Bellanich, C.N. Morrison, Chaos **25**, 043109 (2015)
14. P. Dorey, K. Mersh, T. Romańczukiewicz, Y. Shnir, Phys. Rev. Lett. **107**, 091602 (2011)
15. F.C. Simas, A.R. Gomes, K.Z. Nobrega, J.C.R.E. Oliveira, JHEP **2016**(9), 104 (2016)
16. H. Weigel, J. Phys. Conf. Ser. **482**, 012045 (2014)
17. I. Takyi, H. Weigel, Phys. Rev. D **94**, 085008 (2016)
18. V.A. Gani, A.E. Kudryavtsev, M.A. Lizunova, Phys. Rev. D **89**, 125009 (2014)
19. A.M. Marjaneh, V.A. Gani, D. Saadatmand, S.V. Dmitriev, K. Javidan, JHEP **2017**(7), 028 (2017)
20. E. Belendryasova, V.A. Gani, Commun. Nonlinear Sci. Numer. Simul. **67**, 414 (2019)
21. C.J. Pethick, H. Smith, *Bose-Einstein Condensation in Dilute Gases* (Cambridge University Press, Cambridge, 2002)
22. L.P. Pitaevskii, S. Stringari, *Bose-Einstein Condensation* (Oxford University Press, Oxford, 2003)
23. P.G. Krevrekidis, D.J. Frantzeskakis, R. Carretero-González, *The Defocusing Nonlinear Schrödinger Equation* (SIAM, Philadelphia, 2015)
24. T. Sugiyama, Prog. Theor. Phys. **61**, 1550 (1979)
25. I. Roy, S.V. Dmitriev, P.G. Kevrekidis, A. Saxena, Phys. Rev. E **76**, 026601 (2007)
26. C. Gallo, D. Pelinovsky, Asympt. Anal. **73**, 53 (2011)
27. G. Karali, C. Sourdis, Arch. Rat. Mech. Anal. **217**, 439 (2015)
28. M.P. Coles, D.E. Pelinovsky, P.G. Kevrekidis, Nonlinearity **23**, 1753 (2010)
29. J. Dorignac, J.C. Eilbeck, M. Salerno, A.C. Scott, Phys. Rev. Lett. **93**, 025504 (2004)
30. A. Demirkaya, R. Decker, P.G. Kevrekidis, I.C. Christov, A. Saxena, JHEP **2017**(12), 71 (2017)
31. R.V. Radomskiy, E.V. Mrozovskaya, V.A. Gani, I.C. Christov, J. Phys. Conf. Ser. **798**, 012087 (2017)
32. A. Khare, I.C. Christov, A. Saxena, Phys. Rev. E **90**, 023208 (2014)

Chapter 11
The ϕ^4 Model in Higher Dimensions

Jean-Guy Caputo

Abstract We analyze solutions of the ϕ^4 model in 2 and 3 spatial dimensions. For infinite domains, the solutions are of the bubble type, they collapse in finite time and give rise to transient waves. In inhomogeneous waveguides with Neumann boundary conditions, kinks will accelerate for narrowing widths and can be reflected if they do not have enough energy to cross the defect. In the presence of dissipation, kinks propagate at a constant radial speed. Dissipative kinks will travel in a smooth waveguide but can be trapped by sharp enlargements.

11.1 Introduction

The ϕ^4 potential has a double well and is therefore a simple model of a phase transition. It gives rise to kink solutions that connect the two minima of the potential. The minima are usually chosen equal and the zero energy level is taken there. Then the energy of the kink is concentrated in the transition region. We speak of "localized energy" or "localized wave". In this review, we will consider such waves and describe their dynamics in various contexts. Although extended periodic waves can also exist, they will not be presented herein.

Two types of dynamics occurring with the ϕ^4 potential will be described in this chapter. One is the conservative, ballistic mode where the particle has a mass and obeys Newton's law; a chain of particles will follow a wave equation in the continuum limit. The other type of dynamics is diffusive and yields a reaction-diffusion equation in the continuum limit.

In 1D, kinks in the conservative case have a free parameter: the velocity. In higher dimensions, kinks can be extended to so-called radial kinks where the wave profile only depends on the radial component. This waves are translationally invariant in 1D. However, in 2D, 3D or higher dimensions, radial kinks collapse to 0 and give

J.-G. Caputo (✉)
Laboratoire de Mathématiques, INSA Rouen Normandie,
76801 Saint-Etienne du Rouvray, France
e-mail: caputo@insa-rouen.fr

P. G. Kevrekidis and J. Cuevas-Maraver (eds.), *A Dynamical Perspective on the ϕ^4 Model*, Nonlinear Systems and Complexity 26,
https://doi.org/10.1007/978-3-030-11839-6_11

rise to linear waves. We will describe this collapse and show the effect of setting a minimum r_0 to the radial domain. The onset of collapse is predicted by an energy analysis. Another interesting situation, especially for applications, is the propagation of kinks in a waveguide. Here again, an energy analysis describes the situation. A cousin of the ϕ^4 wave equation is the sG equation which is integrable in 1D, i.e. it has an infinite number of conservation laws and, in particular a localized oscillating traveling wave: the breather (see e.g. for example the book by Eilbeck et al. [1]). In 2D and higher dimensions, both sine-Gordon and ϕ^4 kinks behave in similar ways; we will point out some few differences.

Dissipative kinks have the same analytical expression as conservative kinks except that now the velocity is fixed by the environment. Such dissipative kinks describe for example the propagation of a diffusive combustion front, or the propagation of the nerve impulse in an axon (see the review by [2]). Defects affect such fronts and we will analyze, in 1D, how the front position and width depend on an inhomogeneous reaction term. In 2D, radial fronts are especially interesting as they can be blocked in some waveguides. This occurs in a number of physiological applications (see e.g. the nice discussion by Berestycki et al. [3]). We will give a review of the situation and explain why this blockage occurs.

This chapter is organized as follows. Section 11.1 presents the model and the main features of 1D kinks. The conservative or ballistic case is discussed in Sect. 11.2. Section 11.3 considers the dissipative, so-called Zeldovich reaction-diffusion model, and conclusions are given in Sect. 11.4.

11.2 The Model

11.2.1 Basic Equations

The ϕ^4 PDE is obtained from the Lagrangian density in $\mathbb{R}^d$

$$\mathcal{L} = a\frac{\psi_t^2}{2} - \frac{(\nabla\psi)^2}{2} + \frac{\psi^2}{2} - \frac{\psi^4}{4}, \tag{11.1}$$

where $\psi = \psi(x_1, x_2, \ldots, x_d, t)$ and a is a parameter. The potential, in the interesting bistable case,

$$V(\psi) = -\frac{\psi^2}{2} + \frac{\psi^4}{4} + \frac{1}{4}, \tag{11.2}$$

has two minima at $\psi = \pm 1$ and a maximum at $\psi = 0$. As shown below, the homogeneous solutions $\psi \equiv \pm 1$ are stable while $\psi \equiv 0$ is unstable.

The corresponding Euler–Lagrange equation is

$$a\psi_{tt} - \Delta\psi - \psi + \psi^3 = 0, \tag{11.3}$$

where $\Delta \equiv \partial_{x_1}^2 + \partial_{x_2}^2 + \cdots + \partial_{x_d}^2$ is the usual Laplacian operator.

If dissipation is present, one can add a damping term and obtain the general equation

$$a\psi_{tt} + \epsilon\psi_t + \Delta\psi - \psi + \psi^3 = 0. \tag{11.4}$$

This is the model that is considered in this chapter.

In the two-dimensional case, the Lagrangian is defined as $L = \int \mathrm{d}t\mathrm{d}x\mathrm{d}y\mathcal{L}$. For finite domains, like e.g. an infinite strip in x of width w in y we get a Lagrangian

$$L = \int \mathrm{d}t\mathrm{d}x \int_0^w \mathrm{d}y\mathcal{L}.$$

The Euler–Lagrange equations yield naturally the boundary conditions in y via the integrated terms obtained by integration by parts. Typically we get

$$\delta L = \cdots + \int \mathrm{d}t\mathrm{d}x \left[\frac{\partial\mathcal{L}}{\partial\psi_y}\delta\psi\right]_0^w = 0$$

and this implies

$$\frac{\partial\mathcal{L}}{\partial\psi_y} \equiv \psi_y = 0,$$

at $y = 0$ and $y = w$. We naturally get homogeneous Neumann boundary conditions in y for this strip. In this chapter, we will consider the motion of kinks in 2D infinite domains or in waveguides embedded in 2D. The results can be generalized to $d > 2$.

The sG wave equation, where $V(\psi) = 1 - \cos\psi$, has many similarities with the ϕ^4 model. In 1D, the former is integrable, i.e. it has an infinite number of conservation laws and has a breather solution [1], which does not exist for the latter. However, in higher dimensions integrability is lost and kinks from both models should behave similarly. In fact, it should be expected that all potentials $V(\psi)$ presenting an unstable state per two stable ones (where the latter have equal energy) behave in similar manner. We will see in the chapter that this is partially true.

11.2.2 Stability of Homogeneous States and Kinks

First of all, we examine homogeneous solutions of (11.4), $\psi = \psi_0 + u$, with u small. Linearized equation is

$$au_{tt} + \epsilon u_t - \Delta u - u + 3\psi_0^2 u = 0, \tag{11.5}$$

which can be solved using Fourier analysis; that is, by taking

$$u = \exp[i(\mathbf{k} \cdot \mathbf{x} - \omega t)]$$

one arrives to the dispersion relation

$$-\omega^2 - i\omega + |k|^2 - 1 + 3\psi_0^2 = 0.$$

It is easy to see that for $\psi_0 = 0$ solution, ω can have a negative imaginary part so that this homogeneous solution is unstable. On the contrary, for $\psi_0 = \pm 1$, the imaginary part of ω is always positive and these homogeneous states are stable.

Kinks are solutions connecting the two stable states ± 1. The presence or not of the inertial term ψ_{tt} in (11.4) will considerably modify the dynamics. We will therefore consider $\epsilon = 0$ as the purely ballistic case and $a = 0$ as the purely diffusive case.

11.3 Inertial Case: Dynamics of Fronts and Radial Kinks

11.3.1 The Radial ϕ^4 Equation

In 1D, there is a two-parameter exact solution, see e.g. the book by Eilbeck et al. [1], the kink

$$\psi(x,t) = \pm \tanh\left[\frac{x - vt - x_0}{\sqrt{2(1 - v^2)}}\right], \tag{11.6}$$

where v is the kink velocity and x_0 is an arbitrary position. The dynamics of such kinks has been studied extensively, in particular the kink-antikink collision, see [4]. Note also the problems in describing kink-antikink collisions using collective variables [5].

The radial ϕ^4 equation in d dimensions is

$$\phi_{tt} - \phi_{rr} - (d-1)\frac{\phi_r}{r} + V'(\phi) = 0. \tag{11.7}$$

It can be obtained, for example in a 2D sector, by expanding the whole field ψ onto the azimuthal modes. The calculation done in [6] for the sG equation can be reproduced here for ϕ^4. We give here the main steps of the derivation in the two-dimensional case.

The 2D ϕ^4 equation reads

$$\psi_{tt} - \frac{\psi_{\theta\theta}}{r^2} - \psi_{rr} - \frac{\psi_r}{r} + V'(\psi) = 0. \tag{11.8}$$

As domain we consider the sector $r_0 < r < r_1$, $0 < \theta < \theta_0$ shown in Fig. 11.1. The boundary conditions are homogeneous Neumann so that $\psi_\theta = 0$ for $\theta = 0$ and $\theta = \theta_0$ and $\psi_r = 0$ for $r = r_0$ and $r = r_1$. To reduce the 2D problem it is natural to expand in azimuthal modes using the cosine Fourier series

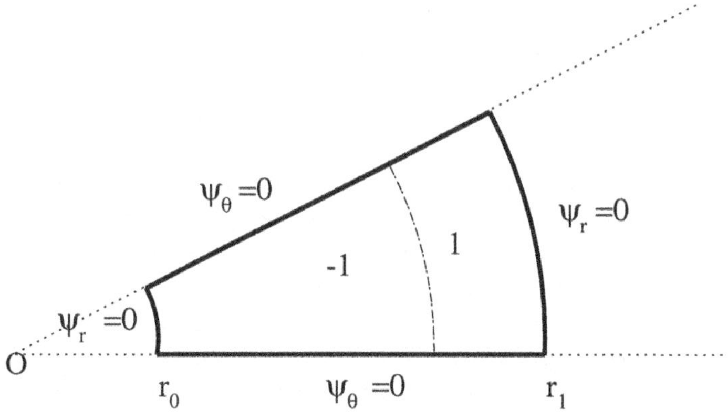

Fig. 11.1 Sketch of the 2d domain for the 2D ϕ^4 equation. The radial kink initial condition is shown as a dotted-dashed line. Adapted from [6], with permission from APS

$$\psi(r, \theta, t) = \sum_{n=0}^{\infty} \phi_n(r, t)\chi_n(\theta), \tag{11.9}$$

where $\chi_n(\theta) = \cos(\frac{n\pi\theta}{\theta_0})$. Plugging the expression (11.9) into (11.8) and projecting onto χ_0 we obtain the evolution of ϕ_0

$$-\phi_{0_{tt}} + \phi_{0_{rr}} + \frac{\phi_{0_r}}{r} = \frac{1}{\theta_0}\int_0^{\theta_0} V'(\phi_0 + \phi_1\chi_1 + \phi_2\chi_2 + \cdots)\mathrm{d}\theta. \tag{11.10}$$

We find that the non zero contribution of the corrections is quadratic. Neglecting it, (11.10) reduces to the radial 1D ϕ^4 equation

$$-\phi_{tt} + \phi_{rr} + \frac{\phi_r}{r} = -V'(\phi). \tag{11.11}$$

This simplification is legitimate because the azimuthal components do not get amplified, see [7].

11.3.2 Review of the Literature

Most studies in higher dimensions have dealt with the sG equation in a semi infinite domain. In 1D, the kink is a topological defect so that if it is present, it cannot disappear. It is then interesting to examine what happens to a 1D kink in a higher dimensional space, where the topological constraint is absent. Christiansen and Lomdahl [8–11] studied radial kinks in 1D. They first noted the return effect where a radial kink with an outgoing velocity will expand up to a maximal radius and then shrink.

Samuelsen explained this using energy arguments [12]; the total energy in the radial case (and factoring out the angular contribution) is

$$E = \int_{r_0}^{r_1} r^{d-1} \mathrm{d}r \left[\frac{\phi_t^2}{2} + \frac{\phi_r^2}{2} + V(\phi) \right]. \tag{11.12}$$

Assume a radial kink, $\phi(r - R(t))$ localized at its position R so E can be approximated as

$$E \approx R^{d-1} E_v^1, \tag{11.13}$$

where E_v^1 is the energy of a 1D kink with velocity v. There exists a radius R_m such that $E = R_m^{d-1} E_0^1$. For that radius, the kink will have lost all its kinetic energy and start to shrink.

A more difficult issue is to find out the behaviour of the kink when it reaches $r = 0$. Geike [13] showed the destruction of the kink at the origin (see also also Bogolubsky and Makhankhov work [14]), and also reported the existence of different behavior between 2D and 3D kinks [13]. Breathers centered at $r = 0$ were shown by Piette et al. to be very long lived [15]. In a recent work [6], we showed that for $d = 2$, the kink disintegrates into a fast breather and for $d \geq 3$, the kink couples to a breather centered at $r = 0$. We also examined the situation for $r_0 > 0$. Although that work was related to the sG equation, the analysis for the ϕ^4 model is very similar.

To finish this review, note that in a recent work [7], we examined the effect of an external potential on the radial sG and ϕ^4 equations.

$$\phi_{tt} - \phi_{rr} - \frac{\phi_r}{r} + V'(\phi)(1 + V_{\text{ext}}(r)) = 0. \tag{11.14}$$

We showed that when V_{ext} is a sufficiently deep potential well centered on $r = 0$, a new kink connecting the state π for sG (0 for ϕ^4) — which is now meta-stable — to the state 2π for sG (1 for ϕ^4) is also a solution of (11.14).

11.3.3 Conservation Laws

Equation (11.7) has the following energy conservation law

$$\frac{\mathrm{d}H}{\mathrm{d}t} = [r^{d-1} \phi_r \phi_t]_{r_0}^{r_1}, \tag{11.15}$$

where the Hamiltonian H is

$$H = \int_{r_0}^{r_1} r^{d-1} \mathrm{d}r \left[\frac{\phi_t^2}{2} + \frac{\phi_r^2}{2} + V(\phi) \right] \equiv \int_{r_0}^{r_1} r^{d-1} \mathrm{d}r \mathcal{H}. \tag{11.16}$$

The flux relation

$$\frac{\mathrm{d}H}{\mathrm{d}t} = {r_1}^{d-1}\phi_r\phi_t|_{r_1}, \tag{11.17}$$

allows to compute the amount of energy escaping from the computational domain at $r = r_1$.

For a kink, the energy is

$$E_{\mathrm{K}}^{1D} = \frac{4}{3}\frac{1}{\sqrt{1-v^2}}. \tag{11.18}$$

Following the same procedure as in [6], we estimate the radial kink energy in d dimensions as

$$E_{\mathrm{K}} = \frac{4}{3}R^{d-1}\frac{1}{\sqrt{1-v^2}}. \tag{11.19}$$

This calculation is correct provided that r is far from $r = 0$, as it relies on the parity of ϕ_r with r. Since the energy is conserved, the expression E_{K} allows to compute the evolution of the radial kink position R as a function of time, being exactly the same as in [16]. In that case, we have that R fulfills the following ODE:

$$\frac{\mathrm{d}R}{\mathrm{d}t} = \pm\sqrt{1-\left(\frac{R}{R_0}\right)^{2(d-1)}}. \tag{11.20}$$

11.3.4 Kink Collisions in Two and Higher Dimensions

Here we recall the situation for the sG equation [6].

Figure 11.2 shows two snapshots of the evolution of a 2D sG kink in a domain $r_0 < r < r_1$, with $r_0 = 1$ and $r_1 = 20$, using as initial condition

$$\psi(r, t = 0) = 4\mathrm{atan}\left(\frac{r - R_0}{\sqrt{1-u_0^2}}\right), \tag{11.21}$$

with $R_0 = 15$ and $u_0 = 0$. On the left panel, which corresponds to the kink at $t = 20$, it is accelerated towards the narrow edge. Notice the absence of radiation and the characteristic overshoot. The right panel presents the solution after collision, at $t = 50$, and shows that the kink has disappeared and only a flat background $\phi \approx 2\pi$ persists with some oscillations. Despite the violence of the collision, all the energy remains in the radial $n = 0$ mode and no azimuthal modes are excited. We expect similar 2D results for the ϕ^4 equation, as shown below.

Figure 11.3 shows snapshots of the ϕ^4 kink colliding with the $r = 0$ boundary for $d = 2$ at times $t = 0, 6.7, 13.3, 20, 26.6, 33.3$ and 40. Clearly the kink is destroyed

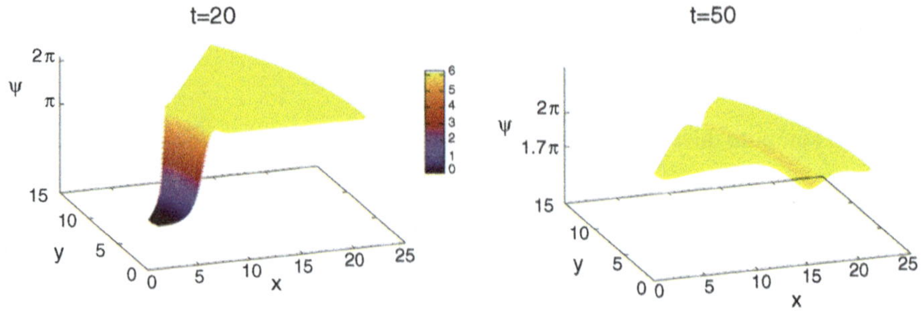

Fig. 11.2 Snapshots of the evolution of a kink for the 2D sG equation in a wedge. The kink is started at $R_0 = 15$ in a domain with $r_0 = 1$, $r_1 = 20$. Left panel shows the solution for $t = 20$ (before collision) and right panel holds for $t = 50$ (after collision). The color range of $\psi(x, y)$ is $[0, 2\pi]$. Adapted from [6], with permission from APS

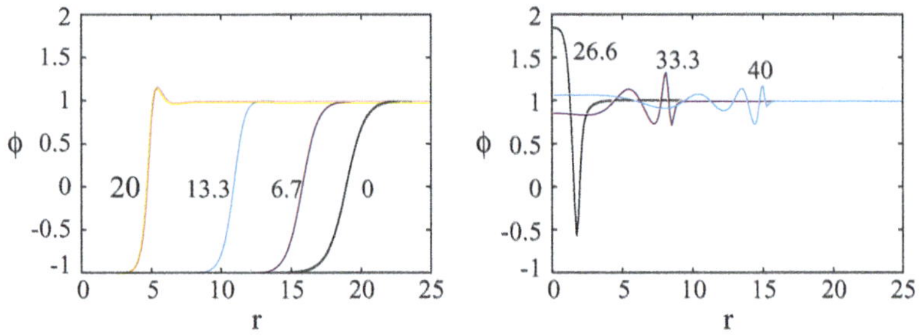

Fig. 11.3 Snapshots of a ϕ^4 kink with $d = 2$ in the r domain $0 \leq r \leq 25$ for different times, before collision (left) and after collision (right). The kink is started at $R = 20$

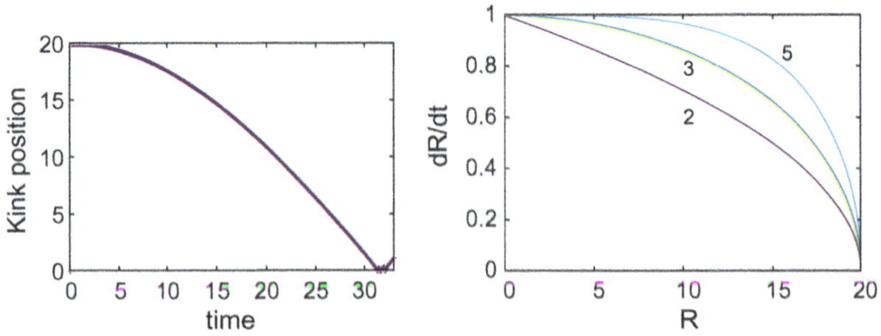

Fig. 11.4 (Left panel) Time evolution of the ϕ^4 kink position $R(t)$ for $d = 2$. (Right panel) Plane $(R, \frac{dR}{dt})$ for a radial kink started with 0 velocity at $R_0 = 20$ for $d = 2$, 3 and 5

in this configuration and only small amplitude radiation remains. This result is similar to the findings of sG equation [6].

The position $R(t)$ of the kink can be computed for the numerical solution before the collision. It is shown in the left panel of Fig. 11.4, together with the estimate

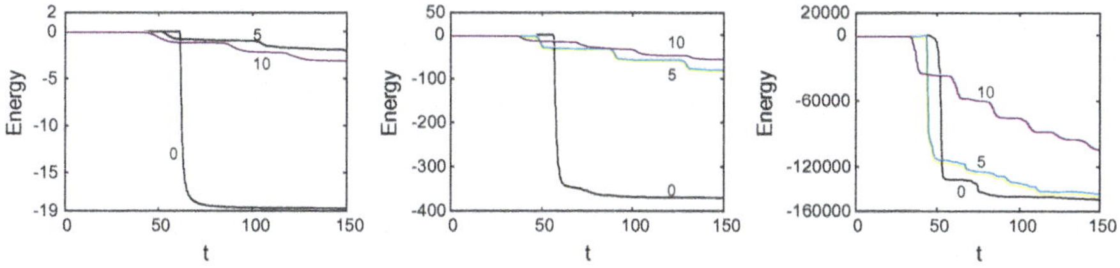

Fig. 11.5 Energy escaping the computational domain at $r = 30$ for a radial ϕ^4 kink started at $r = R_0 = 20$ and $r_0 = 0, 5$ and 10. The spatial dimensions are $d = 2$ (left), $d = 3$ (middle) and $d = 5$ (right)

(11.20). The two curves are indistinguishable. The right panel of the figure shows the evolution of a radial kink in the plane $(R, \frac{\mathrm{d}R}{\mathrm{d}t})$ for $d = 2, 3$ and 5. Notice how the kink becomes relativistic for $d = 3$ even at $R = 10$. Furthermore, there is no difference in the evolution of $R(t)$ for the ϕ^4 and the sG equations (cf. Fig. 4a of [6]).

Note that the acceleration of the kink towards the narrow edge of the domain comes from the term $\frac{\phi_r}{r}$. It decays fastly as R is increased.

11.3.5 Energy Radiated Out at Collision

Figure 11.5 presents the energy escaping the computational domain at $r = 30$ for a radial ϕ^4 kink started at $r = R_0 = 20$ and $r_0 = 0, 5$ and 10 for $d = 2$ (left), $d = 3$ (middle) and $d = 5$ (right) spatial dimensions. To estimate the energy exiting the domain, we use the flux relation (11.17). Here, we see differences with the results for sG dynamics published in [6]. In particular, for sine-Gordon, $d = 3$ gives the fastest decay for $r_0 = 5$.

In the ϕ^4 model, the fastest decay always occurs for $r_0 = 0$. Away from $r = 0$, (11.7) reduces to either 1D sG or 1D ϕ^4. The former has breather solutions which can carry out energy in a coherent fashion as opposed to the latter. This could explain the different behavior between the two systems.

11.3.6 2D Solutions in a Deformed Waveguide

It is interesting to examine the motion of kinks in deformed waveguides in a plane. An immediate application of the sG equation in this context is the physics of long Josephson junctions [17]. These are built from two superposed superconducting films separated by a thin oxide layer (~1 nanometer), enabling the tunneling of Cooper pairs between films. For the ϕ^4 model, this application is absent, nevertheless the situation is interesting from a formal point of view.

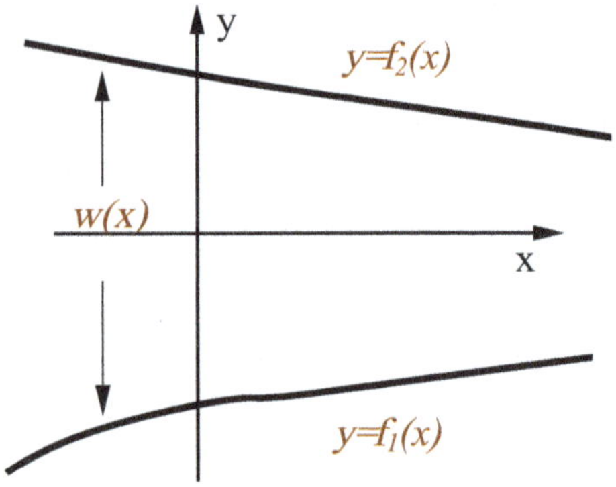

Fig. 11.6 Sketch of an inhomogeneous waveguide

As discussed above, the lateral boundary conditions on the waveguide are homogeneous Neumann. Assuming that the width w varies slowly compared to the wavelength of the solution, we can reduce (11.4) to the following 1D equation

$$\phi_{tt} - \phi_{xx} - \frac{w'(x)}{w(x)}\phi_x + V'(\phi) = 0, \tag{11.22}$$

where x is the curvilinear coordinate along the waveguide [18].

We sketch below the derivation for completeness. Let us consider the 2D ϕ^4 equation

$$\psi_{tt} - (\psi_{xx} + \psi_{yy}) + V'(\psi) = 0, \tag{11.23}$$

in the spatial domain $(x, y) \in \mathbb{R} \times [f_1(x); f_2(x)]$ (see Fig. 11.6).

We assume that f_1 and f_2 vary slowly with x as $f_1(\epsilon x)$ and $f_2(\epsilon x)$ where ϵ is a small parameter. The transverse average is defined as

$$\phi(x) \equiv \frac{1}{w}\int_{f_1}^{f_2} \psi(x, y)\mathrm{d}y, \tag{11.24}$$

where $w = f_2 - f_1$. At the interfaces $y = f_i$, $i = 1, 2$ we have homogeneous Neumann boundary conditions $\nabla\psi \cdot \mathbf{n} = 0$ where $\mathbf{n}$ is the normal. This leads to the following relations

$$-\epsilon f_i'(x)\psi_x(x, f_i) + \psi_y(x, f_i) = 0, \tag{11.25}$$

for $i = 1, 2$. Averaging (11.23) on y, we get

$$\phi_{tt} - \frac{1}{w}\int_{f_1}^{f_2} \psi_{xx}\mathrm{d}y - \frac{1}{w}[\psi_y]_{f_1}^{f_2} + \frac{1}{w}\int_{f_1}^{f_2} V'(\psi)\mathrm{d}y = 0. \tag{11.26}$$

Taking the derivative of (11.24), we have

$$\frac{1}{w}\int_{f_1}^{f_2} \psi_{xx}\mathrm{d}y = \phi_{xx} + \frac{w'(x)}{w(x)}\phi_x - \epsilon f_2'\psi_x(x, f_2) + \epsilon f_1'\psi_x(x, f_1),$$

and the bracket in (11.26) becomes

$$[\psi_y]_{f_1}^{f_2} = \epsilon f_2'\psi_x(x, f_2) - \epsilon f_1'\psi_x(x, f_1)$$

Combining the above two expressions, we get

$$\frac{1}{w}\int_{f_1}^{f_2} \psi_{xx}\mathrm{d}y + \frac{1}{w}[\psi_y]_{f_1}^{f_2} = \phi_{xx} + \frac{w'(x)}{w(x)}\phi_x + O(\epsilon^2),$$

where the terms involving f_1 and f_2 have cancelled. The term $V'(\psi)$ can be approximated by noting that there is y^* such that $\phi(x) = \psi(x, y^*)$. Then

$$\psi(y) = \phi(x) + (y - y^*)\psi_y^* + \cdots .$$

Using this we have

$$V'(\psi) \approx V'(\phi) + V''(\phi)(y - y^*)\psi_y^*$$

so that

$$\frac{1}{w}\int_{f_1}^{f_2} V'(\psi)\mathrm{d}y \approx V'(\phi).$$

Collecting all the terms, we arrive to (11.22).

The radial equation (11.11) is recovered by assuming $w(x) = w^{d-1}$, as expected.

The dynamics of ballistic kinks in such a waveguide is governed by the energy. As in the radial case, we assume that the kink behaves as a collective wave centered at position $X(t)$. In a waveguide of transverse width w, using arguments similar to the ones leading to (11.19), the energy of the ϕ^4 kink is

$$E_{\mathrm{K}} = w(X)\frac{4}{3}\frac{1}{\sqrt{1-(\dot{X})^2}}. \tag{11.27}$$

Since the term $E = \frac{3}{4}E_{\mathrm{K}}$ is constant, we get from (11.27)

$$(\dot{X})^2 = 1 - \frac{w^2(X)}{E^2}. \tag{11.28}$$

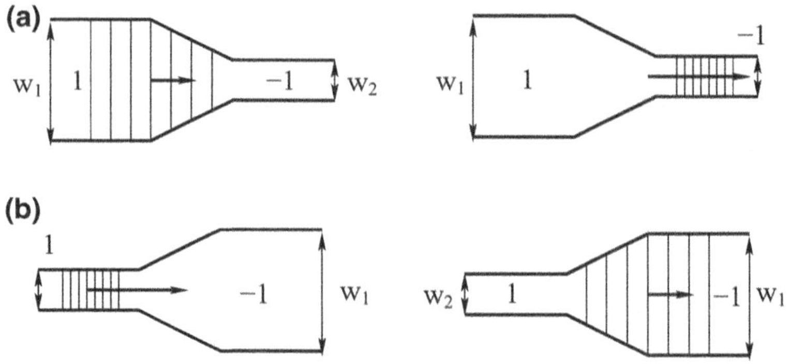

Fig. 11.7 Kink crossing between two waveguides of different width

This predicts that the kink can stop (i.e. $\dot{X}$ becomes 0) for certain widths. After that, it will bounce back. The kink can modify its velocity $\dot{X}$ or even its width as for curved waveguides, see for example [19].

We can predict the evolution of a kink traveling in a guide composed of two homogeneous sections, w_1 and $w_2 < w_1$, as shown in the schematic picture of Fig. 11.7. Energy is conserved so that

$$\frac{w_1}{\sqrt{1-v_1^2}} = \frac{w_2}{\sqrt{1-v_2^2}}. \tag{11.29}$$

If the kink is traveling from the wide section w_1 to the narrow section w_2 as in the top panels (a), expression (11.29) indicates that $v_2 > v_1$ so that the kink will always cross and increase its velocity.

If the kink is traveling from the narrow section w_2 to the wide section w_1 as in panels (b) of Fig. 11.7, the kink will slow down so that there is a threshold velocity v_c

$$v_c = \sqrt{1-\left(\frac{w_2}{w_1}\right)^2}, \tag{11.30}$$

corresponding to have a static kink in guide 1. For $v_2 < v_c$, the kink cannot cross the junction and is reflected. These statements were developed in [18] and tested experimentally by the group of Costabile [20]. A similar reasoning was used recently to describe how sG kinks propagate across forks [21, 22]. ϕ^4 kinks should behave in the exact same way.

11.4 The Zeldovich Reaction-Diffusion Equation

The ϕ^4 potential has also been used in the dissipative (overdamped) context. Zeldovich introduced a reaction-diffusion equation to model the propagation of a combustion front in a solid [23]. This is the kind of combustion that occurs in a candle. The model can also describe nerve impulse propagation along a neuron [2]. This equation, which is typically refereed to as the bistable equation (see [24] for a review), is given by

$$\phi_t = \Delta\phi + \phi(1-\phi)(\phi-\alpha), \tag{11.31}$$

where $0 < \alpha < 1$ is a parameter. In the 1D case, the kink (11.6) is still a solution

$$\phi^K(x,t) = \frac{1}{1+\exp\frac{x-vt}{\sqrt{2}}}, \tag{11.32}$$

however the velocity is now fixed to $v = \frac{1-2\alpha}{\sqrt{2}}$ [2]. Expressions (11.6) and (11.32) match exactly when the nonlinearity of (11.31) is written as in (11.4). The formula (11.32) describes a kink whose value is 1 at $-\infty$ and 0 at $+\infty$ (1-0 kink), traveling from left to right. Note that 0-1 kinks are also a solution that travels from right to left.

In the work of [16], I examined with Benoit Sarels the influence of a modulation $s(x)$ of the nonlinearity. The model is given by

$$\phi_t = \phi_{xx} + s(x)R(\phi). \tag{11.33}$$

It can represent a defect in the combustible material, or for the neuron application, an enlargement of the axon. We assumed a kink ansatz

$$U(z) = \frac{1}{1+\exp z}, \qquad \phi(x,t) = U\left(\frac{x-X(t)}{w(t)}\right), \tag{11.34}$$

and derived collective variable equations for the position X and width w. To this aim, we used (11.33) and

$$\phi\phi_t = \phi\phi_{xx} + s(x)\phi R(\phi), \tag{11.35}$$

integrated them over the whole domain, using the fact that the integrands are localized so that the integrals always existed. We used also the fact that ϕ is a 1-0 kink so that it is flat at both ends. Replacing ϕ by the expressions (11.34), we get two ODEs for X and w

$$\begin{aligned} \dot{X} &= w\int_{-\infty}^{\infty} s(wz+X)R(U(z))\mathrm{d}z, \\ \dot{w} &= \frac{1}{3w} + w\int_{-\infty}^{\infty} s(wz+X)(1-2U(z))R(U(z))\mathrm{d}z. \end{aligned} \tag{11.36}$$

Two different situations can take place, depending on whether the defect $s(x)$ is narrow or wide compared to w. The latter case is the simplest, as the integrals can be calculated adiabatically yielding

$$\begin{aligned} \dot{X} &= \left(\frac{1-2a}{2}\right) ws(X), \\ \dot{w} &= \frac{1}{3w} - \frac{w}{6}s(X). \end{aligned} \tag{11.37}$$

When the defect $s(x)$ is narrow compared to w, we use a delta-function approximation $s(x) = \alpha + \beta\delta(x)$ obtaining

$$\begin{aligned} \dot{X} &= \alpha w \left(\frac{1-2a}{2}\right) + \beta R\left(U\left(\frac{-X}{w}\right)\right), \\ \dot{w} &= \frac{1}{3w} - \alpha\frac{w}{6} + \beta\left(1 - 2U\left(\frac{-X}{w}\right)\right) R\left(U\left(\frac{-X}{w}\right)\right). \end{aligned} \tag{11.38}$$

The solutions of these equations were in excellent agreement with the solutions of the PDE. In particular, they predicted pinning of a wide kink on a narrow defect; inspection of (11.37) shows that no pinning can occur on a wide defect. On the contrary, for β large enough, the RHS of $\dot{X}$ in (11.38) can become zero, implying pinning of the kink. Such a pinning has been observed in other applications like mechanical fractures [25] or the motion of a meniscus in capillaries [26]. It has also been seen in 2D waveguides and this is the object of the rest of the section.

As for the ballistic case, the motion of 2D radial fronts can be inferred from those of 1D kinks. 1-0 kinks expand out to infinity, as expected. More interesting is the propagation in 2D waveguides. This is particularly important since blocking of waves has been observed in different contexts. In physiology, the nerve impulse in a neuron can be stopped by a sudden enlargement of the axon. This could occur in the brain and the heart, see the nice discussion in [3]. The Zeldovich model can also model the propagation of an epidemic. In that context, geographic obstacles like rivers or mountains have been shown to stop the propagation of the front. This occurred in the black death plague in the middle ages (see [27, 28]).

As discussed previously, for the propagation in a waveguide, the transverse boundary conditions are homogeneous Neumann. Juliette Bouhours in her thesis [29] studied systematically the propagation of reaction-diffusion ϕ^4 kinks in waveguides with variable width. Here, as opposed to the ballistic case, the energy plays no role.

A first result of Bouhours [29] is that when the waveguide section decreases, there is always passage of the kink. Blockage will only occur when the section of the waveguide increases. Figure 11.8 shows two types of waveguides. In (a) the width increases rapidly while in (b) it increases gradually. The kink is presented by its contour lines going from 1 to the left to -1 to the right. The numerics of Bouhours [29] show that the kink is blocked for the configuration (a) while it can cross for (b).

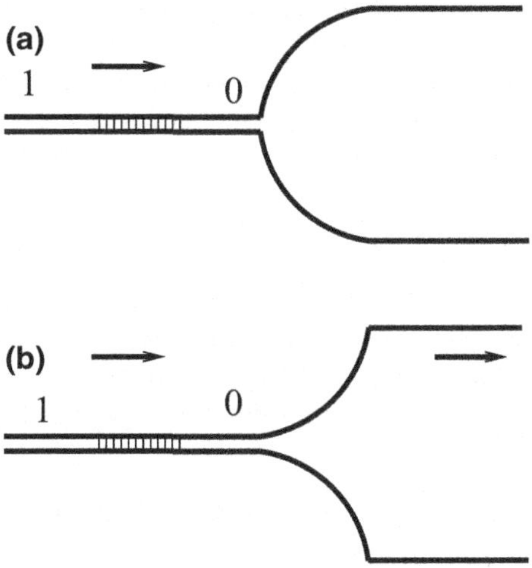

Fig. 11.8 Two waveguides with increasing width. In **a** the kink will be blocked while it will pass in (**b**)

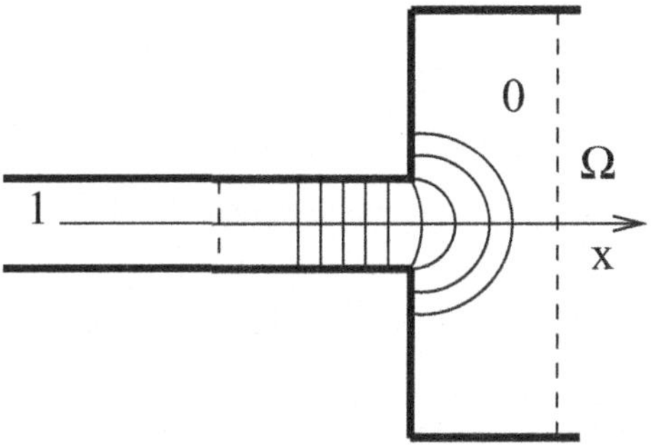

Fig. 11.9 Trapping of a dissipative kink by a sudden enlargement of the waveguide

Berestycki, Bouhours and Chapuisat proved that under a technical condition on the normal, trapping occurs [3].

When the kink gets trapped, it develops into a radial kink on the RHS. This is displayed in Fig. 11.9.

Consider the 2D Zeldovich equation

$$\phi_t = \phi_{xx} + \phi_{yy} + \phi(1-\phi)(\phi - a), \tag{11.39}$$

and that the kink has reached the domain Ω bounded by the dashed lines in Fig. 11.9. Integrating the equation over this domain, we get

$$\partial_t \int_\Omega \phi \mathrm{d}x\mathrm{d}y = \int_{\text{left}} \nabla\phi \cdot \mathbf{n} \mathrm{d}s + \int_{\text{right}} \nabla\phi \cdot \mathbf{n} \mathrm{d}s + \int_\Omega \phi(1-\phi)(\phi - a)\mathrm{d}x\mathrm{d}y,$$

where we have used the fact that the boundary conditions on the exterior of the waveguide are homogeneous Neumann ($\nabla\phi \cdot \mathbf{n} = 0$) and where *left* and *right* indicate the two lateral sides of Ω. In the region on the right, the wave is a radial kink, while it is a 1D kink on the left. We can assume that the line integrals are very small, the first one can be taken sufficiently far so that it is close to 0. The second one is positive but very small because the kink is trapped by the discontinuity of the normal.

The integral

$$I = \int_\Omega \phi(1-\phi)(\phi-a)\mathrm{d}x\mathrm{d}y \equiv \int_\Omega R(\phi)\mathrm{d}x\mathrm{d}y$$

can be computed in the following way. Assume that the region poking into the right domain is for $x = h$ (see Fig. 11.9). Then the integral I can be estimated using a half radial kink ansatz for the right region and a 1D kink for the left region. We have

$$I \approx w_1 \int_{-\infty}^{h} R(\phi(x))\mathrm{d}x + \pi \int_h^\infty R(\phi(r))r\mathrm{d}r. \tag{11.40}$$

Using the estimates of [16], we get

$$R(\phi(r)) = -e^r \frac{-1+a+ae^r}{(1+e^r)^3} \approx -ae^{-r},$$

for large r. Then the second integral is

$$\pi \int_h^\infty R(\phi(r))r\mathrm{d}r = -\pi a(1+h)e^{-h},$$

which is negative. The other term is

$$\int_{-\infty}^{h} R(\phi(x))\mathrm{d}x = \frac{1}{2} - \frac{ae^h}{1+e^h} - \frac{1}{2(1+e^h)^2} \approx \frac{1}{2} - a,$$

which is positive. Combining the two terms of (11.40), we get

$$I \approx w_1(\frac{1}{2} - a) - \pi a(1+h)e^{-h}. \tag{11.41}$$

These terms can balance, making I zero, so that $\partial_t \int_\Omega \phi\mathrm{d}x\mathrm{d}y = 0$ indicating that the kink becomes trapped. Figure 11.10 shows $I(h)$ for $w_1 = 1$ and $w_1 = 3$. For $w_1 = 1$, the expression becomes 0 at a given h while for $w_1 = 3$, $I(h)$ does not change sign. This naive analysis shows that to trap the kink it is crucial to have a radial kink on the right boundary. Note also that w_2 does not explicitly appear. Nevertheless, we need $w_2 > 3w_1$ so that the radial solution can exist.

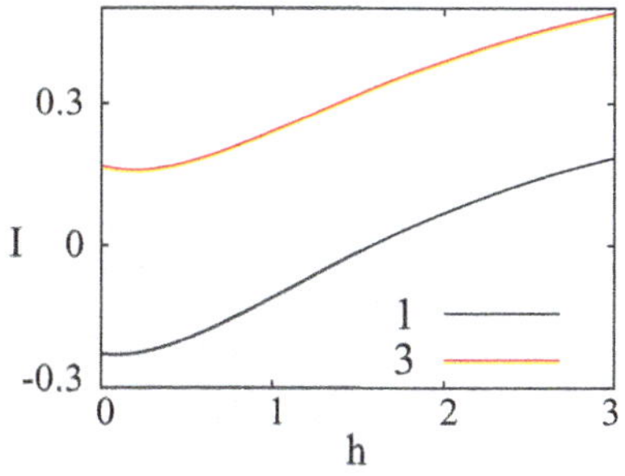

Fig. 11.10 Plot of $I(h)$ for $w_1 = 1$ and $w_1 = 3$

11.5 Conclusion

We analyzed the motion of ϕ^4 kinks in 2D and higher dimensions for a the conservative wave equation and a reaction-diffusion equation.

For the conservative equation, radial kinks collapse to zero. The onset of the collapse is completely controlled by the energy. A simple argument based on the Hamiltonian gives perfect agreement with the numerics. The outcome of this collision depends on the dimension, and whether we consider sG or ϕ^4. The dynamics of such ballistic kinks in a slowly varying waveguide is well described by a simple energy argument. This will predict, for example, the reflection of a kink that does not have enough energy to cross a junction.

Dissipative kinks have a fixed velocity and orientation. For example, 1-0 radial kinks will expand while 0-1 kinks will contract. Using collective coordinates, in 1D, we calculated how a localized defect changes the kink position and width. This predicts pinning of wide kinks on narrow defects. Numerical experiments show that a 2D kink can be pinned by an abrupt widening of the waveguide. We carried out an approximate analysis and it gives the values of the width for which this can happen.

These equations are very interesting as they are simple yet sufficiently general to yield non-trivial effects. Among the perspectives, I would like to mention the effect of an external potential on the waves. In [7], we introduced a potential to stabilize locally the 0 state. It would be interesting to distribute such a potential over the domain and examine the solutions. Another point is that the solutions we have described in the conservative case are transient. It would be interesting to drive the system and examine how to create such solutions.

Acknowledgements The author thanks the CRIANN computing center for the use of its facilities.

References

1. R.K. Dodd, J.C. Eilbeck, J.D. Gibbon, H.C. Morris, *Solitons and Nonlinear Wave Equations* (Academic Press, New York, 1984)
2. A.C. Scott, *Nonlinear Science, Emergence and Dynamics of Coherent Structures* (Oxford University Press, Oxford, 2003)
3. H. Berestycki, J. Bouhours, G. Chapuisat, Calc. Var. Partial Differ. Equ. **55**, 55 (2016)
4. D.K. Campbell, J.F. Schonfeld, C.A. Wingate, Phys. D **9**, 1 (1983)
5. J.-G. Caputo, N. Flytzanis, Phys. Rev. A **44**, 6219 (1991)
6. J.-G. Caputo, M.P. Soerensen, Phys. Rev. E **88**, 022915 (2013)
7. P.G. Kevrekidis, I. Danaila, J.-G. Caputo, R. Carretero, arXiv:1808.02928
8. P.L. Christiansen, O.H. Olsen, Phys. Scr. **20**, 531–538 (1979)
9. P.L. Christiansen, O.H. Olsen, Phys. Lett. A **68**, 185 (1978)
10. P.S. Lomdahl, O.H. Olsen, P.L. Christiansen, Phys. Lett. A **78**, 125 (1980)
11. P.L. Christiansen, P.S. Lomdahl, Phys. D **2**, 482 (1981)
12. M.R. Samuelsen, Phys. Lett. A **74**, 21 (1979)
13. J. Geicke, Phys. Lett. A **98**, 147 (1983)
14. I.L. Bogolubsky, V.G. Makhankov, JETP Lett. **24**, 12 (1976)
15. B. Piette, W.J. Zakrzewski, Nonlinearity **11**, 1103 (1998)
16. J.-G. Caputo, B. Sarels, Phys. Rev. E **84**, 041108 (2011)
17. A. Barone, G. Paterno, *Physics and Applications of the Josephson Effect* (Wiley, New York, 1982)
18. A. Benabdallah, J.-G. Caputo, A.C. Scott, Phys. Rev. B **54**, 16139 (1996)
19. C. Gorria, Yu.B. Gaididei, M.P. Soerensen, P.L. Christiansen, J.-G. Caputo, Phys. Rev. B **69**, 134506 (2004)
20. G. Carapella, N. Martuciello, G. Costabile, Phys. Rev. B **66**, 134531 (2002)
21. J.-G. Caputo, D. Dutykh, Phys. Rev. E **90**, 022912 (2014)
22. D.R. Gulevich, S. Savelev, V.A. Yampolskii, F.V. Kusmartsev, F. Nori, J. Appl. Phys. **104**, 064507 (2008)
23. Y.B. Zeldovich, D.A. Frank-Kamenetsky, Dokl. Akad. Nauk. SSSR **19**, 693 (1938)
24. J. Xin, SIAM Rev. **42**, 161 (2000)
25. D. Dalmas, E. Barthel, D. Vandembroucq, J. Mech. Phys. Solids **57**, 446 (2009)
26. F. Diovatelli, L. Biferale, S. Chibbaro, A. Puglisi, S. Succi, Phys. Rev. E **78**, 036305 (2008)
27. J.D. Murray, *Mathematical Biology II: Spatial Models and Biomedical Applications*, 3rd edn. (Springer, Heidelberg, Berlin, 2001)
28. J. Gaudart, O. Touré, N. Dessay, A. Dicko, S. Ranque, L. Forest, J. Demongeot, O.K. Doumbo, Malar. J. **8**, 61 (2009)
29. J. Bouhours, Ph.D. Thesis, Université Pierre et Marie Curie - Paris VI, 2014. https://tel.archives-ouvertes.fr/tel-01070608

Chapter 12
Higher-Order Field Theories: ϕ^6, ϕ^8 and Beyond

Avadh Saxena, Ivan C. Christov and Avinash Khare

Abstract The ϕ^4 model has been the "workhorse" of the classical Ginzburg–Landau phenomenological theory of phase transitions and, furthermore, the foundation for a large amount of the now-classical developments in nonlinear science. However, the ϕ^4 model, in its usual variant (symmetric double-well potential), can only possess two equilibria. Many complex physical systems possess more than two equilibria and, moreover, the number of equilibria can change as a system parameter (e.g., the temperature in condensed matter physics) is varied. Thus, "higher-order field theories" come into play. This chapter discusses recent developments of higher-order field theories, specifically the ϕ^6, ϕ^8 models and beyond. We first establish their context in the Ginzburg–Landau theory of successive phase transitions, including a detailed discussion of the symmetric triple well ϕ^6 potential and its properties. We also note connections between field theories in high-energy physics (e.g., "bag models" of quarks within hadrons) and parametric (deformed) ϕ^6 models. We briefly mention a few salient points about even-higher-order field theories of the ϕ^8, ϕ^{10}, etc. varieties, including the existence of kinks with power-law tail asymptotics that give rise to long-range interactions. Finally, we conclude with a set of open problems in the context of higher-order scalar fields theories.

A. Saxena
Los Alamos National Laboratory, Theoretical Division and Center for Nonlinear Studies, Los Alamos, NM 87545, USA
e-mail: avadh@lanl.gov

I. C. Christov (✉)
School of Mechanical Engineering, Purdue University, West Lafayette, IN 47907, USA
e-mail: christov@purdue.edu

A. Khare
Department of Physics, Savitribai Phule Pune University, Pune 411007, India
e-mail: khare@physics.unipune.ac.in

P. G. Kevrekidis and J. Cuevas-Maraver (eds.), *A Dynamical Perspective on the ϕ^4 Model*, Nonlinear Systems and Complexity 26,
https://doi.org/10.1007/978-3-030-11839-6_12

12.1 Introduction

The mathematical setting of this chapter is $(1+1)D$ field theories. That is to say, we consider a generic spatiotemporal field, $\phi = \phi(x,t)$ (although, in further sections symbol ϕ may be replaced by the symbol u depending on the context), and its concordant governing evolution equation. Within the context of classical neutral scalar field theories, the evolution of ϕ is determined by a PDE that extremizes the *action* functional (in some appropriate "natural" dimensionless units):

$$S[\phi] = \int \mathrm{d}t \int \mathrm{d}x \left[\frac{1}{2}\phi_t^2 - \frac{1}{2}\phi_x^2 - V(\phi)\right], \tag{12.1}$$

where x and t subscripts henceforth denote partial differentiation. The three terms on the RHS of (12.1) denote, respectively, the kinetic energy of the field, the (opposite of the) potential energy within the field, and V is some to-be-specified term quantifying the field's *self-interaction*. Specifically, we call V "the potential" of the field theory, and it sets the dynamics of the field. As is convention, we term a field theory with a polynomial potential V of degree n as a "ϕ^n field theory." The potential is constructed, derived, modeled or conjectured on the basis of the physical behavior of the system under consideration. We refer the reader to [1–3] for more detailed textbook introductions to the subject, including physical examples of various field theories in high-energy theoretical physics.

The Euler–Lagrange equation extremizing the action S from (12.1) is easily found to be the nonlinear wave (often referred to as KG-type) equation

$$\phi_{tt} - \phi_{xx} + V'(\phi) = 0\,, \tag{12.2}$$

where a prime denotes differentiation with respect to the argument of the function, here ϕ. Note that, on the basis of the Lorentz invariance of (12.2), in this chapter we are only concerned with its static solutions, i.e., $\phi_t = 0$ and $\phi = \phi(x)$ only, with traveling solutions being obtainable from the latter by a boost transformation. Hence, we are interested in how the solutions of the ODE

$$\phi_{xx} = V'(\phi) \tag{12.3}$$

are affected by the choice of V. Specifically, we will (mostly) discuss "simple" polynomials of even degree that possess $\phi \mapsto -\phi$ symmetry (thus endowing the field theory with reflectional, or Z_2, symmetry) as choices for the potential.

In this chapter, we discuss the kink (i.e., domain wall) solutions of (12.3) and their context in the hierarchy of various higher-order field theories, where by "higher-order" we either mean that the potential V is a polynomial of degree greater than four or is non-polynomial. For brevity and clarity, we will often refer the reader to the encyclopedic study of the eighth, tenth- and twelfth-degree field theories (and their kink solutions in the presence of degenerate minima), provided in [4].

12.2 First- and Second-Order Phase Transitions: The Need for Higher-Order Field Theory

The quartic, ϕ^4, potential is the "workhorse" of the Ginzburg–Landau (phenomenological) theory of superconductivity [5–7], taking ϕ as the *order parameter* of the theory (i.e., the macroscopic wave function of the condensed phase). In this context, the third term (i.e., $V(\phi)$) in (12.1) is interpreted as the Landau free energy density, while the combination of the second and third terms (i.e., $\frac{1}{2}\phi_x^2 + V(\phi)$) is the full Ginzburg–Landau free energy density, which allows for domain walls of non-vanishing width and energy to exist between various phases (corresponding to equilibria, i.e., minima of V) in the system. Specifically, a prototypical example of the *continuous* (or second-order) phase transition can be modeled by the classical ϕ^4 (double well) field theory.

To illustrate the second-order phase transition, consider the symmetric quartic (double well) potential

$$V(\phi) = \frac{1}{4}\phi^4 - \frac{\alpha_2}{2}\phi^2 + \frac{1}{4}, \tag{12.4}$$

where α_2 is a parameter that might depend on, e.g., the temperature or pressure of the system in condensed matter physics or the mass of a meson in high-energy theoretical physics. As the temperature or pressure of the system changes, so does α_2, leading to structural changes of the potential in (12.4), as shown in Fig. 12.1a. Note that at $\alpha_2 = 1$, (12.4) can be rewritten as $V(\phi) = \frac{1}{4}(\phi^2 - 1)^2$. Specifically, the *global* minima of this potential, i.e., ϕ_0 such that $V'(\phi_0) = 0$ and $V''(\phi_0) > 0$, are

$$\phi_0 = \left\{\pm\sqrt{\alpha_2}\right\} \quad (\alpha_2 > 0), \tag{12.5}$$

while $\phi_0 = 0$ is a global maximum. As $\alpha_2 \to 0^+$, these two minima smoothly coalesce into a single *global* minimum at $\phi_0 = 0$ ($\alpha_2 \le 0$), as shown in Fig. 12.1c. This smooth process is characteristic of the continuous, i.e., second-order, phase transitions, and is the only type of bifurcation of equilibria that a symmetric double well ϕ^4 potential can exhibit. For $\alpha_2 = 1$ both degenerate minima of the potential satisfy $V(\phi_0) = V'(\phi_0) = 0$, and a domain wall, which solves (12.3) and (12.4), exists connecting the two phases ($\phi_0 = -1$ and $\phi_0 = +1$):

$$\phi_K(x) = \tanh\left(x\middle/\sqrt{2}\right). \tag{12.6}$$

This well-known domain wall, or *kink*, solution [1–3] is illustrated in Fig. 12.1e.

However, in materials science and condensed matter physics, one also observes *discontinuous* (or first-order) phase transitions, or even *successive* series of first-and/or second-order phase transitions. How can those be modeled? One approach is to add degrees of freedom by increasing the degree of the potential V to greater-than-fourth order [8]. For example, first-order transitions can be modeled by sextic, ϕ^6, field theory [9–11]. A triple well potential characteristic of a ϕ^6 field theory naturally arises as a one-dimensional cross-section across a path of strain space passing through the austenite well and two of the martensite wells of the free energy of a two-phase martensitic material with cubic and tetragonal phases [12, Sect. 5.5]. Although another possibility to model a first-order transition is by way of an *asymmetric* double well ϕ^4 potential (e.g., a double well potential in an external field) [13], here we restrict ourselves to symmetric potentials only. Then, in order to capture two or more successive transitions, we must go beyond the ϕ^4 and ϕ^6 field theories to even higher orders [8, 14]. Similarly, higher-order field theories arise in the context of high-energy physics, wherein the availability of more than two equilibria leads to more types of mesons [15, 16], which is indeed necessary for certain nuclear and particle physics models.

To illustrate the first-order phase transition, consider the sextic (triple well) potential

$$V(\phi) = \frac{1}{2}\phi^6 - \phi^4 + \frac{\alpha_2}{2}\phi^2 , \tag{12.7}$$

where α_2 is again a parameter that might depend on, e.g., the temperature or pressure of the system. Varying α_2 leads to structural changes of the potential in (12.7), as shown in Fig. 12.1b. Note that at $\alpha_2 = 1$, (12.7) can be rewritten as $V(\phi) = \frac{1}{2}\phi^2(\phi^2 - 1)^2$. Specifically, the minima of this potential are

$$\phi_0 = 0, \quad \phi_0 = \left\{ \pm \frac{\sqrt{\sqrt{4 - 3\alpha_2} + 2}}{\sqrt{3}} \right\} \quad (\alpha_2 < 4/3) . \tag{12.8}$$

For $\alpha_2 > 1$, the two non-zero minima are *local*, while $\phi_0 = 0$ is the *global* minimum; vice versa for $\alpha_2 < 1$. At $\alpha_2 = 1$, the exchange of global minima is *sudden*, i.e., the global minima at $|\phi_0| \neq 0$ do not coalesce with the one at $\phi_0 = 0$ as in the ϕ^4 example above. This non-smooth process is characteristic of discontinuous, i.e., first-order, phase transitions. In Fig. 12.1d, dashed and solid curves denote the local and global minima values of ϕ_0, respectively.) For $\alpha_2 = 1$ all three minima of the potential become degenerate and satisfy $V(\phi_0) = V'(\phi_0) = 0$, thus domain wall (kink) solutions, which satisfy (12.3) and (12.7), exist connecting a pair of equilibria (either $\phi_0 = -1$ and $\phi_0 = 0$ or $\phi_0 = 0$ and $\phi_0 = +1$):

$$\phi_K(x) = \begin{cases} -\dfrac{1}{\sqrt{e^{2x} + 1}} = -\sqrt{\dfrac{1 - \tanh x}{2}} , & \text{half-kink from } -1 \text{ to } 0, \\[2ex] \dfrac{1}{\sqrt{e^{-2x} + 1}} = \sqrt{\dfrac{1 + \tanh x}{2}} , & \text{half-kink from } 0 \text{ to } +1, \end{cases} \tag{12.9}$$

as illustrated in Fig. 12.1f. Equation (12.9) represents the well-known ϕ^6 *half-kink* [17]. Finally, as $\alpha_2 \to (4/3)^-$, the two non-zero minima disappear entirely (once again suddenly) leaving a single, global minimum at $\phi_0 = 0$, as shown in Fig. 12.1d.

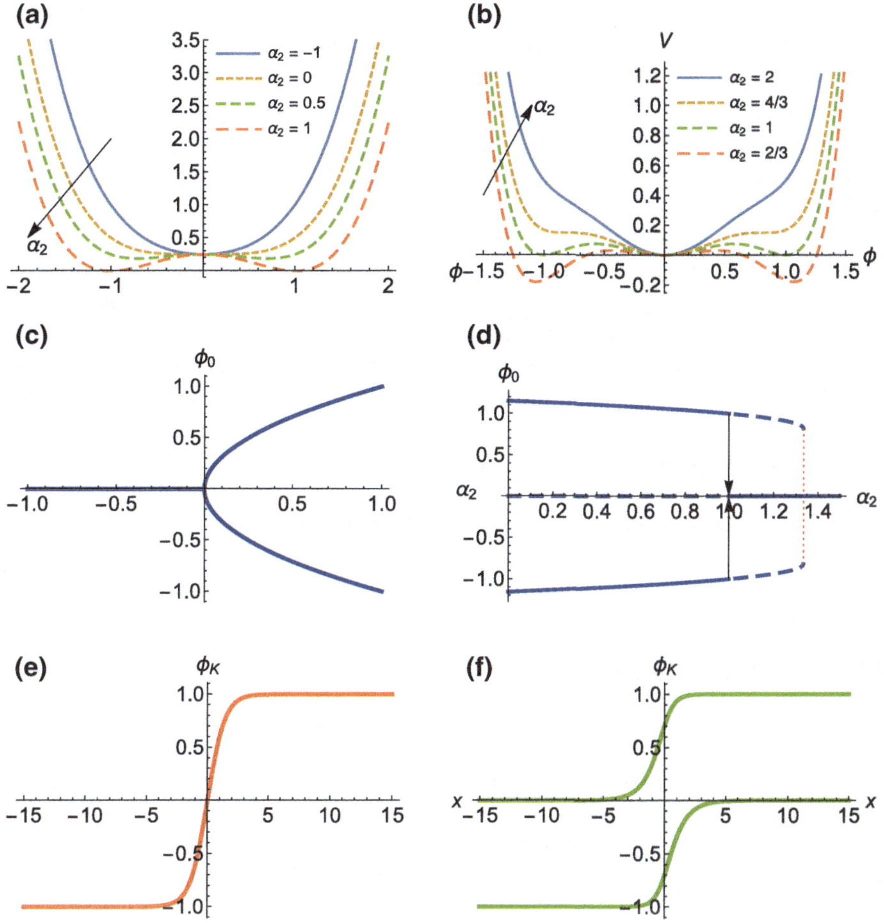

Fig. 12.1 Illustrations of **a** a continuous (second-order) phase transition under the ϕ^4 potential, (12.4) and **b** a discontinuous (first-order) phase transition under the ϕ^6 potential, (12.7). In the literature it is common to shift $V(\phi) \mapsto V(\phi) - V_0(\alpha_2)$ to ensure that the kink solution in (12.6) exists for each α_2 at which there are two degenerate minima, but we have chosen not to do that here in order to not overcomplicate the picture. Panels **c** and **d** show the respective bifurcations in the minima ϕ_0 such that $V'(\phi_0) = 0$ and $V''(\phi_0) > 0$ as a function of the parameter α_2; dashed and solid curves denote the local and global minima, respectively. Panels **e** and **f** show the kink and half-kink solutions, (12.6) and (12.9), respectively, at the critical value $\alpha_2 = 1$ when the potentials in (**a**) and (**b**) have two or three degenerate minima

Beyond these two introductory examples of a second- and a first-order phase transition, similar reasoning can be applied to show that an octic, ϕ^8, field theory can model a second-order transition followed by a first-order transition [4, 8, 18, 19]. Meanwhile, two successive first-order transitions can be modeled by a ϕ^{10} field theory [4, 20]. But, to describe three successive (first- and/or second-order) transitions one must resort to a ϕ^{12} field theory [4, 21, 22]. Continuing in the same vein, for four or more successive transitions, a ϕ^{14} or higher-order (e.g., ϕ^{4n} or ϕ^{4n+2} with $n > 3$) field theory must be employed. So far we have implicitly assumed that, as stated at the beginning of the chapter, we deal with neutral scalar (single-component) field theories. Beyond the scope of this chapter but also relevant is the fact that *multi-component* ϕ^4 or ϕ^6 field theories can also describe successive phase transitions [23].

Higher-order (specifically, higher than sextic) field theories are also needed to capture *all* symmetry-allowed phases in a transition [18, 19]. Certain crystals undergo two successive ferroelastic (i.e., strain as the order parameter) or ferroelectric (i.e., electric polarization as the order parameter) first-order transitions [20]. In particle physics massless mesons interacting via long-range forces are modeled with the ϕ^8 field theory [15]. Additionally, there are examples of isostructural transitions (i.e., the crystal symmetry does not change but the lattice constant changes), which can be described by the ϕ^8 field theory [24]. In biophysics, chiral protein crystallization is modeled via a ϕ^{10} field theory [25]. Similarly, the transitions in certain piezoelectric (i.e., stress-induced polarization) materials with perovskite structure are modeled by the ϕ^{12} field theory [21, 22].

12.3 ϕ^6 Field Theory

As we have just discussed in Sect. 12.2, the location of the global minimum of a triple well ϕ^6 potential abruptly (discontinuously) jumps from $\phi_0 = 0$ to a pair of finite values $|\phi_0| \neq 0$ through the phase transition (as α_2 goes through 1 in the example of (12.7) above). At the phase transition point, the potential has three global minima. This type of phase transition is ubiquitous in nature: from cosmological transitions in the early Universe [26] to solid-solid transformations from one crystal structure to another [18, 19]. Here, it is relevant to mention the significance of the latter from the thermodynamics point of view (see also [27, Sect. 5]), i.e. when the field $\phi(x,t)$ possesses a large number of kinks driven by white noise and balanced by dissipation. At such a discontinuous (first-order) phase transition, described by the ϕ^6 field theory, we expect that the field's self-correlation function will yield finite correlation lengths at the transition temperature, which is associated with latent heat in classical thermodynamics [28].

12.3.1 Exact Kink and Periodic Solutions, Asymptotic Kink Interaction

In the case of three degenerate minima, a ϕ^6 potential can always be factored into the form $\phi^2(\phi^2 - a^2)^2$, up to scaling factors, and then the exact domain-wall solutions are the half-kinks in (12.9). For different values of α_2, different types of exact nonlinear waveforms exist near the "wells" of a ϕ^6 potential [29].

To illustrate these ideas, let us return to (12.3). Multiplying by ϕ_x and forming a complete differential, we may integrate both sides to get the *first integral of motion*:

$$(\phi_x)^2 = 2[V(\phi) - \mathfrak{C}], \tag{12.10}$$

where $\mathfrak{C}$ is a constant of integration. Assuming that $\phi \to \phi_0$ smoothly as $|x| \to \infty$, where ϕ_0 is a degenerate minimum of V such that $V(\phi_0) = V'(\phi_0) = 0$ and $V''(\phi_0) > 0$, fixes the integration constant to $\mathfrak{C} = 0$. Then, a second integration, taking V to be as in (12.7) with $\alpha_2 = 1$, leads to the solution $\phi_K(x)$ in (12.9).

But, what if we do not apply the approach to equilibrium as a boundary condition? Then, what happens when $\mathfrak{C} \neq 0$? To understand this case, note that we may still separate variables in the first-order ODE (12.10) to get the implicit relation:

$$x = \int \frac{\mathrm{d}\phi}{\sqrt{2[V(\phi) - \mathfrak{C}]}}, \tag{12.11}$$

where we have still to specify the limits of integration (hence, no need for a second constant of integration). For clarity, we restrict ourselves to the positive root in (12.11). As discussed in [11, 29], by picking the integration limits to be consecutive zeros of $V - \mathfrak{C}$, with a maximum of V in between, the RHS of (12.11) becomes an *elliptic integral* [30] in the variable $\varphi = \phi^2$, and *Jacobi's elliptic functions* [31] (see also Sect. 5.3) can be used to solve for ϕ. This sets a range of physically admissible choices for $\mathfrak{C}$, namely those between maxima and minima values of $V(\phi)$.

Figure 12.2b summarizes visually these so-called *kink lattice* solutions obtained in [29] by performing the integration in (12.11) with $V(\phi) = \frac{1}{6}\phi^6 - \frac{1}{4}\phi^4 + \frac{\alpha_2}{2}\phi^2$ and inverting the expression in terms of the Jacobi elliptic functions sn and dn:

$$\phi_{KL,1}(x) = \frac{\phi_1}{\sqrt{1 - A^2\,\mathrm{sn}^2(\beta x \mid m)}}, \tag{12.12a}$$

$$\phi_{KL,2}(x) = \frac{\phi_2\,\mathrm{dn}(\beta x, m)}{\sqrt{1 - B^2\,\mathrm{sn}^2(\beta x \mid m)}}, \tag{12.12b}$$

$$\phi_{KL,3}(x) = \frac{\phi_1 C\,\mathrm{sn}(\bar{\beta} x, \bar{m})}{\sqrt{1 - C^2\,\mathrm{sn}^2(\bar{\beta} x \mid \bar{m})}}, \tag{12.12c}$$

where the elliptic moduli $m, \bar{m} \in [0, 1]$, and the constants A, B, C, β and $\bar{\beta}$ are related to the roots $\phi_{1,2,3}$, satisfying $V(\phi_{1,2,3}) = \mathfrak{C}$ (see Fig. 12.2a), as

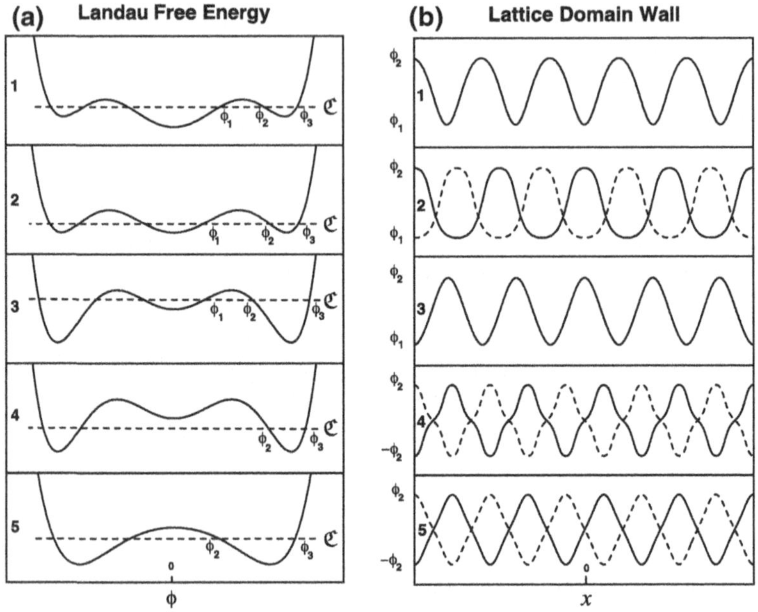

Fig. 12.2 **a** A ϕ^6 potential (i.e., the Landau free energy density of the field theory) at different temperatures (panels 1–5); specifically $V(\phi) = \frac{1}{6}\phi^6 - \frac{1}{4}\phi^4 + \frac{\alpha_2}{2}\phi^2$ for different α_2. $\mathfrak{C}$ is the constant of integration in (12.10), and $\pm\phi_{1,2,3}$ are the solutions to $V(\phi) = \mathfrak{C}$. **b** Illustrations of the corresponding kink lattices: case 1 [$\frac{3}{16} < \alpha_2 < \frac{1}{4}$, $\phi_{KL,1}(x)$], case 2 [$\alpha_2 = \frac{3}{16}$, $\phi_{KL,1}(x)$ as dashed and $\phi_{KL,2}(x)$ as solid], case 3 [$0 < \alpha_2 < \frac{3}{16}$, $\phi_{KL,2}(x)$], cases 4–5 [$\alpha_2 < \frac{3}{16}$ or $\alpha_2 < 0$ but $\pm i\phi_1$ is now a pair of imaginary solutions, $\pm\phi_{KL,3}(x)$ as dashed and solid]. From [M. Sanati and A. Saxena, "Half-kink lattice solution of the ϕ^6 model," J. Phys. A: Math. Gen. 32, 4311–4320 (1999)] © IOP Publishing. Reproduced with permission. All rights reserved

$$A^2 = \frac{\phi_2^2 - \phi_1^2}{\phi_2^2}, \quad B^2 = \frac{\phi_2^2 - \phi_1^2}{\phi_2^3 - \phi_1^2}, \quad C^2 = \frac{\phi_2^2}{\phi_1^2 + \phi_2^2}, \quad \beta^2 = \tfrac{1}{3}\phi_2^2(\phi_3^2 - \phi_1^2),$$
$$\bar{\beta} = \tfrac{1}{3}\phi_3^2(\phi_2^2 + \phi_1^2), \quad m = \frac{\phi_3^2(\phi_2^2 - \phi_1^2)}{\phi_2^2(\phi_3^2 - \phi_1^2)}, \quad \bar{m} = \frac{\phi_2^2(\phi_3^3 + \phi_1^2)}{\phi_3^2(\phi_2^2 + \phi_1^2)}. \tag{12.13}$$

The solutions in (12.12) are exact and *periodic* with period $4K(m)/\beta$, where $K(m)$ is the complete elliptic integral of the first kind [31]. In cases 2, 4 and 5 of Fig. 12.2, these periodic solutions reduce to distinct kinks in the limit of $m \to 1$ (or $\bar{m} \to 1$, as the case might be).

Finally, the asymptotic interaction force between ϕ^6 kinks can be obtained in the usual way via Manton's method [2, 32], from the exponential tail asymptotics as shown in e.g. [29] (the result is also mentioned in [33]).

12.3.2 Linearization About a Kink (Internal Modes) and Linearization About an Equilibrium (Phonon Modes)

Linearizing the field theory about a kink solution, i.e. substituting $\phi(x,t) = \phi_K(x) + \delta e^{i\omega_i t}\chi_i(x) + \text{c.c.}$ into (12.2) and keeping terms up to $O(\delta)$, yields a standard Schrödinger-type eigenvalue problem [27]:

$$\left[-\frac{d^2}{dx^2} + V''\big(\phi_K(x)\big)\right]\chi_i = \omega_i^2 \chi_i, \tag{12.14}$$

where ω_i is the temporal oscillation frequency of the i-th linearization mode, and $\chi_i(x)$ is the eigenfunction giving this mode's spatial structure. The traditional ϕ^4 symmetric double well's kink, e.g. as in (12.6), possesses a translational mode, $\omega_0 = 0$, and an internal mode at an isolated eigenvalue at $\omega_1 = \sqrt{3/2}$ (continuous spectrum begins at $\omega = \sqrt{2}$). In fact, it is even possible to write down $\chi_1(x)$ analytically [27]. Meanwhile, the standard ϕ^6 symmetric triple well's half-kink, e.g. as in (12.9), *does not* possess such an internal mode [17, 33].

However, this issue of whether a single translational mode exists or not, is hardly the whole story in higher-order field theories. As we discuss below, there are ϕ^6 models with *controllably* many internal modes. Meanwhile, much like the "classical" ϕ^4 and ϕ^6 pictures, for a ϕ^8 field theory with four degenerate minima, specifically $V(\phi) = (\phi^2 - a^2)^2(\phi^2 - b^2)$ with $a, b = (\sqrt{3} \mp 1)/2$, which has both full- and half-kink solutions [4], the full-kink has an internal mode ($\omega_1 \approx 1.645$) [34] while the half-kink does not [35]. The possibility of power-law (as opposed to exponential) tails in higher-order field theories adds further complications. The kink of the ϕ^8 model with two degenerate minima, specifically $V(\phi) = (\phi^2 - a^2)^2(\phi^2 - b^2)$ with $a = 4/5$ and $b = 1$, is reported to possess three internal modes ($\omega_1 \approx 2.068$, $\omega_2 \approx 3.192$ and $\omega_3 \approx 3.689$) [35], while kink solutions with power-law tails of other sextic and non-polynomial models possess only the zero mode [36].

Meanwhile, phonon modes, i.e. linear excitations about an equilibrium state ϕ_0, are a simpler matter. Linearizing the field theory about a minimum of V, i.e., substituting $\phi(x,t) = \phi_0 + \delta e^{i(qx - \omega_q t)} + \text{c.c.}$ into (12.2) and keeping terms up to $O(\delta)$, yields

$$\omega_q^2 - q^2 = V''(\phi_0), \tag{12.15}$$

for a phonon mode with temporal frequency ω_q and spatial wave number q. For the example triple well ϕ^6 potential in (12.7), we have $V''(\phi_0) = 15\phi_0^4 - 12\phi_0^2 + \alpha_2$. Substituting the equilibria ϕ_0 from (12.8) into the latter gives us

$$V''(\phi_0) = \left\{\alpha_2,\ \frac{4}{3}\left(2\sqrt{4-3\alpha_2} + 4 - 3\alpha_2\right),\ \frac{4}{3}\left(2\sqrt{4-3\alpha_2} + 4 - 3\alpha_2\right)\right\}, \tag{12.16}$$

where the second and third values, obviously, hold only for $\alpha_2 \le 4/3$ (i.e., as long as those minima exist). In particular, at $\alpha_2 = 1$ for the case of three degenerate minima, we have $V''(\phi_0) = \{1, 4, 4\}$. Since in all cases we have $V''(\phi_0) \neq 0$, then our model ϕ^6 field theory has well-defined phonon modes along an optical branch (i.e., $\omega_q \nrightarrow 0$ as $|q| \to 0$). On the other hand, in certain special cases of higher-than-sixth order field theories (e.g., ϕ^8), a degeneracy occurs and $V''(\phi_0) = 0$, leading to the possibility of *nonlinear phonons*. Nonlinear (or anharmonic) phonons, represent large field excursions of oscillations around the minima of the potential (but do not go over adjacent barriers). Then, in such a case, more terms must be kept in the linearization beyond the vanishing $V''(\phi_0)$ term.

Finally, we note that by Weyl's theorem, the dispersion relation given by (12.15) describes the continuous spectrum for *both* linearization about a uniform equilibrium and for linearization about a coherent structure such as a kink.

12.3.3 Collisional Dynamics of ϕ^6 Kinks and Multikinks

Chapters 2 and 3 discuss the "classical picture" of kink–antikink collisions in the ϕ^4 model as developed/described in the large body of work emanating from [37–39]. In particular, Chap. 3 discusses some of the recently uncovered twists in this classical picture, as far as the collective-coordinate approach is concerned, and how to resolve them. Chapter 4 further delves into the notions of fractal structures in the resonance windows and the finer details of their study under the collective-coordinates (variational) approximation. Thus, in this subsection we simply mention one of the more salient aspects of studying kink collisions in higher-order field theories. Specifically, the availability of multiple stable equilibria in the system, which allows for the existence of half-kinks (recall Fig. 12.1f), opens the possibility of studying collisions between kinks each connecting a *different* pair of equilibria (also called "topological sectors"). Whereas in the prototypical ϕ^4 field theory under the potential in (12.4) (with $\alpha_2 = 1$) we only have a kink (given in (12.6)) connecting -1 to $+1$ or antikink connecting $+1$ to -1, in the example ϕ^6 field theory under the potential in (12.7) (with $\alpha_2 = 1$) we have *two* half-kinks (given in (12.9)) and their corresponding antikinks. Clearly, this key difference between the ϕ^4 and ϕ^6 models gives rise to a potentially far richer phenomenology of kink-kink and kink-antikink collisions.

For example, the collisional dynamics of a "staircase" half-kink+half-kink ansatz, which is formed by superimposing the half-kink from -1 to 0 onto the half-kink from 0 to $+1$, suitably well separated as shown in Fig. 12.3a, were studied by the classical collective coordinate approach in [40], with an updated treatment (resolving certain quantitative discrepancies) given in [41, 42]. These types of kink + kink collisions are obviously not possible in the ϕ^4 model, where one typically studies kink–antikink collisions only. The ϕ^6 collisional phenomenology is, thus, more subtle. Further explorations of multikink configurations, meaning various superpositions of half-kinks in some prescribed arrangements, were presented in [43].

A related possibility in ϕ^6 field theories is exact kink solutions that look like a "bound pair" of kinks (see Fig. 12.3b), similar to the "staircase" kink in Fig. 12.3a

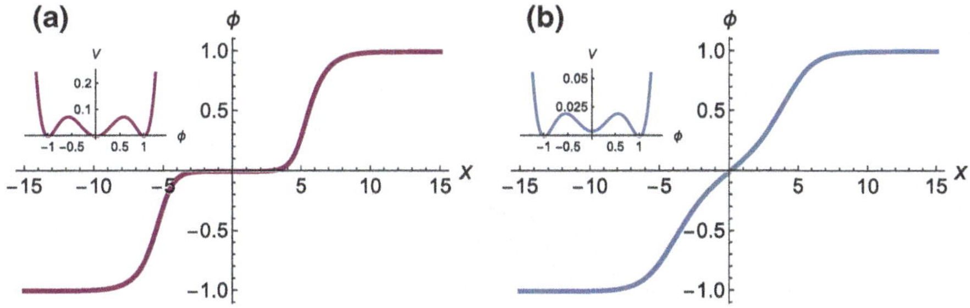

Fig. 12.3 Two types of "staircase" multikink-type ansätze studied in the literature. **a** The example ϕ^6 field theory, (12.7) at $\alpha_2 = 1$ exhibiting three degenerate minima (see inset), allowing for the superposition of two well-separated half-kinks from (12.9). **b** The Christ–Lee model, (12.17) at $\epsilon = 0$ exhibiting two degenerate minima (see inset and note the middle, non-degenerate minimum "lifting off" from the origin) and a "bound pair" exact kink solution given in (12.18)

discussed above. Such kinks can be found in the parametric ϕ^6 model introduced by Christ and Lee [16]; specifically, an example potential (fixing some of the extra parameters from [16]) of this form is

$$V(\phi) = \frac{1}{8(1+\epsilon^2)}(\phi^2 + \epsilon^2)(\phi^2 - 1)^2 . \tag{12.17}$$

The corresponding exact kink solution to (12.3) and (12.17) (see also [44]) is

$$\phi_K(x) = \frac{\epsilon \sinh(x/2)}{\sqrt{1 + \epsilon^2 + [\epsilon \sinh(x/2)]^2}} . \tag{12.18}$$

Notice that as $\epsilon \to 0^+$ or $\epsilon \to \infty$, the potential in (12.17) takes the form, respectively, of the prototypical triple-well ϕ^6 potential (i.e., (12.7) with $\alpha_2 = 1$, suitably normalized) or the prototypical double-well ϕ^4 potential (i.e., (12.4) with $\alpha_2 = 1$, suitably normalized), discussed above. The context of the Christ–Lee model is not condensed matter physics or phase transitions, but rather it was introduced in high-energy theoretical physics as a "bag" model in which the role of quarks within a hadron is played by the domain wall solutions of the field theory.

For the Christ–Lee model, with potential given by (12.17), studying the collisional dynamics of the kink solutions from (12.18) yields highly nontrivial results (as compared to the "classical picture" of ϕ^4 kink–antikink collisions). Specifically, as the parameter ϵ is tuned in the Christ–Lee model, one can *control* the number of internal modes (i.e., non-zero isolated eigenvalues of (12.14)) of the staircase-like kink. Although it has long been posited [37] that the internal mode of the kink's linearization (recall Sect. 12.3.2) sets, to a large extent, the collisional dynamics, recent results using the ϕ^6 model [33] have proposed an additional mechanism unrelated to the internal mode. After the work in [33], it was further shown in [44] that the resonance window structure exhibits quite counterintuitive behaviors as the number

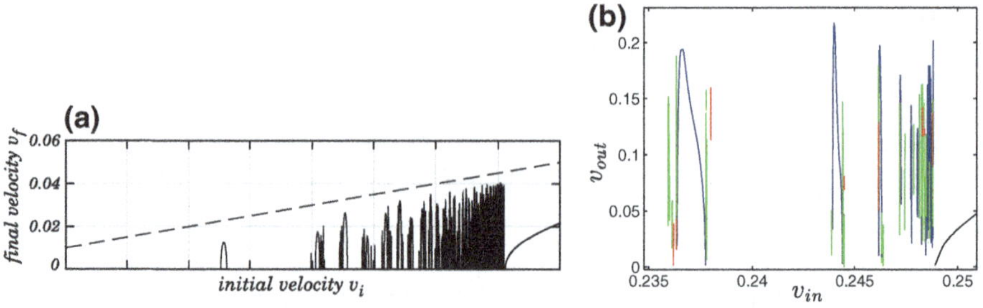

Fig. 12.4 Resonance window maps, based on direct numerical simulation of (12.2), of the final kink velocity (v_{f} or v_{out}) upon a collision of prescribed initial kink velocities (v_{i} or v_{in}). **a** The traditional ϕ^6 model (i.e., (12.7) with $\alpha_2 = 1$) and no internal modes for the half-kink (i.e., (12.9)). **b** The parametric ϕ^6 model (i.e., (12.17) with $\epsilon = 0.5$) and four internal modes for the staircase-like kink (i.e., (12.18)); colors indicate how many bounces it took to escape (black for one, blue for two, green for three, red for four). Panel **a** is reprinted with permission from [Patrick Dorey, Kieran Mersh, Tomasz Romańczukiewicz, and Yasha Shnir, Physical Review Letters, 107, 091602, 2011] [33] © The American Physical Society. Panel **b** is reprinted (without modification) from [44]. © 2017 The Authors of [44], under the CC BY 4.0 license

of internal modes in the Christ–Lee model under (12.17) is tuned. Specifically, this increase in the number of internal modes does *not* lead to more complex resonance structures of ever more multi-bounce windows. Instead, for a wider range of collision velocities, the staircase-like kinks simply scatter elastically off to infinity.

Figure 12.4 shows a comparison between (a) the "classical" ϕ^6 resonance window (kink with one internal mode) structure of kink collisions and (b) the parametric ϕ^6 theory under (12.17) with $\epsilon = 0.5$ (i.e., four internal modes of the staircase-like kink). The study of ϕ^4 kink interactions and resonances is a time-honored subject that has led to elegant demonstrations of Hamiltonian dynamics and even mechanical demonstrations of the two-bounce windows [45]. Following [46], given initially symmetrically located kinks with equal and opposite velocities v_{in}, a direct numerical simulation of (12.2) is performed, colliding the kinks. If they "escape" the collision going off to infinity, the escape velocity v_{out} is recorded and plotted. Clearly, only for some ranges of v_{in} is there a computable v_{out}. The ranges in which v_{out} does not exist are termed *resonance* windows in which the kinks continue to bounce back-and-forth forming a bound pair of sorts. Counterintuitively, the structure of these resonance windows in the absence of an internal mode, Fig. 12.4a, is far more complex than in the presence of four internal modes, Fig. 12.4b. We will not delve into this mystery further here because there remain many open problems about kink interactions in ϕ^6 (and even higher-order) field theories.

12.3.4 Statistical Mechanics of the ϕ^6 Theory, Including Quasi-exactly Solvable Results

Equation (12.3) subject to the ϕ^6 potential, e.g. as given in (12.7), represents a highly anharmonic system. Therefore, the number of nonlinear (e.g., soliton and breather) and linear (e.g. phonons) elementary excitations is thermally controlled. In order to determine the thermal density of these excitations, and their individual contribution to correlation functions (and other thermodynamic quantities such as specific heat and entropy), one must investigate their statistical mechanics. In one dimension, entropy considerations dictate the presence of kinks. Thus, the interactions between kinks and phonons and possibly other excitations play a crucial role in the overall thermodynamics of the process. This question has been of significant interest in condensed matter physics for the past 40 years [27, Sect. 5]. The latter can be studied using a probability density function (PDF), which can be calculated either analytically via the path-integral approximation scheme [47, 48] or numerically by way of Langevin dynamics [49]. In these ways, one can obtain equilibrium properties; and, not just the PDF but also the presence of heterophase fluctuations in the vicinity of a phase transition, the field configuration(s), the average total kink-number density, correlation functions, structure factors, specific heat, internal energy and entropy.

The ϕ^4 model and its attendant kink field have been extensively studied in the literature using techniques such as the path integral formalism [47, 48]. As discussed in Chap. 5, Langevin dynamics were also developed for computing the thermodynamic quantities of a ϕ^4 field theory [50–53]. For higher-order field theories, on the other hand, not much is known beyond the very preliminary results regarding ϕ^6 in [54]. In general, we expect a much richer phenomenology in terms of the possible kink structures and their interactions, under higher-order field theories.

An important departure of the ϕ^6 model (and, indeed, all higher-order field theories of the form ϕ^{4n+2}) from the ϕ^4 model, is that it leads to a *quasi-exactly solvable* (QES) problem [55] for the PDF of the kink field. This result was shown in [9, 56] for ϕ^6, then some further exact PDFs were obtained for ϕ^{10} in [4]. Let us illustrate the basic idea of this approach. Via the path-integral (transfer operator) formalism [47, 48, 50, 51] (see also [57, Sect. 10.5]), one can reduce the statistical mechanics problem of finding the PDF to solving, once again, a Schrödinger-type eigenvalue problem:

$$\left[-\frac{1}{2\beta^2}\frac{\mathrm{d}^2}{\mathrm{d}\phi^2} + V(\phi)\right]\Psi_k = E_k\Psi_k, \tag{12.19}$$

where β is an inverse temperature, (Ψ_k, E_k) is the sought after eigenpair and V is the model potential. For $V(\phi)$ given in (12.7), (12.19) is a well-known QES eigenvalue problem [58]. Specifically, one posits one solution to (12.19) (out of the infinite number of possible ones) in the form

$$\Psi_0(\phi) = \exp\left\{-\frac{1}{2}\phi^2\left(\phi^2 - K\right)\right\}, \tag{12.20}$$

where E_0 and K are still to be determined. Upon substituting (12.20) for the wavefunction $\Psi_0(\phi)$ and (12.7) for $V(\phi)$ into (12.19) and requiring that equality holds, one obtains the consistency conditions:

$$\alpha_2 = -\frac{1}{2}, \quad K = 2, \quad E_0 = -\frac{1}{4}, \quad \beta = 2\,. \tag{12.21}$$

Thus, for the specific ϕ^6 potential in (12.7) with $\alpha_2 = -1/2$ and at the precise (inverse) temperature $\beta = 2$, (12.20) represents an *exact* ground state PDF (i.e., the wavefunction has no nodes) for the ϕ^6 field theory, as long as $E_0 = -1/4$ and $K = 2$. Finally, the PDF for the field is just the normalized squared ground state wave function Ψ_0^2 from (12.20).

This is but one example for an exact solution, many other ansätze that would conceivably lead to further exact PDF solutions are provided in [58], potentially including excited states. The exactness of these solutions (and the accuracy of the path-integral formalism) can subsequently be verified by Langevin simulations [49, 51]. Other examples of QES *non-polynomial* field theories that have both exact kink solutions and quasi-exactly solvable thermodynamics are discussed in [59–61]. Finally, we emphasize once more that the PDF for the ϕ^4 model *can only be obtained numerically* (or approximately using certain Gaussian fits) [50, 51].

12.4 ϕ^8 Field Theory

12.4.1 Successive Phase Transitions

A ϕ^8 field theory can be used to describe a first-order transition followed by a second-order phase transition. That is to say, as the coefficients of the potential are varied, it is possible to observe coalescence (continuously) and global/local exchange (discontinuously) of minima. A comprehensive discussion is given in [4, Sect. II-A]. Let us now illustrate, through Fig. 12.5 and its discussion, how a succession of a first-order and a second-order phase transition can be described using the octic potential

$$V(\phi) = \phi^8 - 4\phi^6 + \frac{9}{2}\phi^4 - \alpha_2\phi^2 + \frac{1}{16}, \tag{12.22}$$

where α_2 is a free parameter that can be varied to observe the successive phase transitions (e.g., it can be considered a function of the system's temperature). The coefficient of ϕ^8 in V can be taken to be as unity, without loss of generality, by an appropriate rescaling of the x-coordinate.

First, note that, as $\alpha_2 \to 0^+$, the potential in (12.22) has an absolute minimum at $\phi_0 = 0$ into which two global minima at

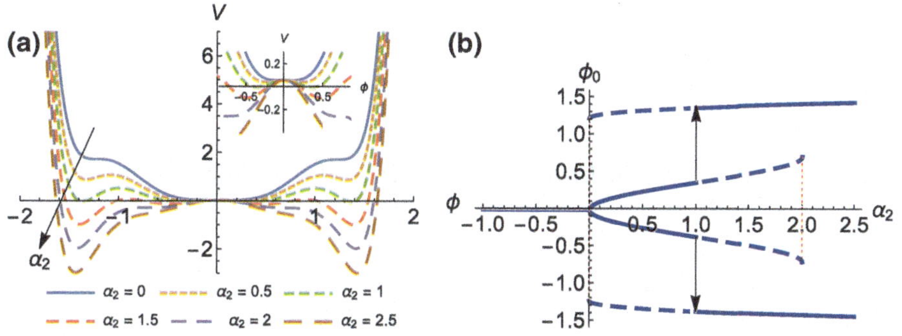

Fig. 12.5 **a** Structure of the ϕ^8 potential in (12.22) for different α_2 (i.e., values of the coefficient of the quadratic term), showing the various phases and phase transitions in this field theory; inset shows zoom near the origin. **b** Bifurcations in the minima ϕ_0 such that $V'(\phi_0) = 0$ and $V''(\phi_0) > 0$ as a function of the parameter α_2; dashed and solid curves denote the local and global minima values, respectively

$$\phi_{0,\text{inner}} = \pm\frac{1}{2}\sqrt{\mathrm{i}\sqrt{3}\left(r - \frac{1}{r}\right) - r - \frac{1}{r} + 4}\,, \quad r = \sqrt[3]{\sqrt{(\alpha_2 - 2)\alpha_2} + \alpha_2 - 1}\,, \tag{12.23}$$

have coalesced. Note that $\phi_{0,\text{inner}}$ are actually real numbers (thus, exist) only for $0 \leq \alpha_2 \leq 2$. Meanwhile the two local minima,

$$\phi_{0,\text{outer}} = \pm\sqrt{1 + \frac{r}{2} + \frac{1}{2r}} \qquad (\alpha_2 \geq 0)\,, \tag{12.24}$$

where r is as given in (12.23), have become the inflection points $\phi_0 = \pm\sqrt{3/2}$ at $\alpha_2 = 0$. This behavior is analogous to the ϕ^4 scenario illustrated in Fig. 12.1a. Hence, $\alpha_2 = 0$ corresponds to a second-order phase transition.

Second, note that for $\alpha_2 = 1$, the potential in (12.22) has four (the maximum number of) degenerate global minima, and can be factored into the form $V(\phi) = (\phi^2 - a^2)^2(\phi^2 - b^2)^2$ with $a = \sqrt{\frac{1}{2}\left(2 - \sqrt{3}\right)}$ and $b = \sqrt{\frac{1}{2}\left(2 + \sqrt{3}\right)}$. As α_2 passes through the value of 1, the inner pair of minima and the outer pair of minima suddenly exchange their local/global nature. Hence, $\alpha_2 = 1$ corresponds to a first-order phase transition temperature of the system. This behavior is analogous to the ϕ^6 scenario illustrated in Fig. 12.1b.

Going further, for $\alpha_2 = 2$, the potential in (12.22) has absolute minima at $\phi_{0,\text{outer}} = \pm\sqrt{2}$, a maximum at $\phi_0 = 0$ and inflection points at $\phi_{0,\text{inner}} = \pm\sqrt{2}/2$. Meanwhile for $1 < \alpha_2 < 2$, the potential in (12.22) has global minima at $\phi_{0,\text{outer}}$, as given by (12.24), local minima at $\phi_{0,\text{inner}}$, as given by (12.23), and three maxima, including one at $\phi_0 = 0$. Then, for $0 < \alpha_2 < 1$, the situation is reversed and the global minima are at $\phi_{0,\text{inner}}$, while the local minima are at $\phi_{0,\text{outer}}$; there are still three maxima, including one at $\phi_0 = 0$. For $\alpha_2 < 0$, the example potential has only a single minimum at $\phi_0 = 0$ and no other extrema.

12.4.2 Exact Kink Solutions: "The Rise of the Power-Law Tails"

A classification and enumeration of kink solutions to ϕ^8 field theories with degenerate minima can be found in [4, Sect. II]. First, we note that, given the extra degrees of freedom, an octic potential can have up to four simultaneous degenerate minima, at the first-order phase transition. In this case a kink *and* a half-kink are possible, each one with different energy [4]. Next, let us now summarize the most salient feature of these kink solutions: the possibility of *algebraic* ("slow") decay of the kinks' shapes $\phi_K(x)$ towards the equilibria ϕ_0 as $|x| \to \infty$, i.e. *power-law* tail asymptotics.

Consider an octic potential with two degenerate minima (equilibria) at $\phi_0 = \pm a$ (i.e., a double-well potential), specifically $V(\phi) = \frac{\lambda^2}{2}(\phi^2 - a^2)^4$, which has an exact, *implicit* kink solution of (12.3) [4, (32)]:

$$x(\phi) = \frac{2a\phi}{\gamma_1(a^2 - \phi^2)} + \frac{1}{\gamma_1} \ln\left(\frac{a+\phi}{a-\phi}\right), \tag{12.25}$$

where $\gamma_1 = 4\lambda a^3$. The implicit relation for $x(\phi)$ in (12.25) can be easily inverted to give $\phi_K(x)$ using, e.g., MATHEMATICA. From (12.25), the asymptotics of the tails of this kink are found to be algebraic (and symmetric) [4, (33)]:

$$\phi_K(x) \simeq \mp a \left(1 \pm \frac{1}{\gamma_1 x}\right), \quad x \to \mp\infty. \tag{12.26}$$

Next, consider an octic potential with three degenerate minima (equilibria) at $\phi_0 = 0, \pm a$ (i.e., a triple-well potential), specifically $V(\phi) = \frac{\lambda^2}{2}\phi^4(\phi^2 - a^2)^2$, which has an exact half-kink solution [4, (23)] of (12.3) given implicitly by

$$x(\phi) = -\frac{2a}{\gamma_2 \phi} + \frac{1}{\gamma_2} \ln\left(\frac{a+\phi}{a-\phi}\right), \tag{12.27}$$

where $\gamma_2 = 2\lambda a^3$ [see also [15, (67)]; but it should be noted that there is a typographical error therein that is evident upon comparison with (12.27)]. From expanding (12.27) perturbatively as $|x| \to \infty$, the "tails" of the kink can be shown to be of mixed algebraic/exponential (asymmetric) type [4, (24)]:

$$\phi_K(x) \simeq a \times \begin{cases} -\dfrac{2}{\gamma_2 x}, & x \to -\infty, \\ 1 - 2\mathrm{e}^{-\gamma_2 x - 2}, & x \to +\infty. \end{cases} \tag{12.28}$$

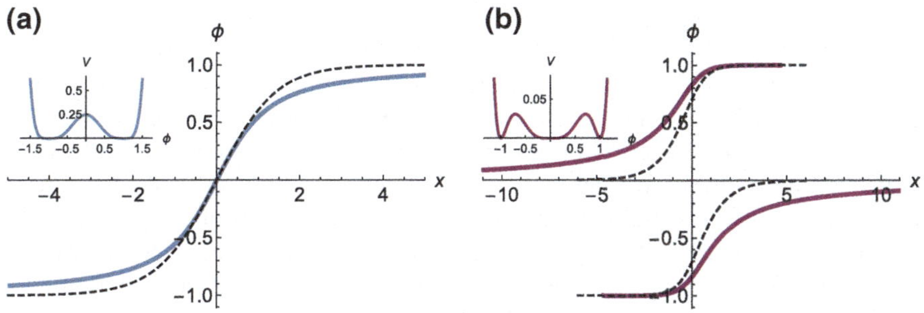

Fig. 12.6 Kink solutions of octic field theories with power-law tail asymptotics. **a** The kink from (12.25) ($a = 1$ and $\lambda = 1/\sqrt{2}$) and the corresponding double well ϕ^8 potential as an inset. To illustrate the slow tail decay, the ϕ^4 kink from (12.6) is superimposed as a dashed curve. **b** The half-kink from (12.27) ($a = 1$ and $\lambda = 1$) and the corresponding triple well ϕ^8 potential as an inset. To illustrate the slow tail decay, the ϕ^6 half-kinks from (12.9) are superimposed as dashed curves

The tail asymptotics highlighted by (12.26) and (12.28) (illustrated in Fig. 12.6) are in *stark* contrast to the exponentially decaying kinks and half-kinks of the ϕ^4 and ϕ^6 models, respectively. Of course, these are not the only examples of double and triple well ϕ^8 potentials. Other cases are discussed in [4, Sect. II], including kink solutions with the "usual" exponential tail asymptotics. Furthermore, the slow (algebraic) decay of the tails is indicative of *long-range* interactions of kinks [62]. It is noteworthy that kinks with algebraic tail asymptotics can also be obtained in certain sextic potentials [36, 63]. Some initial forays into the excitation spectra of kinks with power-law tails (i.e., linearization about a kink, along the lines of Sect. 12.3.2) were presented in [34, 35].

12.4.3 Collisional Dynamics and Interactions of ϕ^8 Kinks

Very little is known about kink collisions under the ϕ^8 (or any higher-order) polynomial field theory, beyond some preliminary results [64, 65]. The main challenge in studying such collisions is that an ansatz of superimposed single-kink solutions must be used as initial conditions. Thus, while cases of kinks with exponential decay may be studied along the same lines as the ϕ^4 and ϕ^6 theories (see Chap. 2 and also recall the discussion and references above), the case of power-law tails is not so simple. In particular, due to the slow algebraic decay of power-law tails, it is neither clear how to quantify the condition of initially "well separated" kinks, nor how to decide the truncation length of the finite computational domain.

For example, even though a ϕ^8 kink–antikink pair appears to show a weakly *repulsive* character under certain discretizations, resonance windows typical of *attractive* interactions are observed [64, 65]. At this time, this counterintuitive result remains poorly understood, and it is not known how the numerical discretization of the slowly decaying tails affects it. Further mysteries (specifically, unexplained quantitative discrepancies) arise when comparing Manton's [2, 32] method for estimating the kink–antikink force of interaction to results from the collective-coordinate approach (see [62] wherein the kink–antikink force of interaction was estimated to decay as the fourth power of their separation). The issue of how to numerically discretize kinks with power-law tails, and how to quantify whether they are "well separated" initially, is equally thorny under the collective-coordinate approach [66].

Our current understanding of this subject is evolving. Recent developments suggest that direct numerical simulation approaches that prepare an initial condition for kink–antikink collisions via "standard" superpositions (summation or product) of kinks and antikinks do not accurately account for "long" (algebraically) decaying tails. As a result, a number of unexpected and, to some degree, unwarranted results arise from collision simulations based on such ansätze. To uncover the key physics of kink–antikink collisions in the presence of long-range interactions (power-law tails), the first step is, thus, to determine the proper superposition to be employed in constructing initial conditions. This topic is the subject of ongoing research [66].

12.4.4 Statistical Mechanics of the ϕ^8 Field Theory and Phonons

Field theories of the ϕ^8 type are *not* QES so their statistical mechanics can only be studied by Langevin simulations [50–53] or the "double-Gaussian" variational approximation [50, 51]. In principle, one can obtain the lowest energy state numerically e.g. by Langevin dynamics and use it to calculate the PDF and the concordant thermodynamic quantities. Likewise, the eigenvalues of (12.19) can be computed numerically and used in the transfer operator approach. Finally, there exist special cases of the ϕ^8 field theories with two and three degenerate minima having $V''(\phi_0) = 0$ [4, Table I], again leading to the possibility of nonlinear phonon modes. The impact of the latter on the field thermodynamics is, as of now, unexplored.

12.5 Beyond

There is a veritable zoology of (kink and other) exact solutions in higher-order field theories, depending on the specified potential. Here, we make no attempt to systematically classify or organize these solutions as such an endeavor would constitute a book on its own. Instead, we highlight some (a) interesting and (b) novel aspects of kinks in higher-order field theories "beyond" ϕ^8.

12.5.1 Brief Overview of the ϕ^{10} Field Theory

12.5.1.1 Successive Phase Transitions and Kink Solutions

As in Sect. 12.4.1, one can design a specific ϕ^{10} potential, in which varying the coefficient of the ϕ^2 term leads to a succession of *two first-order* phase transitions [4, Sect. III]; for brevity, we do not include the latter discussion here. From amongst the many features that ϕ^{10} kinks can exhibit, we summarize the following from [4]: (a) in the case of five degenerate minima, four quarter-kinks of different energy, e.g., a pair connecting 0 to $+a$ (or $-a$ to 0) and a pair connecting $+a$ to $+b$ (or $-b$ to $-a$), for some a and b, exist; (b) kinks are generally asymmetric; (c) kinks with power-law tails exist, with a variety of possible decays in the case of three degenerate minima.

12.5.1.2 Statistical Mechanics of the ϕ^{10} Theory, Including Quasi-exactly Solvable Results

As mentioned above, ϕ^{10} is the next example of a QES field theory after ϕ^6. Following the approach in Sect. 12.3.4, we posit the following generalization of the ansatz in (12.20):

$$\Psi_0(\phi) = \exp\left\{-\frac{\sqrt{2}}{6}\phi^2\left(\phi^2 - K\right)^2\right\}, \tag{12.29}$$

which clearly has three maxima at $\phi = 0$ and $\phi = \pm\sqrt{K}$; at these ϕ values, $\Psi_0(\phi) = 1$, while at all other ϕ values, $0 < \Psi_0(\phi) < 1$. Upon substituting (12.29) for the wavefunction $\Psi_0(\phi)$ and a generic tenth-order potential, namely $V(\phi) = \phi^{10} - \alpha_8\phi^8 + \alpha_6\phi^6 - \alpha_4\phi^4 + \alpha_2\phi^2$ as in [4], into (12.19) and requiring that equality holds, one obtains two sets of consistency conditions:

$$\alpha_8 = \frac{8K}{3}, \quad \alpha_6 = \frac{22K^2}{9}, \quad \alpha_4 = \frac{16K^3 + 45\sqrt{2}}{18}, \quad \alpha_2 = \frac{K^4 + 18\sqrt{2}K}{9},$$
$$E_0 = \frac{K^2}{3\sqrt{2}}, \quad \beta = 1\,. \tag{12.30}$$

Hence, as long as the potential is of the generic form above, with coefficients $\alpha_{8,6,4,2}$ depending on K as in (12.30), then (12.29) is an exact ground-state wavefunction (no nodes) of the ϕ^{10} field theory at the (inverse) temperature of $\beta = 1$ with eigenvalue $E_0 = K^2/(3\sqrt{2})$. The PDF is the normalized squared wavefunction, Ψ_0^2. Figure 12.7 shows a visual comparison between the exact PDFs obtained herein for the ϕ^6 and ϕ^{10} field theories. Once again, employing different ansätze (from e.g. [58]) can yield exact excited-state PDFs, as also shown in [4]. As discussed above (see also Chap. 2), the exactness of these PDFs can be verified via Langevin simulations.

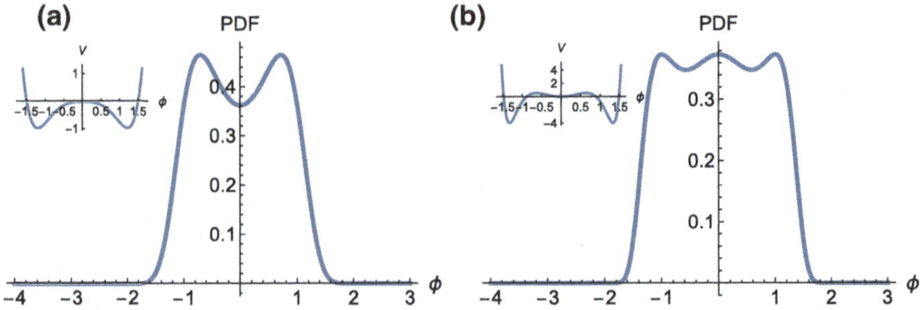

Fig. 12.7 **a** Exact PDF for a specific ϕ^6 potential (i.e., (12.7) with $\alpha_2 = -1/2$, shown as an inset), based on the wavefunction in (12.20) with $K = 2$. **b** Exact PDF for a specific ϕ^{10} potential (i.e., the generic potential defined by the coefficients $\alpha_{8,6,4,2}$ in (12.30), shown as an inset), based on the wavefunction in (12.29) with $K = 1$

12.5.2 ϕ^{4n+2} Field Theories with Three Degenerate Minima

Generalizing the result in [67] on ϕ^{2n+2} field theories, let us consider a special family of ϕ^{4n+2} field theories with three degenerate minima under a potential of the form

$$V(\phi) = \frac{\lambda^2}{2}\phi^2(\phi^{2n} - a^2)^2, \quad n = 1, 2, 3, \ldots . \tag{12.31}$$

By standard methods, it can be shown that these field theories have *explicit* exact half-kink solutions (connecting $-a^{1/n}$ and 0 or 0 and $+a^{1/n}$) given by

$$\phi_K(x) = \mp\{A[1 \mp \tanh(\beta x)]\}^{1/(2n)} , \tag{12.32}$$

provided that $A = a^2/2$ and $\beta = \lambda n a^2$. For $a = n = 1$, (12.32) reduces to (12.9).

12.5.3 Complex, $\mathcal{PT}$-Invariant Solutions of the ϕ^4 Field Theory

Since the introduction of the concept of $\mathcal{PT}$-symmetry in the late 1990s, a host of new physical insights (see, e.g., [68] and the references therein) have emerged, resulting in the rapid growth of research on open systems with balanced loss and gain. Here, $\mathcal{P}$ stands for parity symmetry $\{x, t\} \mapsto \{-x, t\}$, while $\mathcal{T}$ stands for time-reversal symmetry $\{x, t, \mathrm{i}\} \mapsto \{x, -t, -\mathrm{i}\}$. Then, the combined $\mathcal{PT}$-symmetry stands for $\{x, t, \mathrm{i}\} \mapsto \{-x, -t, -\mathrm{i}\}$.[1] Recently, novel complex periodic as well as hyperbolic kink solutions with $\mathcal{PT}$-eigenvalue -1 have been derived for a number

[1]Here, $\mathrm{i} = \sqrt{-1}$.

of real nonlinear equations, including the ϕ^4 and ϕ^{4n+2} models, and several higher-order non-polynomial field theories such as sG, double-sine-Gordon and double-sine-hyperbolic-Gordon, etc. [69, 70]. But, while kinks of $\mathcal{PT}$-symmetric nonlinear field theories are not affected by loss/gain, their stability critically depends on the loss/gain profile [71].

In this section, let us consider a model ϕ^4 theory: (12.3) with $V(\phi) = -\frac{a}{2}\phi^2 + \frac{b}{4}\phi^4$. Taking $a, b > 0$, $V(\phi) = \mathfrak{C}$ has real solutions $\phi_{1,2,3}$. Then, similarly to how the periodic solutions for ϕ^6 were constructed in Sect. 12.3.1 (recall Fig. 12.2), it can be shown that this ϕ^4 theory has a kink lattice solution

$$\phi_{KL}(x) = A\sqrt{m}\,\mathrm{sn}(\beta x\,|\,m)\,, \tag{12.33}$$

provided that $A = \sqrt{2\beta^2/b}$ and $\beta = \sqrt{a/(1+m)}$. As before, sn, cn and dn are Jacobi's elliptic functions with modulus $m \in [0, 1]$ [31]. Equation (12.33) reduces to (12.6) for $a = b = m = 1$. Remarkably, this same field theory also admits two complex, $\mathcal{PT}$-invariant periodic kink lattice solutions with $\mathcal{PT}$-eigenvalue -1:

$$\phi_{cKL,1}(x) = A\sqrt{m}\,[\mathrm{sn}(\beta x\,|\,m) \pm \mathrm{i}\,\mathrm{cn}(\beta x\,|\,m)]\,, \quad \beta = \sqrt{2a/(2-m)}\,, \tag{12.34a}$$

$$\phi_{cKL,2}(x) = A\left[\sqrt{m}\,\mathrm{sn}(\beta x\,|\,m) \pm \mathrm{i}\,\mathrm{dn}(\beta x\,|\,m)\right], \quad \beta = \sqrt{2a/(2m-1)}\,, \tag{12.34b}$$

provided that $A = \sqrt{\beta^2/(2b)}$. Notice that, unlike the solution in (12.33), $a < 0$ is allowed here if $m < 1/2$ in (12.33).

In the limit $m \to 1^-$, (12.34a) and (12.34b) both reduce to the complex, $\mathcal{PT}$-invariant kink solution

$$\phi_{cK}(x) = A[\tanh(\beta x) \pm \mathrm{i}\,\mathrm{sech}(\beta x)]\,, \tag{12.35}$$

with $A = \sqrt{\beta^2/(2b)}$ and $\beta = \sqrt{2a}$. While the width, $1/\beta$, of the complex, $\mathcal{PT}$-invariant kink in (12.35) is half of the width of the real kink (i.e., (12.33) with $m = 1$), their amplitudes are the same. As described in [72], the existence of a complex, $\mathcal{PT}$-invariant kink solution can be traced back to translational invariance: if $\tanh(\beta x)$ is a solution, then so is $\tanh(\beta x + x_0)$. Now, take $x_0 = \mathrm{i}\pi/4$ and observe that $\tanh(\beta x \pm \mathrm{i}\pi/4) = \tanh(2\beta x) \pm \mathrm{i}\,\mathrm{sech}(2\beta x)$, which immediately substantiates the existence of complex, $\mathcal{PT}$-invariant kink solution with half of the width. Clearly, this argument applies to any model that admits a kink solution of the form $\tanh x$.

Unfortunately, however, a similar argument for the existence of complex, $\mathcal{PT}$-invariant periodic solutions such as those in (12.34a) and (12.34b) is lacking. The obvious generalization would be to argue that if $\mathrm{sn}(\beta x\,|\,m)$ is a solution, then so is $\mathrm{sn}(\beta x + x_0\,|\,m)$ due to translational invariance. To this end, take $x_0 = \mathrm{i}K'(m)/2$, and, on using the addition theorem for sn, one finds that

$$\mathrm{sn}\left(\xi \pm \tfrac{1}{2}\mathrm{i}K'(m)\,|\,m\right) = \frac{(1+\sqrt{m})\,\mathrm{sn}(\xi\,|\,m) \pm \mathrm{i}\,\mathrm{cn}(\xi\,|\,m)\,\mathrm{dn}(\xi\,|\,m)}{m^{1/4}[1+\sqrt{m}\,\mathrm{sn}^2(\xi\,|\,m)]}\,, \tag{12.36}$$

where, recalling that $K(m) = K'(1-m)$, one has used the fact that (see [31])

$$\operatorname{sn}\left(\tfrac{1}{2}\mathrm{i}K'(m)\,|\,m\right) = \frac{\mathrm{i}}{m^{1/4}}, \quad \operatorname{cn}\left(\tfrac{1}{2}\mathrm{i}K'(m)\,|\,m\right) = \frac{\left(1+\sqrt{m}\right)^{1/2}}{m^{1/4}},$$
$$\operatorname{dn}\left(\tfrac{1}{2}\mathrm{i}K'(m)\,|\,m\right) = \left(1+\sqrt{m}\right)^{1/2}. \tag{12.37}$$

Inspired by the identity in (12.36), recently two of us asked [73] if there is a more general complex, $\mathcal{PT}$-invariant periodic solution. To this end, consider the ansatz

$$\phi_{cKL,3}(x) = \frac{A\,\operatorname{sn}(\beta x\,|\,m) \pm \mathrm{i}B\,\operatorname{cn}(\beta x\,|\,m)\,\operatorname{dn}(\beta x\,|\,m)}{1 + D\,\operatorname{sn}^2(\beta x\,|\,m)}, \tag{12.38}$$

where A, B, D and β have to be determined in terms of a, b and m. After a lengthy calculation, we find that (12.38) is a complex, $\mathcal{PT}$-invariant periodic solution, if

$$A = \sqrt{(2/b)(D+1)(D+m)\beta^2}, \quad \beta = \sqrt{a/(m+1)}, \quad bB^2 = 2D\beta^2. \tag{12.39}$$

Unlike the real or the complex periodic kink solutions discussed above, the periodic kink solution in (12.38) exists even if $b < 0$. In particular, if $-1 < D < -m$, then $b < 0$, which shows that this is a distinct periodic kink solution. In the special case

$$A = \frac{(1+\sqrt{m})}{m^{1/4}}F, \quad B = \frac{F}{m^{1/4}}, \quad D = \sqrt{m}, \tag{12.40}$$

the solution in (12.38) takes the form

$$\phi_{cKL,3}(x) = \frac{F[(1+\sqrt{m})\,\operatorname{sn}(\beta x\,|\,m) \pm \mathrm{i}\,\operatorname{cn}(\beta x\,|\,m)\,\operatorname{dn}(\beta x\,|\,m)]}{m^{1/4}[1+\sqrt{m}\,\operatorname{sn}^2(\beta x\,|\,m)]}, \tag{12.41}$$

while the conditions in (12.39) become $F = \sqrt{2m\beta^2/b}$ and $\beta = \sqrt{a/(m+1)}$, which coincide *exactly* with the condition under which the solution in (12.33) exists.

In the limit $m \to 1$, (12.38) leads to a more general, complex kink solution:

$$\phi_{cK,3}(x) = \frac{A\tanh(\beta x) \pm \mathrm{i}B\,\operatorname{sech}^2(\beta x)}{1 + D\tanh^2(\beta x)}, \tag{12.42}$$

provided that $A = \sqrt{(2/b)(D+1)^2\beta^2}$, $\beta = \sqrt{a/2}$, and $bB^2 = 2D\beta^2$. However, (12.42) does not represent a new kink solution. Specifically, as argued above, translational invariance means that given the "standard" kink solution $\hat{A}\tanh(\beta x)$, another kinks solution is $\hat{A}\tanh(\beta x + \mathrm{i}x_0)$. Then, it is easily shown that (12.42) and the standard kink solution $\hat{A}\tanh(\beta x)$ are related via

$$A = (1 + D)\hat{A}\,, \quad B = \sqrt{D}\hat{A}\,, \quad D = \frac{1 - \cos(2x_0)}{1 + \cos(2x_0)}\,. \tag{12.43}$$

Summarizing: while the complex $\mathcal{PT}$-invariant periodic kink is a new solution, the complex $\mathcal{PT}$-invariant hyperbolic kink is not. Using addition theorems for cn and dn, more general complex, $\mathcal{PT}$-invariant pulse solutions (similar in structure to (12.38) but with $\mathcal{PT}$-eigenvalue $+1$) have also been recently obtained [73]. Determining the stability of these new solutions is an open problem.

Another observation in the recent work [73] is that there exist remarkable connections between the complex solutions of various real scalar field theories. For example, consider a general field under (12.3) with $V(\phi) = \frac{a}{2}\phi^2 + \frac{b}{2n+2}\phi^{2n+2} + \frac{c}{4n+2}\phi^{4n+2}$, where n is a positive integer. It is amusing to note that, for given a, b, c, and some complex solution $\phi = \phi_1(x)$, then $\phi = \pm i\phi_1(x)$ is also a solution for the same values of the parameters a, b, c if n is an even integer (i.e., $n = 2, 4, 6, \ldots$), or with the same a and c *but* with $-b$ if n happens to be an odd integer (i.e., $n = 1, 3, 5, \ldots$). As special cases, $c = 0$ and $n = 1$ yields a ϕ^4 field, while $c \neq 0$ and $n = 1$ yields a ϕ^6 field.

12.6 Conclusion

In this chapter, we confined our attention to one-dimensional, higher-than-fourth-order field theories. Just as the ϕ^4 model has served as a prototype for describing second-order phase transitions and their attendant kinks (or domain walls) as well as breathers, the ϕ^6 model is a prototype for exploring first-order transitions with a richer phenomenology and different types of kinks. In particular, we discussed exact kink solutions of the ϕ^6 model, collisional dynamics of various kinks and statistical mechanics of this field theory. However, the ϕ^6 model is incapable of describing two or more (first- or second-order) successive phase transitions, and we must resort to ϕ^8 or even higher-order field theories. In this context, we discussed exact kink solutions and their interaction in the ϕ^8 model and interestingly elucidated the possibility of kinks with power-law tail asymptotics, which is quite different from the exponential tails in the ϕ^4 and ϕ^6 field theories. We also briefly considered ϕ^{10} as well as a general ϕ^{4n+2} field theory with degenerate minima and discussed their kink solutions. Finally, we explored complex, $\mathcal{PT}$-invariant kink solutions of polynomial field theories, and in particular ϕ^4.

12.6.1 Open Problems

Beyond the ϕ^6 model, we merely scratched the surface of a number of open questions for higher-order field theories, their kink-solution collisional dynamics, their statistical mechanics, and their connection to other nonlinear science models, etc.

The thermodynamic limiting case of infinite-order (continuous) phase transitions is an exciting area in this vein. The behavior of topological excitations in two (and possibly three) dimensional higher-order field theories is an entirely open issue as well. Nonlocal higher-order field theories, coupled higher-order models and the kink solutions that they harbor remain topics for future investigation.

One of the major open problems in higher-order field theories of the type discussed here is the kink collisional dynamics. Not only is there a far richer phenomenology of kinks in higher-order field theories (including kinks with power-law tails, the difficulties associated with studying their collisions having been mentioned in Sect. 12.4.2), but there are also many more possibilities for pair-wise interaction. The coexistence of kinks with pure power-law (or pure exponential tail) asymptotics with kinks with mixed tail asymptotics (i.e., power-law as $x \to -\infty$ but exponential as $x \to +\infty$) is possible in ϕ^{12} field theories with five and four degenerate minima [4, Sects. IV-B.2, IV-C.2]. What is the nature of these distinct kink-kink interactions? Can we generalize Manton's approach [2, 32] for calculating kink-(anti)kink effective force of interaction to power-law tails?

Furthermore, kinks with different energies can co-exist, as is the case described in [4, (9)–(17)]. To summarize, consider the octic potential with four degenerate minima: $V(\phi) = (\phi^2 - a^2)^2(\phi^2 - b^2)^2$. It possesses an exact kink solution connecting $-a$ to $+a$ (also found in [15]) with energy (rest mass) $E_{K,1} = \frac{4\sqrt{2}}{15}a^3(5b^2 - a^2)$. There is also an exact half-kink solution connecting a to b (or $-b$ to $-a$) with energy (rest mass) $E_{K,2} = \frac{2\sqrt{2}}{15}(b-a)^3(b^2 + 3ab + a^2)$. In [4], it was shown that $E_{K,1} \gtreqless E_{K,2}$ if $b/a \lesseqgtr 2/(3 - \sqrt{5})$. In particular, for $b/a = 2/(3 - \sqrt{5})$, the kinks and half-kinks have equal energies. So far, in lower-order field theories, the kinks (and antikinks) being scattered necessarily have the same energy because the field theory can have only two (ϕ^4 and ϕ^6) or three (ϕ^6) degenerate minima. Here, for the first time, two kinks of the same type as well as two kinks of different types can exist having equal or unequal energies. Thus, the question to be addressed is: how does the ratio b/a affect kink scattering dynamics? A similar situation occurs in the ϕ^{12} field theory with six degenerate minima [4, (113)].

Another branch of open questions relates to the fact that the ϕ^{4n+2} field theories with three degenerate minima mentioned in Sect. 12.5.2 offer a *parametrized* way to "turn up" the order of the field theory while maintaining the basic exact half-kink structure in (12.32). Thus, we ask: how does n affect kink scattering dynamics, starting with the known case of the ϕ^6 field theory [33] for $n = 1$? Furthermore, what is the number of internal modes that the half-kink structure in (12.32) possesses, and does this number depend on n?

It is also worth investigating the stability of the complex $\mathcal{PT}$-symmetric periodic kink solutions of the ϕ^4 field theory discussed in Sect. 12.5.3.

Finally, we inquire: Have we understood all the connections between the solutions of non-polynomial field theories like sine-Gordon, double-sine-Gordon, double-sine-hyperbolic-Gordon and the solutions of higher-order polynomial field theories? The former present infinite-order (both periodic and non-periodic) potentials, and some of the early motivation for studying higher-order field theories originated from truncating infinite-order periodic potentials to obtain polynomial field theories [15] (see also [74–76] and [4, Sect. V]). Specifically, it would be of interest to find out how the nature of kink interactions in non-polynomial theories differs from the corresponding one under a truncated higher-order field theory.

Acknowledgements I.C.C. acknowledges the hospitality of the Center for Nonlinear Studies and the Theoretical Division at Los Alamos National Laboratory (LANL), where the authors' collaboration on higher-order field theory was initiated. We acknowledge the support of the U.S. Department of Energy (DOE): LANL is operated by Triad National Security, L.L.C. for the National Nuclear Security Administration of the U.S. DOE under Contract No. 89233218CNA000001. I.C.C. also thanks V.A. Gani and P.G. Kevrekidis for many insightful discussions on kinks, collisions, collective coordinates, Manton's method and ϕ^8 field theory. A.K. is grateful to INSA (Indian National Science Academy) for the award of INSA Senior Scientist position.

References

1. R. Rajaraman, *Solitons and Instantons: An Introduction to Solitons and Instantons in Quantum Field Theory* (North Holland, Amsterdam, 1982)
2. N. Manton, P. Sutcliffe, *Topological Solitons* (Cambridge University Press, Cambridge, 2004)
3. T. Vachaspati, *Kinks and Domain Walls* (Cambridge University Press, Cambridge, 2006)
4. A. Khare, I.C. Christov, A. Saxena, Phys. Rev. E **90**, 023208 (2014)
5. L.D. Landau, Zh Eksp, Teor. Fiz. **7**, 19 (1937)
6. V.L. Ginzburg, L.D. Landau, Zh Eksp, Teor. Fiz. **20**, 1064 (1950)
7. M. Tinkham, *Introduction to Superconductivity* (McGraw-Hill, New York, 1996)
8. Y.M. Gufan, E.S. Larin, Sov. Phys. Dokl. **23**, 754 (1978)
9. S.N. Behera, A. Khare, Pramana **15**, 245 (1980)
10. F. Falk, Acta Metall. **28**, 1773 (1980)
11. F. Falk, Z. Phys, B **51**, 177 (1983)
12. R. Abeyaratne, J.K. Knowles, *Evolution of Phase Transitions: A Continuum Theory* (Cambridge University Press, Cambridge, 2006)
13. M. Sanati, A. Saxena, Am. J. Phys. **71**, 1005 (2003)
14. Y.M. Gufan, *Structural Phase Transitions* (Nauka, Moscow, 1982) (in Russian)
15. M.A. Lohe, Phys. Rev. D **20**, 3120 (1979)
16. N.H. Christ, T.D. Lee, Phys. Rev. D **12**, 1606 (1975)
17. A. Khare, Lett. Math. Phys. **3**, 475 (1979)
18. J.-C. Tolédano, P. Tolédano, *The Landau Theory of Phase Transitions* (World Scientific, Singapore, 1987) (Chapter IV)
19. P. Tolédano, V. Dmitriev, *Reconstructive Phase Transitions* (World Scientific, Singapore, 1996)
20. B. Mroz, J.A. Tuszynski, H. Kiefte, M.J. Clouter, J. Phys. Condens. Matter **1**, 783 (1989)
21. D. Vanderbilt, M.H. Cohen, Phys. Rev. B **63**, 094108 (2001)
22. I.A. Sergienko, Y.M. Gufan, S. Urazhdin, Phys. Rev. B **65**, 144104 (2002)
23. E.B. Sonin, A.K. Tagantsev, Ferroelectrics **98**, 291 (1989)
24. S.V. Pavlov, M.L. Akimov, Crystall. Rep. **44**, 297 (1999)
25. A.A. Boulbitch, Phys. Rev. E **56**, 3395 (1997)
26. A. Vilenkin, E.P.S. Shellard, *Cosmic Strings and Other Topological Defects* (Cambridge University Press, Cambridge, 2000)
27. A.R. Bishop, J.A. Krumhansl, S.E. Trullinger, Phys. D **1**, 1 (1980)
28. L.E. Reichl, *A Modern Course in Statistical Physics*, 4th edn. (Wiley, New York, 2016)
29. M. Sanati, A. Saxena, J. Phys. A Math. Gen. **32**, 4311 (1999)

30. P.A. Byrd, M.D. Friedman, *Handbook of Elliptic Integrals for Engineers and Physicists* (Springer, Berlin, 1954)
31. M. Abramowitz, I.A. Stegun (eds.), *Handbook of Mathematical Functions* (Dover Publications, New York, 1964)
32. N.S. Manton, Nucl. Phys. B **150**, 397 (1979)
33. P. Dorey, K. Mersh, T. Romańczukiewicz, Y. Shnir, Phys. Rev. Lett. **107**, 091602 (2011)
34. V.A. Gani, V. Lensky, M.A. Lizunova, E.V. Mrozovskaya, J. Phys. Conf. Ser. **675**, 012019 (2016)
35. V.A. Gani, V. Lensky, M.A. Lizunova, JHEP **2015**(8), 147 (2015)
36. D. Bazeia, R. Menezes, D.C. Moreira, J. Phys. Commun. **2**, 055019 (2018)
37. D.K. Campbell, J.F. Schonfeld, C.A. Wingate, Phys. D **9**, 1 (1983)
38. P. Anninos, S. Oliveira, R.A. Matzner, Phys. Rev. D **44**, 1147 (1991)
39. T.I. Belova, A.E. Kudryavtsev, Phys. Usp. **40**, 359 (1997)
40. V.A. Gani, A.E. Kudryavtsev, M.A. Lizunova, Phys. Rev. D **89**, 125009 (2014)
41. H. Weigel, J. Phys. Conf. Ser. **482**, 012045 (2014)
42. I. Takyi, H. Weigel, Phys. Rev. D **94**, 085008 (2016)
43. A. Moradi Marjaneh, V.A. Gani, D. Saadatmand, S.V. Dmitriev, K. Javidan, JHEP **2017**(7), 028 (2017)
44. A. Demirkaya, R. Decker, P.G. Kevrekidis, I.C. Christov, A. Saxena, JHEP **2017**(12), 071 (2017)
45. R.H. Goodman, A. Rahman, M.J. Bellanich, C.N. Morrison, Chaos **25**, 043109 (2015)
46. R.H. Goodman, R. Haberman, SIAM J. Appl. Dyn. Syst. **4**, 1195 (2005)
47. D.J. Scalapino, M. Sears, R.S. Ferell, Phys. Rev. B **6**, 3409 (1972)
48. J.A. Krumhansl, J.R. Schrieffer, Phys. Rev. B **11**, 3535 (1975)
49. J.R. Morris, R.J. Gooding, Phys. Rev. Lett. **65**, 1769 (1990)
50. F.J. Alexander, S. Habib, Phys. Rev. Lett. **71**, 955 (1993)
51. F.J. Alexander, S. Habib, A. Kovner, Phys. Rev. E **48**, 4284 (1993)
52. L.M.A. Bettencourt, S. Habib, G. Lythe, Phys. Rev. D **60**, 105039 (1999)
53. S. Habib, G. Lythe, Phys. Rev. Lett. **84**, 1070 (2000)
54. S. Habib, A. Saxena, in *Nonlinear Evolution Equations & Dynamical Systems NEEDS'94*, ed. by V.G. Makhankov, A.R. Bishop, D.D. Holm (World Scientific, Singapore, 1995) p. 310
55. P.G.L. Leach, Phys. D **17**, 331 (1985)
56. A.D. Bruce, Adv. Phys. **29**, 111 (1980)
57. T. Dauxois, M. Peyrard, *Physics of Solitons* (Cambridge University Press, Cambridge, 2006)
58. A.G. Ushveridze, *Quasi-exactly Solvable Models in Quantum Mechanics* (Institute of Physics Publishing, Bristol and Philadelphia, 1994) (Sec. 1.4)
59. A. Saxena, S. Habib, Phys. D **107**, 338 (1997)
60. A. Khare, S. Habib, A. Saxena, Phys. Rev. Lett. **79**, 3797 (1997)
61. S. Habib, A. Khare, A. Saxena, Phys. D **123**, 341 (1998)
62. R.V. Radomskiy, E.V. Mrozovskaya, V.A. Gani, I.C. Christov, J. Phys. Conf. Ser. **798**, 012087 (2017)
63. A.R. Gomes, R. Menezes, J.C.R.E. Oliveira, Phys. Rev. D **86**, 025008 (2012)
64. E. Belendryasova, V.A. Gani, J. Phys. Conf. Ser. **934**, 012059 (2017)
65. E. Belendryasova, V.A. Gani, Commun. Nonlinear Sci. Numer. Simulat. **67**, 414 (2019)
66. I.C. Christov, R. Decker, A. Demirkaya, V.A. Gani, P.G. Kevrekidis, R.V. Radomskiy, Phs. Rev. D. **99**, 016010 (2019)
67. F. Cooper, L.M. Simmons Jr., P. Sodano, Phys. D **56**, 68 (1992)
68. C.M. Bender, Rep. Prog. Phys. **70**, 947 (2007)
69. A. Khare, A. Saxena, Phys. Lett. A **380**, 856 (2016)
70. A. Khare, A. Saxena, J. Phys. A Math. Theor. **51**, 175202 (2018)
71. A. Demirkaya, D.J. Frantzeskakis, P.G. Kevrekidis, A. Saxena, A. Stefanov, Phys. Rev. E **88**, 023203 (2013)
72. J. Cen, A. Fring, J. Phys. A Math. Theor. **49**, 365202 (2016)

73. A. Khare, A. Saxena, Connections between complex PT-invariant solutions and complex periodic solutions of several nonlinear equations, arXiv:1808.02159
74. D. Bazeia, M.A. González León, L. Losano, J. Mateos Guilarte, Phys. Rev. D **73**, 105008 (2006)
75. D. Bazeia, M.A. González León, L. Losano, J. Mateos Guilarte, EPL **93**, 41001 (2011)
76. D. Bazeia, M.A. González León, L. Losano, J. Mateos Guilarte, J.R.L. Santos, Phys. Scr. **87**, 045101 (2013)

Chapter 13
Effects of Discrete Breathers on Thermal Transport in the ϕ^4 Lattice

Daxing Xiong and Sergey V. Dmitriev

Abstract In this chapter we discuss thermal transport in the ϕ^4 lattice which bears the feature of nonconserved momentum, and therefore its thermal transport behavior is usually suggested to be normal, i.e. following the empirical Fourier's law. However, a full understanding of the mechanism behind this normal behavior is still incomplete. Here, we present evidence about this mechanism stemming from the interactions of phonons with discrete breathers.

13.1 Introduction

Understanding the macroscopic irreversibility of physical systems from the knowledge of their microscopic deterministic dynamics is one of the central tasks in statistical physics. Among them, one of the most vivid examples is the anomalous thermal transport in low-dimensional systems. This topic has attracted significant interest over the last twenty years (see the review articles [1, 2] for a brief introduction). Generally, the thermal transport properties can be investigated by introducing a temperature gradient in the system. Fourier's law tells us that the heat current $\mathbf{J}$ is proportional to the temperature gradient ∇T according to $\mathbf{J} = -\kappa \nabla T$, with the thermal conductivity κ being a constant at the given averaged temperature. During the past 200 years, this celebrated empirical law has witnessed a great success in the three-dimensional (3D) bulk materials, and thus transport behavior following Fourier's law is usually known as "normal heat transport". Nevertheless, nowadays, with the rapid development of

D. Xiong (✉)
Department of Physics, Fuzhou University, Fuzhou 350108, Fujian, China
e-mail: phyxiongdx@fzu.edu.cn

S. V. Dmitriev
Institute for Metals Superplasticity Problems of RAS, Khalturin St. 39,
450001 Ufa, Russia
e-mail: dmitriev.sergey.v@gmail.com

S. V. Dmitriev
National Research Tomsk State University, Lenin Avenue 36,
634050 Tomsk, Russia

P. G. Kevrekidis and J. Cuevas-Maraver (eds.), *A Dynamical Perspective on the ϕ^4 Model*, Nonlinear Systems and Complexity 26,
https://doi.org/10.1007/978-3-030-11839-6_13

experimental and computing technologies for low-dimensional materials, more and more examples of "anomalous thermal transport" defying Fourier's law have been observed [3–7]. Usually, this anomaly lies in the fact that the thermal conductivity κ is not a material constant, rather it depends on the system size L. In particular, in one-dimensional (1D) case κ diverges with L according to the power-law, i.e., $\kappa \sim L^{\alpha}$ with $0 \leq \alpha \leq 1$ [7]. This means that Fourier's law demands $\alpha = 0$, and thus $\alpha > 0$ indicates an anomalous heat conduction. In particular, the case of $\alpha = 1$ corresponds to the so-called ballistic heat transport.

Indeed, such an anomalous scaling law has now been observed experimentally in a number of low-dimensional materials, such as nanowires, thin films, carbon nanotubes, holey silicon, and others [5, 8–12]. Furthermore, the theoretical design of thermal transistors [13, 14], thermal logic gates [15–17], and thermal diodes [18–22] has arisen in the rapidly growing field of *phononics* [23, 24]. Therefore, a better understanding of the underlying mechanism of this low-dimensional anomalous thermal transport, in particular, to explore the microscopic ingredients necessary for the validity (or failure) of Fourier's law, is of both theoretical and practical interest.

Up to now, many conditions for Fourier's law to be valid have been theoretically checked, such as the role of integrability [25], chaotic [6, 7] and stochastic [26] dynamics, presence of defects, etc. Integrable systems always demonstrate ballistic heat conduction [27–33]. Roughness of graphene nanoribbon edges can suppress thermal conductivity by two orders of magnitude [34]. An isolated carbon nanotube shows anomalous heat transport, while a nanotube on a substrate displays normal thermal conductance [35]. Thermal conductivity of silicene decreases upon doping by isotopes [20].

On the theoretical side, application of an on-site potential, which destroys the conservation of momentum [36–38], has now been recognized as one of the key factors affecting the type of thermal conductance, i.e., it is generally believed that a system without (with) a momentum conservation property should obey (disobey) Fourier's law. One such models where momentum conservation is absent is the ϕ^4 lattice (see [38, 39] and the model presented in Sect. 13.2). On the one hand, the observed Fourier's diffusive heat transport in this model has been attributed to the scattering of the energy carriers by the on-site potential [38], which causes an exponential decay of the correlation of momentum. On the other hand, the analytic results have indicated that this model supports the temporally periodic and exponentially localized in space oscillatory modes, termed as discrete breathers (DBs) [39]. DB solutions can be stable or unstable [40], movable or immobile, and they can have a very long lifetime. It is of interest to relate the heat transport to DB properties, in particular, to DB mobility.

In this Chapter we demonstrate the possible mechanism for Fourier's law observed in the ϕ^4 lattice. We will discuss the recently established close relationship between heat transport and DBs [41]. Such a relation might provide insight into our understanding of macroscopic thermal transport behavior from the microscopic DB dynamics. In addition, scattering of phonons by DBs will be addressed. Toward this end, in what follows we will describe the most essential results of the work [41] and add new results on phonon-DB interactions.

13.2 Discrete ϕ^4 Model

13.2.1 Description of the Model

The Hamiltonian of the ϕ^4 lattice is usually written as

$$H = \sum_k H_k, \; H_k = \frac{p_k^2}{2} + V(r_{k+1} - r_k) + U(r_k), \tag{13.1}$$

where r_k is the displacement of the kth particle from its equilibrium position and p_k is its momentum; $V(\xi) = \frac{\xi^2}{2}$ is the harmonic, nearest-neighbor inter-particle potential; $U(\xi) = k_2\frac{\xi^2}{2} + k_4\frac{\xi^4}{4}$ is the external pinning (or on-site) ϕ^4 potential. As the anharmonic part is of importance to the ϕ^4 potential, throughout the chapter we fix $k_2 = 1$ and vary $k_4 > 0$ to control the strength of the anharmonicity in U. It is worth noting that a positive k_4 ensures the potential U to be a single-well, hard type, which is able to support the stable standing and moving DBs with frequencies above the linear phonon band [39]. For the soft-type U, we note that recently the stable DBs with frequencies below the linear phonon band were discussed [42].

13.2.2 Phonon Dispersion Relation

To explore both the thermal transport and DB properties, it is relevant to consider the phonon dispersion relation. For the linearized form of Hamiltonian (13.1), it can be found that

$$\omega_q = \sqrt{4\sin^2\left(\frac{q}{2}\right) + 1}, \tag{13.2}$$

where q is the wave number and ω_q is the corresponding phonon frequency. This indicates that the phonon group velocity is

$$v_{\mathrm{g}} = \frac{\mathrm{d}\omega_q}{\mathrm{d}q} = \frac{\sin q}{\sqrt{4\sin^2(q/2) + 1}} = \frac{\sin q}{\omega_q}. \tag{13.3}$$

Figure 13.1 presents both ω_q (a) and v_q (b) versus q. The spectrum is within $\omega_q^{\min} = 1$ and $\omega_q^{\max} = \sqrt{5}$. In the vicinity of $q = 0$ (long waves) and $q = \pm\pi$ (short waves) the group velocity v_{g} is close to zero. On the other hand, the fast phonons with $v_{\mathrm{g}}^{\max}$, which mostly contribute to thermal transport, are located close to $q^{\max} \simeq 1.179$ with $v_{\mathrm{g}}^{\max} \simeq 0.618$.

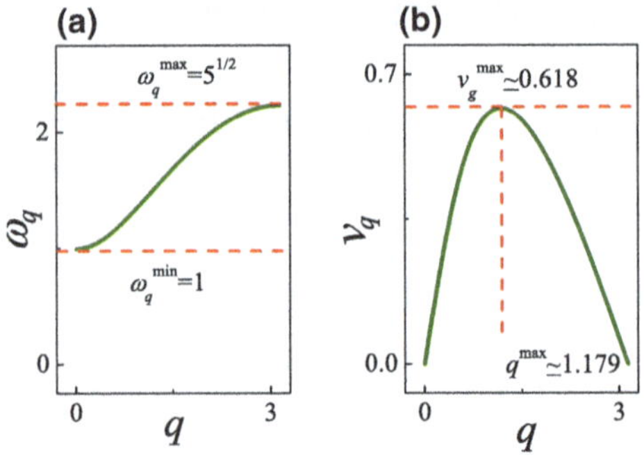

Fig. 13.1 The phonon dispersion relation (**a**) and the relevant group velocity (**b**) for the ϕ^4 lattice. In **a** the horizontal dashed lines denote $\omega_q^{\min} = 1$ and $\omega_q^{\max} = \sqrt{5}$; in **b** the vertical (horizontal) dashed line represents $q^{\max} \simeq 1.179$ ($v_g^{\max} \simeq 0.618$)

13.3 Numerical Techniques

13.3.1 Thermal Transport

There are various theories [43–48] and three main numerical approaches that can be used to study thermal transport properties. Most of the theoretical models predicted that the momentum-nonconserving ϕ^4 lattice will follow the normal heat transport, but did not give an underlying microscopic explanation based on its relation to DBs. Here we mainly deploy numerical techniques to illustrate this point.

The first numerical technique is usually referred to as the direct nonequilibrium molecular dynamics simulation [7]. It is similar to the usual estimate of thermal conductivity in real experiments in 3D bulk materials. In simulations, here one first connects two ends of the system (with size L) to two heat baths with a small temperature difference ΔT for a long time to obtain a nonequilibrium steady state. Then, the heat current J flowing across the system can be measured and finally the thermal conductivity κ can be derived according to

$$\kappa = JL/\Delta T. \tag{13.4}$$

The second numerical technique is based on the Green–Kubo formula [49–52]. Here, one focuses on the system's equilibrium heat current correlation function

$$C(t) = \langle J_{\text{tot}}(t) J_{\text{tot}}(t+\tau)\rangle_{\text{eq}}, \tag{13.5}$$

where J_{tot} is the total energy current, $\langle\cdot\rangle_{\text{eq}}$ denotes the ensemble average. In simulations, we often resort to the time average as long as the considered systems are chaotic and ergodic. With this, the Green–Kubo formula tells us that

$$\kappa = \frac{1}{k_B T^2} \lim_{t\to\infty} \lim_{L\to\infty} \frac{1}{L} \int_0^t C(\tau)\mathrm{d}\tau, \tag{13.6}$$

where k_B is the Boltzmann constant and usually set to unity. In 1D cases, however, if the power-law scaling $\kappa \sim L^{\alpha}$ is assumed, actually one does not need to calculate the Green–Kubo integration, instead, we always focus on the time decay of the equilibrium current correlation by

$$C(t) \sim t^{-\delta}; \tag{13.7}$$

then a slow decay ($0 < \delta < 1$) of $C(t)$ leads to the diverging thermal conductivity κ through the Green–Kubo formula, and we naturally have

$$\alpha = -\delta + 1. \tag{13.8}$$

The third numerical technique is related to the energy or heat diffusion process, since one usually regards the possible origin of heat transport as the diffusion processes of the relevant conserved quantities. To apply this idea for energy diffusion, one resorts to the analysis of evolution of local energy (heat) perturbations. Eventually, the spread of these perturbations defines the overall information of heat transport [53].

At finite temperature, a natural method to study energy diffusion is to excite a wave packet with energy much higher than the equilibrium energy density, then observe the relaxation process of this energy spot before it reaches an equilibrium with its environment [54]. However, it is not straightforward to identify the wave packet after it completely immerses into the environment, and this will cause great statistical fluctuations of the estimate [53]. To avoid the emerging statistical fluctuations, a novel equilibrium fluctuation-correlation method, which focuses on the lattice site-site energy correlation function, is proposed by Zhao [55]. It has been first used for studying the total energy fluctuations spreading. To make the results comparable to the prediction of hydrodynamics theory [44, 46, 47], the method has been further extended to be applicable in investigating the space-space fluctuation spreading, by considering the coarse-grained space-space heat correlation function [56]. In the latter case, the correlation function can be expressed as

$$\rho_Q(m,t) = \frac{\langle \Delta Q_{i+m}(t) \Delta Q_i(0) \rangle_{\mathrm{eq}}}{\langle \Delta Q_i(0) \Delta Q_i(0) \rangle_{\mathrm{eq}}}, \tag{13.9}$$

where

$$Q_i(t) \equiv E_i(t) - \frac{(\langle E \rangle_{\mathrm{eq}} + \langle F \rangle_{\mathrm{eq}}) g_i(t)}{\langle g \rangle_{\mathrm{eq}}} \tag{13.10}$$

is the corresponding conserved local heat energy density, at a coarse-grained bin space with index i and time t. This expression for heat is derived from the basic thermodynamics in conventional hydrodynamic theory [46]. It indicates that the heat energy density (in the bin i) is closely related to the associated particle (within the bin i) density $g_i(t) \equiv \sum_k g_k(t)$, energy density $E_i(t) \equiv \sum_k E_k(t) = \sum_k p_k^2/2 + U(r_k) + [V(r_{k+1} - r_k) + V(r_k - r_{k-1})]/2$, and pressure density $F_i(t) \equiv \sum_k F_k(t)$, where k numbers particles within ith bin. For the pressure, since our system is one dimensional, $F_k(t)$ for each particle is equal to the force and can be calculated from the gradient of the potential. Now in simulations, if one obtains $Q_i(t)$, then its fluctuation is $\triangle Q_i(t) = Q_i(t) - \langle Q \rangle_{\text{eq}}$, and the correlation function can be finally computed.

Next, one usually regards $\rho_Q(m, t)$ as the probability distribution function (PDF) of the corresponding heat diffusion process. Then its mean square displacement (MSD), which is denoted by $\langle \Delta x_Q^2(t) \rangle_{\text{eq}}$, suggests a scaling behavior

$$\langle \Delta x_Q^2(t) \rangle_{\text{eq}} = \sum_{x=-\infty}^{\infty} x_Q^2 \rho_Q(x, t) \sim t^{\mu}, \tag{13.11}$$

characterizing this process. In particular, the superdiffusive scaling exponent $1 < \mu < 2$ is thus related to the diverging thermal conductivity $\kappa \sim L^{\alpha}$ $(0 < \alpha < 1)$ by

$$\alpha = \mu - 1, \tag{13.12}$$

based on the Lévy walk assumption [55]. However, since (13.11) involves a sum to infinity, in practice, the Lévy walk theory [53] tells us that it is equivalent to explore the scaling behavior of the PDFs

$$t^{1/\gamma} \rho_Q(m, t) \simeq \rho_Q(t^{-1/\gamma} m, t). \tag{13.13}$$

From this, in the superdiffusive regime, we have

$$\gamma = 3 - \mu = 2 - \alpha, \tag{13.14}$$

which is also connected to the system size scaling of the thermal conductivity.

13.3.2 Discrete Breathers: The Method

Study of DBs at a finite temperature equilibrium system is still a challenge. At present, one can only rely upon numerical computations. To directly demonstrate the existence of DBs in the thermalized systems, one often adopts the absorbing boundary conditions approach [57, 58], i.e., we first thermalize the system to a sufficiently large temperature when the nonlinear effects contribute to the dynamics, then the thermostat is turned off, and we next add the absorbing boundary conditions to the two ends of the system for a long enough time. This strategy will cause all the mobile excitations, such as phonons and solitons, to be absorbed by the boundaries, and finally the immobile excitations like DBs can persist, if they exist. Finally, from these residual oscillations, we are able to recover the DB properties in the focused equilibrium state, e.g., to identify the DBs frequencies and compare them with the phonon frequencies.

In addition to exploring DBs at a finite temperature, investigation of DBs at zero temperature is also very suggestive. In a nonzero temperature equilibrium state, the effects of the thermostat can be decomposed into a series of kicks. A special high energy kick then is suggested to excite a single DB. Thus, study of the dynamics of a single DB also provides additional information in relation to transport.

First of all we note that, as suggested, e.g., by Khomeriki et al. [59], an approximate moving DB solution can be found analytically by inserting into the equations of motion the ansatz

$$r_k(t) = (-1)^n[\psi(x,t)\exp(i\omega_{\max}t) + \psi^{\star}(x,t)\exp(-i\omega_{\max}t)], \tag{13.15}$$

where x is a continuous variable corresponding to n and $\psi(x,t)$ is a slowly varying function of x and t. This will result in the NLS equation with the well-known moving soliton solution [60]. Restoring the initial variables a moving DB solution for the ϕ^4 discrete equation can be obtained. This solution gives a very good result for DBs with not very large amplitude. Here, however, we aim to obtain DB solutions with machine accuracy. For this we use the analytical form of the solution as the ansatz and find the relation between DB parameters numerically.

For the standing DB we employ the following ansatz

$$r_k(0) = \frac{(-1)^k A_{\mathrm{DB}}}{\cosh[\theta(k-x_0)]}, \qquad \frac{\mathrm{d}r_k(0)}{\mathrm{d}t} = 0, \tag{13.16}$$

where A_{DB} denotes the DB's amplitude, θ parameterizes its inverse width, x_0 represents the position of the DB, and if we set $x_0 = L/2$, DB is centered on a particle, while for $x_0 = (L+1)/2$, it is located at the bond center. With these initial conditions, in practice, one can first choose a DB's amplitude A_{DB}, then find θ by using the trial and error method [61], which in turn minimizes the oscillations of A_{DB}. This provides a simulation approach to find the suitable inverse width for a given amplitude of the DB. Finally, after θ has been obtained, we then can analyze the DB's properties.

For moving DBs we use the following ansatz

$$r_k(t) = \frac{(-1)^k A_{\mathrm{DB}}\cos[\omega_{\mathrm{DB}}t + \nu(k-x_0)]}{\cosh[\theta(k-x_0)]}. \tag{13.17}$$

The parameter ν here defines the DB velocity, which is zero if $\nu = 0$. We use (13.17) as initial condition to initiate a DB for chosen parameters A_{DB} and ν and measure its actual velocity numerically. The inverse width θ is also a function of A_{DB}.

13.4 Normal Thermal Transport

13.4.1 $k_4 = 1$

For the ϕ^4 lattice, to observe the normal heat transport behavior, one usually sets the on-site potential U with $k_4 = 1$ to introduce a relatively strong anharmonicity. To this aim, we first use the direct nonequilibrium molecular dynamics simulation to present the system size L dependence of the thermal conductivity κ. This gives us a first intuition of the normal heat transport conforming to Fourier's law. This is achieved, for each size system, by adopting fixed boundary conditions and connecting the two ends of the system to the Nosé–Hoover thermal reservoirs [62] with temperatures $T_+ = 0.6$ and $T_- = 0.4$ to produce a small temperature difference. After a long enough time of applying both thermal reservoirs, a nonequilibrium steady state will be reached, and this temperature difference will induce a temperature gradient, which gives the heat current defined by [1]

$$J = \langle J_k \rangle_{\text{eq}} = \left\langle p_k \frac{\partial V(r_{k+1} - r_k)}{\partial r_k} \right\rangle_{\text{eq}}. \tag{13.18}$$

In practice, we have evolved the system at least for a time up to $t = 4 \times 10^8$, then checked the temperature profiles. The typical temperature profile curves for different size L is shown in Fig. 13.2. As can be seen, such curves indicate certain scaling property, which suggests that one can estimate the thermal conductivity κ by $\kappa = JL/(T_+ - T_-)$ [1]. Through this, we plot κ as a function of L (from $L = 160$ to $L = 10240$) in Fig. 13.3. It shows that $\kappa(L)$ will attain a constant limiting value of about $\kappa \simeq 11.1$ for $L \geq 2560$. This suggests a finite thermal conductivity with the increase of the system size, and we thus infer that for $L \to \infty$, κ will be a constant independent of L, implying the validity of Fourier's law.

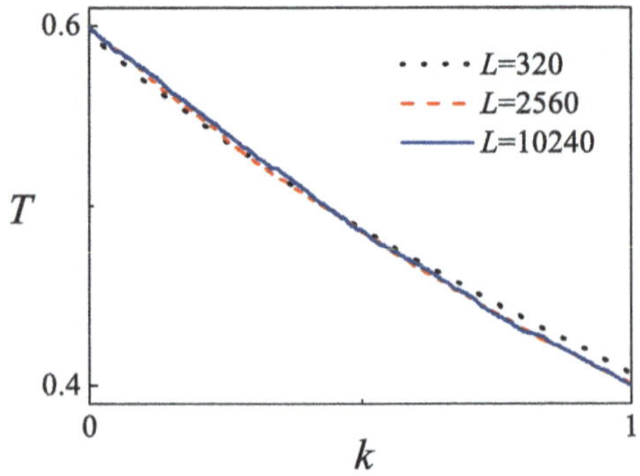

Fig. 13.2 Temperature profiles for the focused system with $k_4 = 1$ for three measured system size $L = 320$, $L = 2560$, and $L = 10240$, respectively. To see the scaling behavior, the x-axis has been rescaled by L

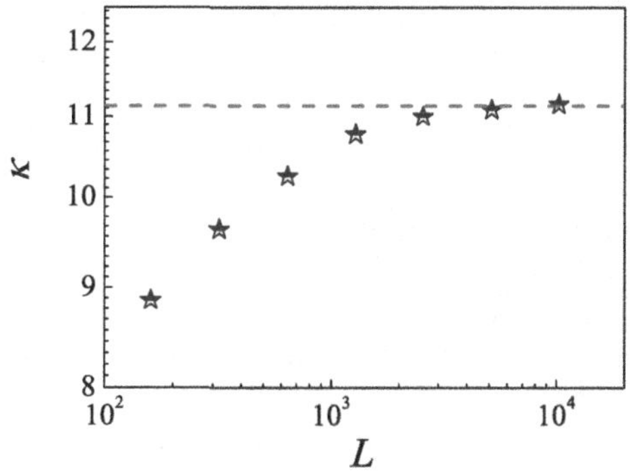

Fig. 13.3 The system size dependence of κ for the focused system with $k_4 = 1$. Here, the error bars are smaller than the symbol size, and the horizontal dashed line gives the converging value of κ for a large L

13.4.2 Phenomenological Picture of Heat Spread in Momentum Non-conserving Lattices

The normal transport behavior in momentum non-conserving systems has been discussed phenomenologically in terms of phonon-lattice interactions [37]. The phonons are strongly scattered by the on-site potential which destroys the conservation of momentum. This scattering then results in the thermal resistance, eventually leading the transport to follow Fourier's law. However, this phenomenological picture does not provide a detailed microscopic picture for the relevant scattering. One should answer by what excitations the phonons are scattered. In addition, this phenomenological picture does not explain ballistic heat conduction in the case of harmonic Hamiltonian (13.1) with $k_4 = 0$ [63], where momentum is also not conserved. Any harmonic lattice demonstrates the ballistic heat conduction [25, 63]. In the harmonic case, there is no temperature gradient and thus no thermal resistance, resulting in the fact that phonons are propagating ballistically. So here the phonon-lattice interactions should be absent, suggesting that to introduce the interplay of phonons and the on-site potential certain anharmonicity in U is necessary. This suggests a strategy to vary the strength of the anharmonicity of U by adjusting the value of k_4. This allows to explore the detailed crossover from the ballistic to normal heat transport.

13.5 Heat Spread

We next use the equilibrium fluctuation-correlation method to present more detailed information of heat transport. Before doing that, we note that Piazza and Lepri [48] have adopted the notion of temperature perturbations to investigate the heat wave propagation in the ϕ^4 lattice ($k_4 = 1$) and found that the telegraph equation might be suitable for quantitatively describing this system's heat wave propagation at the macroscopic level.

We are particularly interested in the heat spread analysis upon variation of the on-site potential anharmonicity. For this purpose, for each k_4, the system is initially placed in contact with a Langevin heat bath [1, 2] of temperature $T = 0.5$, which is the averaged temperature of $T_+ = 0.6$ and $T_- = 0.4$ adopted above, and thus we can make a comparison between them. This heat bath will enable us to prepare a thermalized equilibrium state under the given temperature. With this equilibrium system, we then apply the Runge–Kutta integration algorithm of 7–8th order with a time step of $h = 0.05$ to evolve the system. During the evolution, one finally samples the relevant data and calculates the correlation function $\rho_Q(m, t)$.

More specifically, we consider the system with periodic boundary conditions and $L = 4001$ particles. To make the particle number equal to the system size L, we set both the equilibrium distance between particles and the lattice constant to unity. This also leads the averaged pressure to always be zero, since both the interparticle interaction V and the on-site potential U are even. Recall that pressure is equal to the average force acting on any reference particle in the positive or negative direction of the chain, $\langle F_i \rangle_{\rm eq} = \frac{1}{t} \int_0^t F_i(\tau) {\rm d}\tau$. To calculate the space-space correlation similarly to the description of hydrodynamics, we fix the number of bins to be $(L-1)/2$, so each bin has a size of 2. With this, we prepare each equilibrium system by evolving the system for a long enough time (at least $>10^7$ time units) from initially random states, and use the ensembles of about 8×10^9 data points to compute the correlation.

Figure 13.4 depicts the shapes of $\rho_Q(m, t)$ (for three typical long times) variation of k_4, from small to large. On the one hand, we see that for each time, for a relatively small k_4, $\rho_Q(m, t)$ is a U shape with some oscillations on the fronts (see Fig. 13.4a). This U shape is typically observed in the ballistic classical and quantum random walks [64], and thus it might correspond to the ballistic heat transport. This is coincident with the estimate of heat conduction in [63]. In fact, for $k_4 = 0$, with the phonon dispersion relation ω_q [see (13.2)] at hand, one can predict this U shape by [64]

$$\rho_Q(m, t) \simeq \left| \frac{1}{2\pi} \int_{-\pi}^{\pi} e^{i(mq - \omega_q t)} {\rm d}q \right|^2 . \tag{13.19}$$

On the other hand, for a strong enough anharmonicity, we do see a Gaussian-like profile of $\rho_Q(m, t)$ (see Fig. 13.4e, f), which implies the diffusive normal heat transport obeying Fourier's law, in good agreement with the numerical results in [38] and our estimate in Fig. 13.3 of normal heat conduction.

More interesting things take place in the middle range of k_4, where as k_4 increases, there are two clear transition processes: at first, $\rho_Q(m, t)$'s front peaks are damped, then after this attenuation process is almost completed, the central peak of $\rho_Q(m, t)$ becomes more and more humped. This apparently gives more detailed evidence of the transition, and suggests a crossover from ballistic to normal heat spread in the system.

Next, to characterize the corresponding transport types, it is convenient to regard $\rho_Q(m, t)$ as a PDF and analyze its scaling behavior. Figure 13.5 presents $\rho_Q(m, t)$ as follows from the scaling formula (13.13). In practice, to extract the scaling expo-

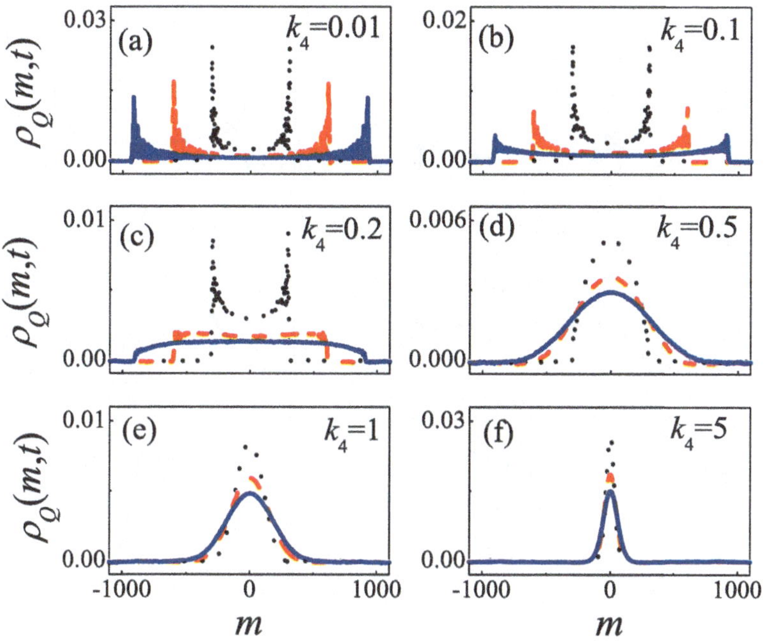

Fig. 13.4 The measured $\rho_Q(m,t)$ for several typical k_4 from small to large, where the dotted, dashed, and solid lines represent the result of $t = 500$, $t = 1000$, and $t = 1500$, respectively. Adapted from [41]. ©2018 APS, under the copyright license

nent γ, we measure the time scaling behavior of the height H_c^Q of the central peak of $\rho_Q(m,t)$ and find that its long-time asymptotics will follow $H_c^Q \sim t^{-1/\gamma}$ [53]. Figure 13.6 presents two typical examples of this time scaling behavior. Now as long as the scaling exponent γ has been obtained, (13.13) will be applied, and the rescaled results of $\rho_Q(m,t)$ are plotted in Fig. 13.5. Note that in general, in the superdiffusive transport regime, such a scaling formula with $1 < \gamma < 2$ only applies to the central part of the PDF, but for the ballistic ($\gamma = 1$) and diffusive ($\gamma = 2$) transports, it is validated for the whole density. Then, Fig. 13.5 indicates that, indeed, under small (large) k_4, the thermal transport is close to ballistic (normal) (see Fig. 13.5a, f), and in the middle range of k_4, there is a crossover (see Fig. 13.5b–e).

It is thus of interest to present a complete picture of γ versus k_4, as shown in Fig. 13.7. Therein six data points are extracted from Fig. 13.5 and others are calculated additionally in the same fashion. This plot suggests that the transition point between ballistic and diffusive transport is between $k_4 = 0.1$ and $k_4 = 0.4$. It is reasonable to take $k_4 = 0.2$ as the transition point, which is also in agreement with the results of Fig. 13.4, where in the case of $k_4 = 0.2$ the front peaks of $\rho_Q(m,t)$ are absent for a long time.

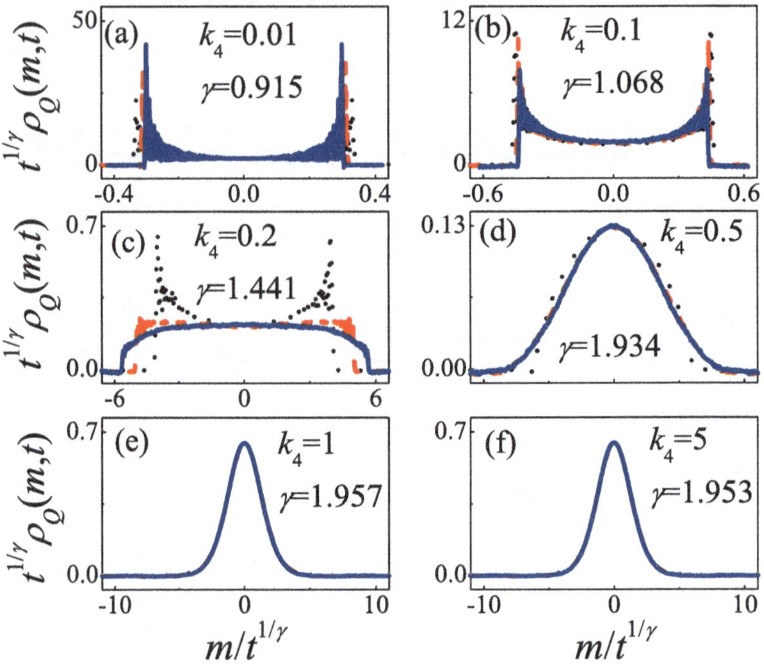

Fig. 13.5 Rescaled $\rho_Q(m, t)$ for the estimate in Fig. 13.4

Fig. 13.6 Two typical results ($k_4 = 0.01$ and $k_4 = 5$) of the height H_c^Q of the central peaks of $\rho_Q(m, t)$ versus t indicating $H_c^Q \sim t^{-1/\gamma}$, which enables us to extract the scaling exponent γ

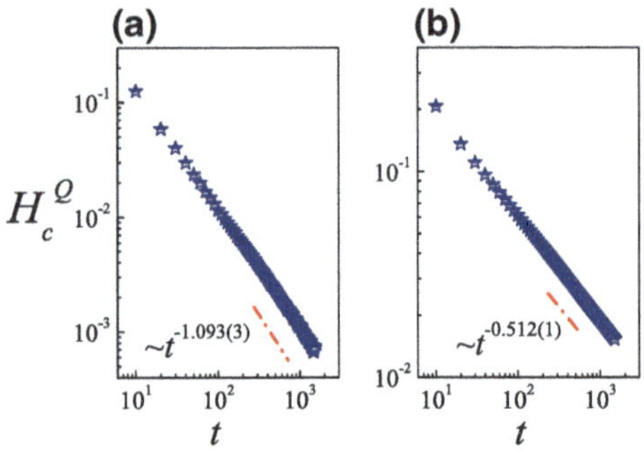

13.6 Discrete Breathers Dynamics

13.6.1 Finite Temperature Case

Given the above k_4 dependent crossover behavior of the thermal transport, we now explore the DB properties. We first focus on the non-zero temperature case ($T = 0.5$, same as above). We first look at the initial power spectrum $P_{\text{Th}}(\omega)$ of a thermalized system. To see this, we consider a relatively short size system of $L = 200$ for facilitat-

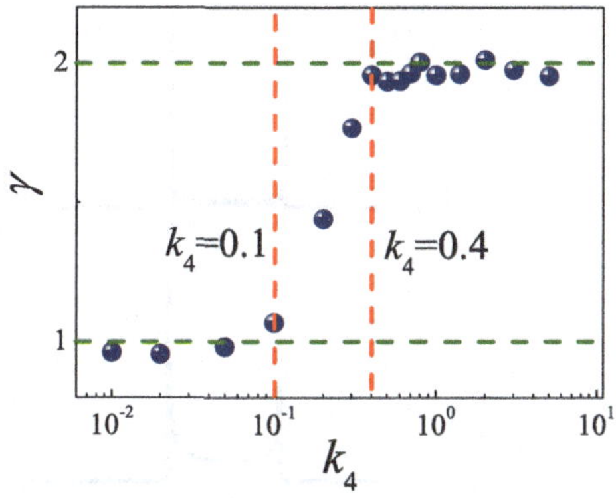

Fig. 13.7 γ as a function of k_4, where the horizontal lines denote $\gamma = 1$ and $\gamma = 2$, corresponding to the ballistic and the Fourier's heat transport, respectively; the vertical lines indicate $k_4 = 0.1$ and $k_4 = 0.4$, respectively. Adapted from [41]. ©2018 APS, under the copyright license

ing the computation. We use the Langevin heat baths [1, 2] to thermalize the system, and then turn off the baths and calculate the power spectrum $P_{\mathrm{Th}}(\omega)$. Generally, it can be computed by applying a frequency analysis of the particles velocity/momentum (since the particle mass is unity):

$$P_{\mathrm{Th}}(\omega) = \lim_{\tau \to \infty} \frac{1}{\tau} \int_0^{\tau} p(t) \exp(-i\omega t) \mathrm{d}t. \tag{13.20}$$

Figure 13.8 presents $P_{\mathrm{Th}}(\omega)$ as a function of ω for different k_4. As can be seen, for a relatively small k_4, $P_{\mathrm{Th}}(\omega)$ shows a band structure with two asymmetric sharp peaks on the edges (see Fig. 13.8a). The origin of these peaks can be explained from the corresponding density of phonon states. In a harmonic system with on-site potential, the phonon group velocity vanishes at the edges of the phonon band and that is why the density of phonon states has two peaks near the edges. As k_4 increases, due to the hard type anharmonicity in the on-site potential, the blue shift of the spectrum $P_{\mathrm{Th}}(\omega)$ is observed. This is an evidence that the DBs with frequencies above the linear phonon band are excited. On the other hand, excited DBs play the role of pins and the long-wave phonons disappear resulting in the depletion of the spectrum in the long-wavelength region, which is close to the lower edge of the phonon spectrum. These changes in the spectrum $P_{\mathrm{Th}}(\omega)$ become noticeable for $k_4 \geq 0.2$. Note that this threshold value of k_4 is in line with the results shown in Fig. 13.7, suggesting that the crossover from ballistic to normal transport happens in the range from $k_4 = 0.1$ to $k_4 = 0.4$.

With this evidence, we then implement the absorbing boundary conditions for a long enough time (usually simulation time of $t = 2 \times 10^6$ is sufficient to absorb all the movable excitations), we finally compute the power spectrum $P_{\mathrm{Re}}(\omega)$ of these residual oscillations, according to (13.20).

Figure 13.9 depicts $P_{\mathrm{Re}}(\omega)$ versus ω for the same values of k_4 after the application of the absorbing boundary conditions. It shows that (i) at a relatively small k_4, there are two peaks left, corresponding to the phonons with frequencies close to the lower and upper boundaries of the band having nearly zero group velocity (see Fig. 13.9a).

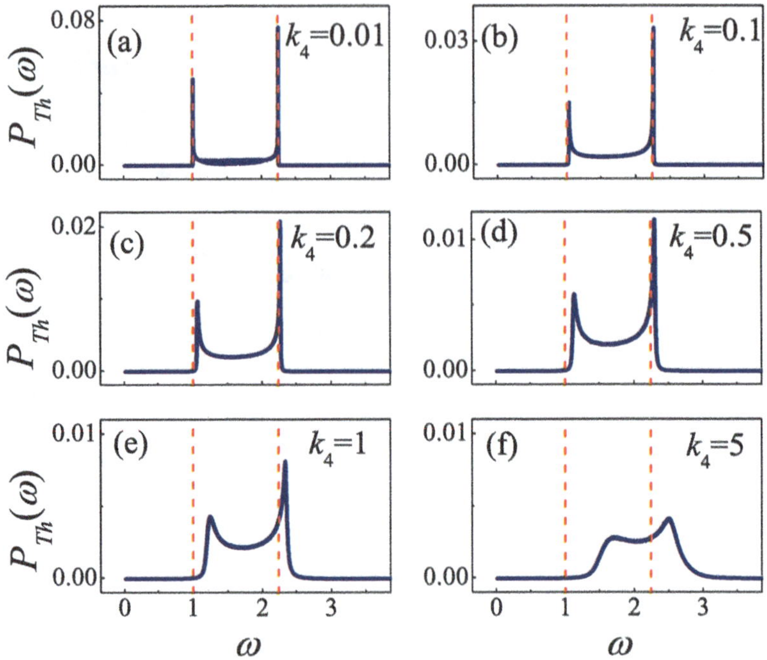

Fig. 13.8 The initial power spectrum $P_{\mathrm{Th}}(\omega)$ of the equilibrium thermalized systems before adding the absorbing sources for different strengths of anharmonicities: **a** $k_4 = 0.01$, **b** $k_4 = 0.1$, **c** $k_4 = 0.2$, **d** $k_4 = 0.5$, **e** $k_4 = 1$, and **f** $k_4 = 5$, where the vertical dashed lines denote the lower and upper boundaries of the linear phonon band, respectively. Adapted from [41]. ©2018 APS, under the copyright license

(ii) With the increase of k_4, the lower frequency peak becomes weaker, but only a slight change of the upper component occurs (see Fig. 13.9b–d). This is because DBs, that appear at higher values of k_4, operate as pinning centers. The longest phonon wavelength is now defined not by the length of the chain but by the average distance between the DBs. Shorter waves can be absorbed by the boundaries. The wave packets with shortest wavelength survive between DBs and they are not absorbed by the boundaries being practically immobile. (iii) Finally, if we look at the case $k_4 = 5$, only the components with frequencies higher than the linear phonon band can be seen (see Fig. 13.9f). This is the indication of presence of immobile DBs in the system. It might have the effect of making the Gaussian peak of $\rho_Q(m, t)$ narrower (see Fig. 13.4f).

We then turn to the analysis of the profiles of these residual excitations. For the cases (i)–(iii), we give one typical shape of the profile from one initial choice of the corresponding equilibrium states, respectively, and show them in Fig. 13.10. As can be seen, when k_4 is large enough, highly localized DB modes can be discerned (see Fig. 13.10c). They correspond to the so-called Sievers–Takeno (ST) modes [65], which are the standing DBs. This gives the power spectrum shown in the case (iii),

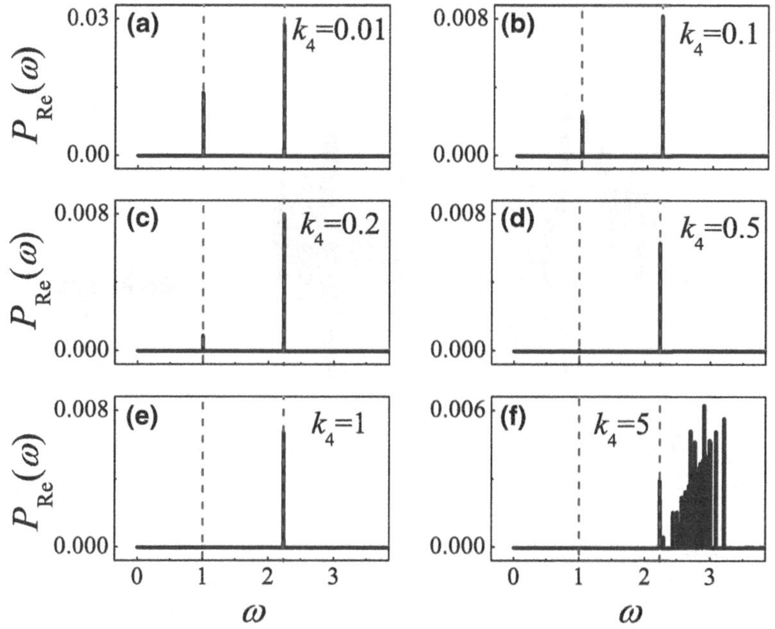

Fig. 13.9 The power spectrum $P_{\text{Re}}(\omega)$ of the equilibrium thermalized systems after implementing the absorbing sources for different strengths of anharmonicity: **a** $k_4 = 0.01$, **b** $k_4 = 0.1$, **c** $k_4 = 0.2$, **d** $k_4 = 0.5$, **e** $k_4 = 1$, and **f** $k_4 = 5$, where the vertical dashed lines denote the lower and upper boundaries of the linear phonon band, respectively. Adapted from [41]. ©2018 APS, under the copyright license

i.e., Fig. 13.9f. As k_4 goes down, the standing ST modes transform into mobile DBs and they are absorbed by the boundaries. That is why, only short-wavelength and long-wavelength phonons, whose group velocity is nearly zero, remain in the system, see Fig. 13.10b–e. A theoretical consideration of excitation of DBs in thermal equilibrium can be found in [66].

13.6.2 Zero Temperature Case

To use the ansatz described by (13.17), we should first know the relevant parameter of a standing DB. For this, as mentioned, one can use (13.16) as initial conditions to produce one ST DB mode. This standing DB can be stably excited when a trial and error method [61] is applied. This method gives a balanced relation between the DB's amplitude A_{DB} and the inverse width θ. After both parameters have been obtained, one then can evolve the DB with the increase of time, and calculate its frequency ω_{DB} and total energy E_{DB}, i.e., the sum of kinetic and potential energy.

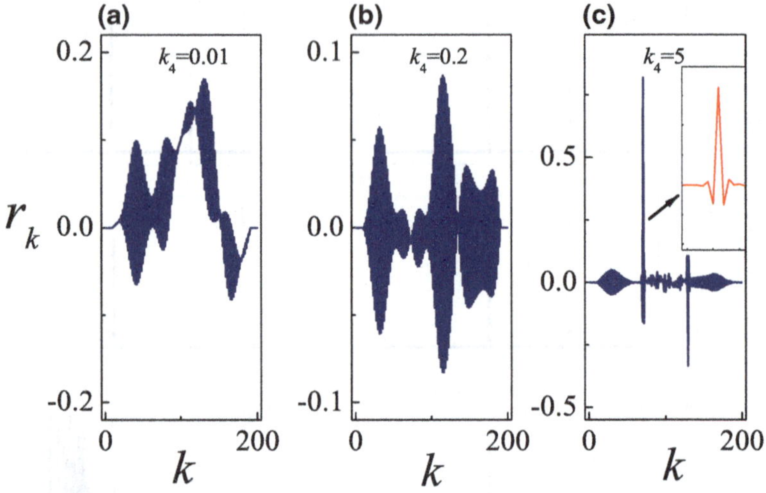

Fig. 13.10 Profiles of the residual thermal fluctuations after applying a long-time absorbing boundary conditions for three k_4 values: **a** $k_4 = 0.01$, **b** $k_4 = 0.2$, and **c** $k_4 = 5$. The inset in **c** is used for demonstrating the Sievers–Takeno DB mode

Table 13.1 presents the results of θ, ω_{DB}, and E_{DB} for a set of A_{DB} for $k_4 = 1$ and $k_4 = 5$, respectively. Among them, $k_4 = 5$ is used for demonstrating the immobile ST DB mode, while $k_4 = 1$ is as a comparison for analyzing the mobile DB. Before going on, Table 13.1 provides various trends, i.e., for a given k_4, as A_{DB} increases, all of θ, ω_{DB}, and E_{DB} increase. However, with the increase of k_4, if we fix the amplitude, both θ and ω_{DB} will increase indicating that as the anharmonicity becomes stronger, the DB will be more localized and with higher frequency. Let us now produce a single moving DB. This, of course, will not be an exact moving DB solution but, as it will be shown numerically, for not very large propagation velocities the DB demonstrates a very long lifetime practically radiating no energy. We adopt (13.17) with a set of ν for setting the initial conditions, and take other DB parameters from Table 13.1. Similarly to the standing DB, with the evolution of time of the stable solution, one then can measure the DB velocity v_{DB}. Figure 13.11 depicts v_{DB} versus ν for two k_4 values for different A_{DB}. It shows that the DB mobility depends on its amplitude; large amplitude DBs are immobile. While for a relatively small amplitude DB is mobile, v_{DB} is nonzero and nearly proportional to ν within the range of $|\nu| \leq 0.5$. In the intermediate range of A_{DB}, v_{DB} increases with ν slower than linearly. This indicates that under such A_{DB}, this moving DB will radiate energy and thus its velocity will gradually decrease. Further analysing the DB's parameters shown in Table 13.1, one might recognize that the mobility/immobility transition takes place at about $\theta \simeq 0.7$. Indeed, from Table 13.1, we have $\theta = 0.639$ for $A_{\mathrm{DB}} = 1$ ($k_4 = 1$), and $\theta = 0.748$ for $A_{\mathrm{DB}} = 0.5$ ($k_4 = 5$).

Table 13.1 The relevant parameters for a set of DBs with different amplitudes for $k_4 = 1$ and $k_4 = 5$. Adapted from [42]. ©2018 APS, under the copyright license

A_{DB}	θ	ω_{DB}	E_{DB}
$k_4 = 1$			
0.25	0.154	2.240	2.028
0.5	0.309	2.257	4.091
0.75	0.474	2.282	6.069
1.0	0.639	2.319	8.147
1.25	0.871	2.354	9.487
1.5	1.135	2.410	10.744
$k_4 = 5$			
0.25	0.349	2.262	0.909
0.5	0.748	2.332	1.751
0.75	1.360	2.451	2.318
1.0	1.960	2.687	3.570
1.25	2.441	3.009	6.091
1.5	2.814	3.380	10.344

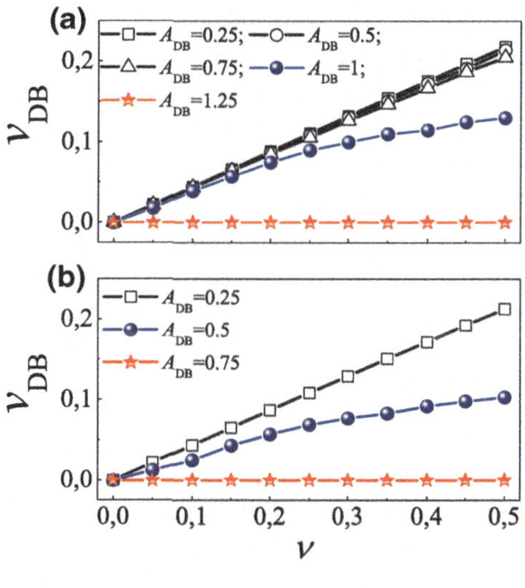

Fig. 13.11 The measured v_{DB} as a function of the parameter ν in the ansatz (13.17) for $k_4 = 1$ (**a**) and $k_4 = 5$ (**b**). Adapted from [41]. ©2018 APS, under the copyright license

13.7 Scattering of Phonons by Movable and Immobile DBs

Our main conjecture is that phonons in the ϕ^4 lattice can interact inelastically with the thermally excited DBs resulting in the normal heat transport. It is thus important to see how phonons of different frequency and amplitude interact with DBs.

Recall that DBs with relatively small amplitude are movable, while those with the amplitude above a threshold valued are immobile (see Sect. 13.6.2). For $k_4 = 1$

considered in that section, the threshold value is about $A_{\mathrm{DB}} = 1$ (see Fig. 13.11a). We thus consider the cases of $A_{\mathrm{DB}} = 0.5$ and $A_{\mathrm{DB}} = 1.25$, representing the movable and immobile DBs, respectively. Single standing DB is excited at the center of the system denoted by $k_c = (L+1)/2$ with the help of the ansatz (13.16) by taking the values of parameters from Table 13.1.

To excite the phonon wave packet, one can apply ac driving to a particle far away from the DB, at location $k^* = k_c - 500$, i.e.,

$$r_{k^*}(t) = A_{\mathrm{P}} \sin(\omega_{\mathrm{P}} t). \tag{13.21}$$

Here A_{P} and ω_{P} denote the driving amplitude and frequency, respectively. The ac driving is applied within the time $0 \le t \le 1500$, and then it is turned off. In this way, two identical phonon wave packets moving from the driven particle in opposite directions are excited. The one moving to the left is used to calculate the energy emitted by the driven particle during the time of driving, which gives the energy of incident phonon wave, E_I, interacting with the DB. The phonon wave packet emitted to the right interacts with the DB and, generally speaking, it splits into reflected and transmitted wave packets having energies E_R and E_{Tr}, respectively. The energies E_I, E_R, E_{Tr}, and the DB energy are calculated when the phonon wave packets move away and separate from the DB, see Fig. 13.13. The time of energy calculation is estimated by knowing the phonon group velocity v_{g} given by (13.3). After the external driving is stopped at $t = t_1$, the phonon wave packet needs a time S/v_{g} to pass through the DB, which is $S = 500$ sites away. To achieve sufficient separation of the phonon wave packets from the DB we double this time and calculate the energies at time $t_1 + 2S/v_{\mathrm{g}}$. The wave packet has energy $A_{\mathrm{P}}^2\omega_{\mathrm{P}}^2/2$ per particle. We take $A_{\mathrm{P}} = 0.04$ and ω_{P} within the phonon band, i.e., $\omega_{\mathrm{P}} \sim 1$. Then the energy per particle in the emitted phonon wave packet is of the order of 10^{-3}. The particles at the ends of the wave packets having energies less than 10^{-6} are not included in the summation.

After we obtain E_I, E_R and E_{Tr}, the reflection and transmission coefficients are defined by

$$R = \frac{E_R}{E_I}; \quad Tr = \frac{E_{Tr}}{E_I}, \tag{13.22}$$

respectively. If the DB does not radiate energy during the interaction with the phonon wave then $R + Tr = 1$. In fact, the DB's properties can change as a result of interaction with the phonon wave. That is why we calculate the energy, E_{DB}^*, and velocity, v_{DB}^*, of the DB after the interaction. Recall that before the interaction we have $E_{\mathrm{DB}} = 4.091$ for $A_{\mathrm{DB}} = 0.5$ and $E_{\mathrm{DB}} = 9.487$ for $A_{\mathrm{DB}} = 1.25$, and $v_{\mathrm{DB}} = 0$ in both cases.

13.7.1 Case of Small Amplitude Phonons

Interaction of phonons with DBs has been studied for the KG chains [67, 68] and for the NLS equation [69–71]. It has been shown that for vanishing phonon amplitude the scattering is elastic [67], i.e., the DB does not radiate its energy. In the case when the DB internal mode is taken into account, the DB always radiates its energy interacting with a phonon. Thus, a DB can be transparent or opaque for the small-amplitude phonons [67].

Here we first analyze the case of relatively small-amplitude phonon wave packets by setting $A_P = 0.04$ in (13.21) and consider various driving frequencies lying in the linear phonon band.

In Fig. 13.12 we plot the transmission coefficient Tr, the DB energy E_{DB} and velocity v_{DB} as a function of the phonon frequency ω_P after the interaction with the phonon wave packet. As mentioned, we obtain Tr for the phonon wave packet by measuring E_R and E_{Tr}. For the DB, E_{DB} is computed in the same way as E_R and E_{Tr}, i.e., we count the associated particles' energy for the DB (for the particles with oscillations on the edges, only the energies larger than 10^{-6} are included).

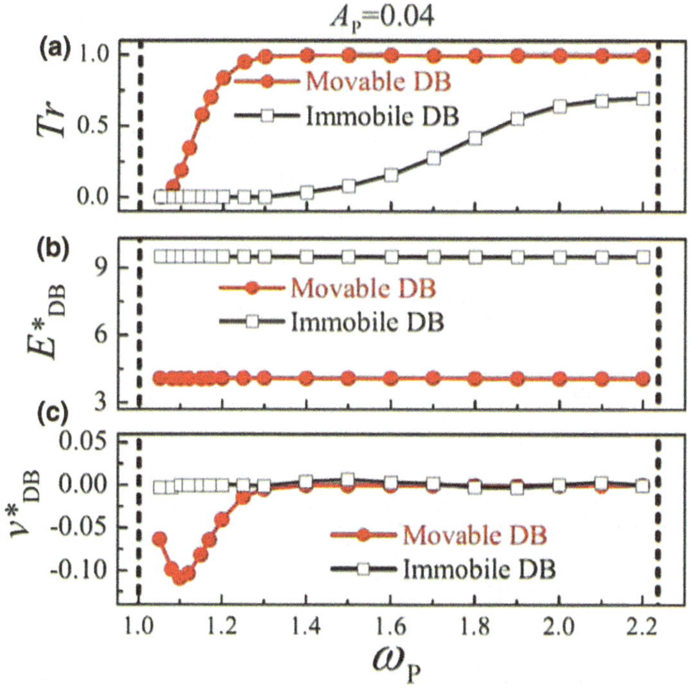

Fig. 13.12 The transmission coefficient Tr (**a**), the DB's energy E^*_{DB} (**b**), and the DB's velocity v^*_{DB} (**c**) as a function of the incident phonon frequency ω_P for $A_P = 0.04$ and $k_4 = 1$, where the circles and squares indicate the results of movable and immobile DBs, respectively

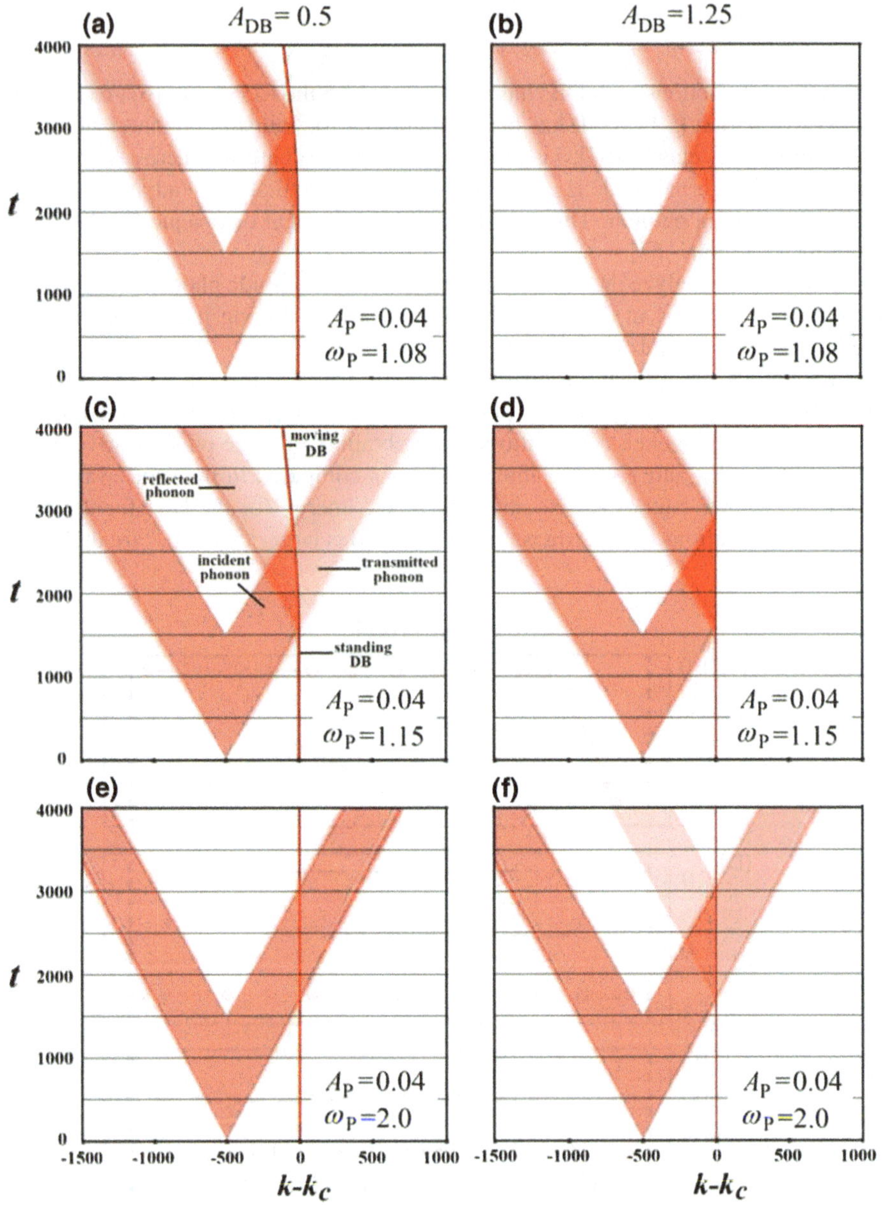

Fig. 13.13 Some typical results ($k_4 = 1$) of interaction of the small-amplitude phonons with DBs: **a**, **c**, **e** for the movable DB ($A_{DB} = 0.5$) and **b**, **d**, **f** for the immobile DB ($A_{DB} = 1.25$). To see these, particles with higher total energy are shown on the (k, t) plane with denser colors. All the DBs are initially at rest at $k - k_c = 0$, and after the interaction with the phonon wave, they either move to the left or still keep standing. The source of the phonon wave packets is located at $k - k_c = -500$ and it operates from $t = 0$ till $t = 1500$. So, the excited phonon wave packet will move to the right interacts with the DB. This causes three parts of phonon wave packets before and after the scattering, i.e., the incident, the reflected, and the transmitted phonons, as indicated for example in (**c**)

v_{DB} is estimated by considering the speed of the central particle of the DB. The results for the movable and immobile DBs are shown by circles and squares, respectively. Vertical dashed lines indicate the edges of the phonon band. Phonons with frequencies very close to the edges of the phonon band are not considered since they have very small group velocity. From Fig. 13.12a it can be seen that the movable DB, which has relatively small amplitude, does not interact with the majority of the phonon spectrum, being transparent for the phonons with $\omega_P > 1.3$. On the other hand, the immobile, large-amplitude DB is an efficient scatterer of the phonons. It completely reflects phonons with $\omega_P < 1.4$. The transmission coefficient increases monotonically with increasing ω_P above this value, but it does not exceed $Tr = 0.7$ even for the short-wavelength phonons having frequencies close to the upper edge of the phonon spectrum. From Fig. 13.12b one can see that the DB energy does not change as a result of the interaction with the small-amplitude phonon, and this is true for both movable and immobile DBs. Finally, according to the results presented in Fig. 13.12c, the long-wavelength phonons with $\omega_P < 1.4$ accelerate the movable DB toward the incident phonon wave. The immobile DB is not accelerated by small-amplitude phonons of any frequencies.

Several examples of phonon-DB interaction in the case of small-amplitude phonons excited with $A_P = 0.04$ are given in Fig. 13.13, where the left (right) panels are for the movable (immobile) DB. The total energy of particles is shown on the (k, t) plane with denser color. Three kinds of phonon frequencies, i.e., $\omega_P = 1.08$ in Fig. 13.13a, b, $\omega_P = 1.15$ in Fig. 13.13c, d, and $\omega_P = 2.0$ in Fig. 13.13e, f, are considered. Since all the figures are plotted in a similar manner, we only indicate the incident, reflected, and transmitted phonon wave packet, as well as the DB in Fig. 13.13c for an example. As shown, the DB is initially at rest at $k - k_c = 0$, but after the interaction with the phonon wave it moves to the left. The source of the phonon wave packets is located at $k - k_c = -500$ and it works from $t = 0$ till $t = 1500$. This causes the phonon wave packets moving to the right to interact with the DB. The movable DB is accelerated by the long-wavelength phonons (with small ω_P), as shown in Fig. 13.13a, c. In these cases, after interaction with the phonon wave packets, the DB starts to move toward the incident phonon. For the phonons with $\omega_P > 1.3$, the movable DB is transparent as illustrated in Fig. 13.13e. The immobile DB fully reflects phonons with $\omega_P < 1.4$, see Fig. 13.13b, d. But for a larger frequency case, phonons are partly reflected and partly transmitted after interaction with the immobile DB, as indicated in Fig. 13.13f.

13.7.2 Effects of Phonon Amplitude

In Sect. 13.7.1, the scattering of small-amplitude phonons by DBs for different phonon frequencies was examined. Here we consider the effect of phonon amplitude on this scattering. Two values of DB frequency are considered: $\omega_P = 1.1$ and $\omega_P = 1.2$. Our main results are presented in Figs. 13.14 and 13.15, where the transmission coefficient Tr, the DB energy E_{DB}, and the DB velocity v_{DB} after the inter-

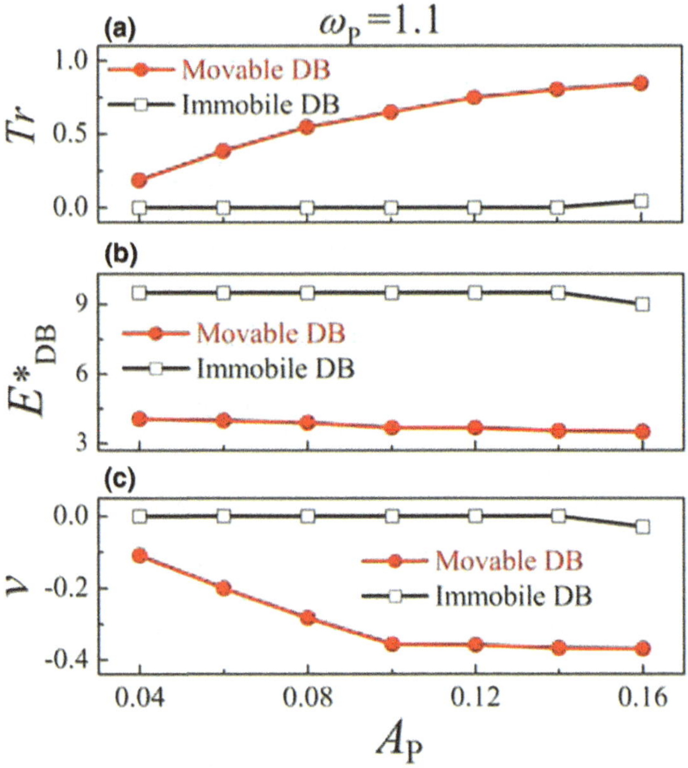

Fig. 13.14 The transmission coefficient Tr (**a**), the DB's energy $E^*_{\rm DB}$ (**b**), and the DB's velocity $v^*_{\rm DB}$ (**c**) as a function of the incident phonon amplitude $A_{\rm P}$ for $\omega_{\rm P} = 1.1$ and $k_4 = 1$, where the circles and squares indicate the results of movable and immobile DBs, respectively

action with the phonon wave as a function of phonon amplitude are plotted. The results for the movable and immobile DBs are shown by circles and squares.

Figure 13.14a shows that in the case of $\omega_{\rm P} = 1.1$, the transparency of the movable DB increases with the increase of the phonon amplitude, but this is not the case for the immobile DB. The latter remains reflective in the whole studied range of phonon amplitudes. Such increase of the mobile DB's transparency is reasonable and it is related to the acceleration of the DB during the interaction with the phonon wave packets (see Fig. 13.14c). It turns out that compared to the standing DB, the moving DB is more transparent. Furthermore, Fig. 13.14b shows that the movable DB loses a part of its energy due to the scattering with the phonon wave packet and this effect is stronger for a larger phonon amplitude. However, for the immobile DB, the case of loss of energy only happens when the largest phonon amplitude investigated is applied.

In the case of $\omega_{\rm P} = 1.2$ (see Fig. 13.15), for the movable DB we have trends similar to those observed in Fig. 13.14. Namely, the movable DB is always accelerated by

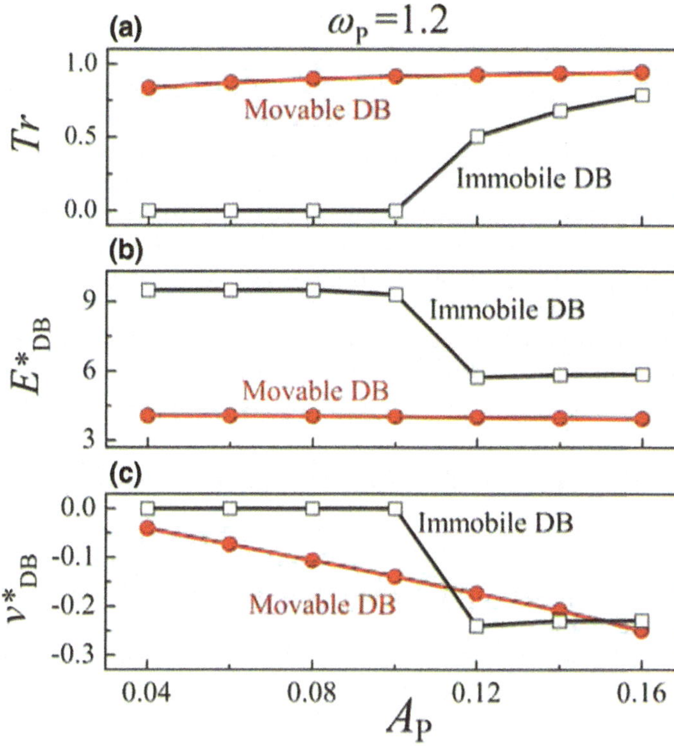

Fig. 13.15 The same as in Fig. 13.14, but for the phonon frequency $\omega_P = 1.2$

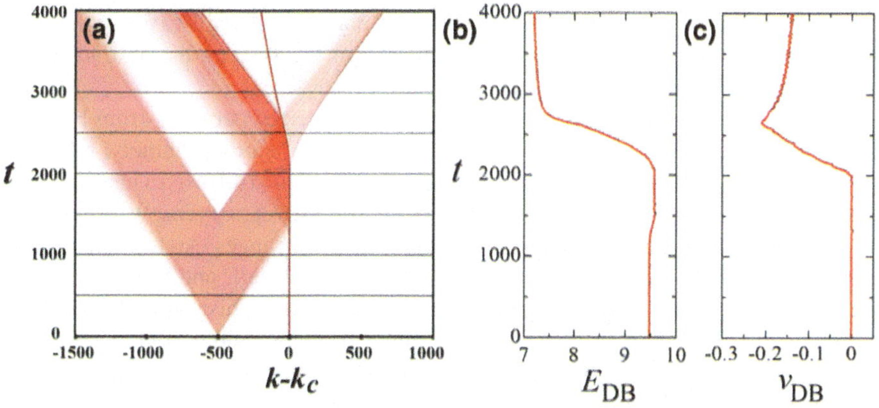

Fig. 13.16 Interaction of the phonon wave packet ($A_P = 0.11$ and $\omega_P = 1.2$) with the immobile DB ($A_{DB} = 1.25$). **a** Energy distribution in (k, t) plane. **b** and **c** show, respectively, the time evolution of the DB's energy (**b**) and velocity (**c**). All these results are under $k_4 = 1$

the incident phonon, and after the scattering, the movable DB loses a part of its energy. Both effects become stronger when the phonon amplitude is larger. However, for the immobile DB, a completely different picture emerges. For $A_P \le 0.1$ (see Fig. 13.15), the immobile DB completely reflects the phonon, i.e., it always keeps its energy conserved and has zero velocity after interaction with the phonon wave packet. However, for $A_P \ge 0.11$, the DB loses a substantial part of its energy, becomes mobile and hence, more transparent for the phonons. Figure 13.16 presents an example for $A_P = 0.11$, where the time evolution of the energy distribution, the DB energy, and the DB velocity are plotted in (a)–(c), respectively. As it can be seen, the phonon wave packet reaches the location of the DB at about $t = 1500$. Between $t = 1500$ and $t = 2000$ the DB is at rest keeping its energy (see Fig. 13.16b, c). Only for $t \ge 2000$, the DB starts to move and radiates energy being partly transparent for the phonon.

13.8 Relating Discrete Breather Properties to Heat Spread

We summarize below the results on DB properties and relate them to thermal transport. In the ϕ^4 lattice, from our numerical analysis, we know: (i) the DB's mobility depends on its degree of spatial localization, i.e., the parameter θ in the ansatz (13.17). The DB is movable when θ is below a threshold value, $\theta \approx 0.7$, and becomes immobile for larger θ. (ii) θ is related to DB's amplitude A_{DB}, see Table 13.1. For increasing anharmonicity parameter k_4, the threshold value of θ is achieved for smaller DB amplitudes. For instance, for $k_4 = 1$, DBs are immobile when $A_{DB} > 1$, while for $k_4 = 5$ they are immobile already for $A_{DB} > 0.5$. (iii) The movable DBs scatter only long-wavelength phonons and they are accelerated by such phonons, while the immobile DBs are very efficient scatterers for phonons of any wavelengths. This scattering will influence the transport of energy in the system, as the transport is mainly carried by the phonon waves.

Coming back to the heat spread in a system at finite temperature, from Fig. 13.7 one can identify three ranges of the nonlinearity parameter k_4, namely, relatively small (large) k_4 corresponds to range I (III), where ballistic (diffusive) heat spread is observed, while in the intermediate range II of k_4, a crossover happens. With the above knowledge of DBs, one can understand why these three ranges exist. In the range I, almost no DBs are excited since the anharmonicity is small, and thus the phonon waves can propagate ballistically without any scattering. This naturally results in the ballistic U shaped heat spread. In the range II, as the anharmonicity becomes stronger, DBs emerge, but these DBs are mainly of the mobile type. For k_4 increasing within the range II, the number of mobile DBs in the chain increases and they scatter phonon waves more efficiently, which then changes the scaling behaviors as shown in Fig. 13.7. Surely, this phonon-DBs scattering process will eventually induce the diffusive heat spread as long as the time is long enough (with a long enough system size). That is why we see the diffusive transport at $k_4 = 1$. Further increase of k_4 suppresses the DB mobility. Immobile DBs can localize energy and scatter phonons efficiently and their appearance in the system leads to diffusive transport; that is the reason why the Gaussian-like heat spreading density now becomes narrower and narrower with the increase of k_4 (as shown in Fig. 13.4f) providing the mechanism for the heat transport obeying Fourier's law, as a result of multiple phonon scattering.

13.9 Summary and Perspectives

To summarize, this chapter presents a possible microscopic origin of the normal thermal transport in the ϕ^4 lattice with sufficiently large anharmonicity parameter (for fixed temperature) or at sufficiently large temperature (for fixed anharmonicity parameter). In the present study the temperature was fixed at $T = 0.5$ and the value of the anharmonicity parameter was varied.

With a relatively strong anharmonicity (the on-site potential is anharmonic, while the inter-particle bonds are harmonic), one can clearly identify the system size independence of the thermal conductivity. This is a general piece of evidence for the diffusive thermal transport obeying Fourier's law. Nevertheless, the well-known argument that nonconservation of momentum is the reason for normal heat transport in the ϕ^4 model, does not give much information about the underlying microscopic mechanism. Motivated by this, we have suggested to investigate the heat spread in the system by varying the strength of the anharmonicity, k_4. In general, we find that under a given temperature and with the increase of k_4, the heat spread undergoes a crossover from ballistic to normal transport, characterized by the corresponding scaling behavior. In particular, such a crossover process is accompanied by some rich details, i.e., if one looks at the shapes of the heat spreading densities, at a small k_4 we always see a U shape [64], while with the increase of k_4, this ballistic U shape is destroyed starting from its fronts. The damping of the fronts then causes the central peak of the density to be humped, and finally a shape close to the Gaussian like density is observed. This Gaussian density will become narrower on further increase of k_4. Such rich details provide a more complete picture to our understanding of the thermal transport of the system.

Bearing such a whole picture in mind, we have further shown that this crossover process seems to be closely related to the system's DB properties. Both movable and immobile DBs can be excited in the systems at zero and finite temperatures, depending on the DB amplitude and anharmonicity strength (k_4). For large enough k_4, the system supports immobile DBs. These DBs can localize energy and scatter phonons very efficiently, and thus contribute to the narrower and narrower Gaussian heat spreading densities. For a relatively small k_4, one has less chance to observe DB formation. In this case, the transport is carried by phonon waves which propagate ballistically along the system, therefore resulting in the U shaped ballistic energy density. Between the mentioned two cases (in an intermediate range of k_4), the movable DBs are present in the system. This kind of DBs scatter long-wavelength phonon waves, thus contributing to the crossover process.

Finally, both of the above macroscopic heat spreading behavior and microscopic DBs dynamics imply that the anharmonicity here can play a critical role. We should note that the ϕ^4 lattice here only can represent one kind of systems with nonconserved momentum, as the anharmonicity lies only in the on-site potential. Quite recently, in the work of [72, 73] it has been found that, applying the anharmonicity into the interparticle interactions and keeping the on-site potential harmonic makes the system bearing momentum-nonconserving feature as well but following a quite unusual heat spreading behavior, i.e., for the Toda system with a harmonic pinning, still, the ballistic-type heat spread, similarly to those found in the integrable systems will be observed [72]. For the β-Fermi–Pasta–Ulam systems along with harmonic on-site potential, the superdiffusive heat spread has been shown [73]. Such recent results are quite unexpected in light of the general belief that the diffusive Gaussian-like heat spread occurs in the momentum-nonconserving systems. Therefore, it is of interest to explore DB properties in such systems to further extend this chapter's conclusions. Another very important case to analyze in the future is the model with soft-type anharmonicity coupling and without on-site potential. In this case DBs cannot exist and affect thermal conductivity. There exists a class of discrete ϕ^4 models showing no static Peierls–Nabarro barrier [74–77]. Analysis of DB properties and heat transport in such systems is of great interest as well.

Acknowledgements D.X. is supported by the NNSF (Grant No. 11575046) of China; the NSF (Grant No. 2017J06002) of Fujian provice of China; the Qishan scholar research fund of Fuzhou university of China. S.V.D. is supported by the Russian Science Foundation, grant No. 16-12-10175 and by The Tomsk State University competitiveness improvement programme.

References

1. S. Lepri, R. Livi, A. Politi, Phys. Rep. **377**, 1 (2003)
2. A. Dhar, Adv. Phys. **57**, 457 (2008)
3. C.W. Chang, D. Okawa, H. García, A. Majumdar, A. Zettl, Phys. Rev. Lett. **101**, 075903 (2008)
4. A.A. Balandin, S. Ghosh, W.Z. Bao, I. Calizo, D. Teweldebrhan, F. Miao, N. Lau, Nano Lett. **8**, 902 (2008)
5. V. Lee, C.-H. Wu, Z.-X. Lou, W.-L. Lee, C.-W. Chang, Phys. Rev. Lett. **118**, 135901 (2017)
6. G. Casati, J. Ford, F. Vivaldi, W.M. Visscher, Phys. Rev. Lett. **52**, 1861 (1984)
7. S. Lepri, R. Livi, A. Politi, Phys. Rev. Lett. **78**, 1896 (1997)
8. C.-W. Chang, in *Thermal Transport in Low Dimensions*, ed. by S. Lepri. Lecture Notes in Physics, vol. 921 (Springer, Berlin, 2016), p. 305
9. T.-K. Hsiao, B.-W. Huang, H.-K. Chang, S.-C. Liou, M.-W. Chu, S.-C. Lee, C.-W. Chang, Phys. Rev. B **91**, 035406 (2015)
10. T.-K. Hsiao, H.-K. Chang, S.-C. Liou, M.-W. Chu, S.-C. Lee, C.-W. Chang, Nat. Nanotechnol. **8**, 534 (2013)
11. J. Lee, J. Lim, P. Yang, Nano Lett. **15**, 3273 (2015)
12. Y.R. Koh, M. Shirazi-Hd, B. Vermeersch, A.M.S. Mohammed, J. Shao, G. Pernot, J.-H. Bahk, M.J. Manfra, A. Shakouri, Appl. Phys. Lett. **109**, 243107 (2016)
13. B. Li, L. Wang, G. Casati, Appl. Phys. Lett. **88**, 143501 (2006)
14. K. Joulain, J. Drevillon, Y. Ezzahri, J. Ordonez-Miranda, Phys. Rev. Lett. **116**, 200601 (2016)
15. L. Wang, B. Li, Phys. Rev. Lett. **99**, 177208 (2007)
16. A. Fornieri, C. Blanc, R. Bosisio, S. D'Ambrosio, F. Giazotto, Nat. Nanotechnol. **11**, 258 (2016)

17. S. Murad, I.K. Puri, Appl. Phys. Lett. **102**, 193109 (2013)
18. B. Li, L. Wang, G. Casati, Phys. Rev. Lett. **93**, 184301 (2004)
19. B. Liu, J.A. Baimova, C.D. Reddy, S.V. Dmitriev, W.K. Law, X.Q. Feng, K. Zhou, Carbon **79**, 236 (2014)
20. B. Liu, C.D. Reddy, J. Jiang, H. Zhu, J.A. Baimova, S.V. Dmitriev, K. Zhou, J. Phys. D Appl. Phys. **47**, 165301 (2014)
21. S. Hu, M. An, N. Yang, B. Li, Small **13**, 1602726 (2017)
22. Z. Duan, D. Liu, G. Zhang, Q. Li, C. Liu, S. Fan, Nanoscale **9**, 3133 (2017)
23. N. Li, J. Ren, L. Wang, G. Zhang, P. Hanggi, B. Li, Rev. Mod. Phys. **84**, 1045 (2012)
24. M. Madovan, Nature **503**, 209 (2013)
25. Z. Rieder, J.L. Lebowitz, E. Lieb, J. Math. Phys. **8**, 1073 (1967)
26. G. Basile, C. Bernardin, S. Olla, Phys. Rev. Lett. **96**, 204303 (2006)
27. A.A. Sokolov, A.M. Krivtsov, W.H. Muller, Phys. Mesomech. **20**, 305 (2017)
28. V.A. Kuzkin, A.M. Krivtsov, Phys. Solid State **59**, 1051 (2017)
29. V.A. Kuzkin, A.M. Krivtsov, Dokl. Phys. **62**, 85 (2017)
30. M.B. Babenkov, A.M. Krivtsov, D.V. Tsvetkov, Phys. Mesomech. **19**, 282 (2016)
31. A.M. Krivtsov, Dokl. Phys. **60**, 407 (2015)
32. V.A. Kuzkin, A.M. Krivtsov, Phys. Status Solidi B **252**, 1664 (2015)
33. V.A. Kuzkin, A.M. Krivtsov, J. Phys. Condens. Matter **29**, 505401 (2017)
34. A.V. Savin, Y.S. Kivshar, B. Hu, Phys. Rev. B **82**, 195422 (2010)
35. A.V. Savin, B. Hu, Y.S. Kivshar, Phys. Rev. B **80**, 195423 (2009)
36. T. Prosen, D.K. Campbell, Phys. Rev. Lett. **84**, 2857 (2000)
37. B. Hu, B. Li, H. Zhao, Phys. Rev. E **57**, 2992 (1998)
38. B. Hu, B. Li, H. Zhao, Phys. Rev. E **61**, 3828 (2000)
39. D. Chen, S. Aubry, G.P. Tsironis, Phys. Rev. Lett. **77**, 4776 (1996)
40. P.G. Kevrekidis, J. Cuevas-Maraver, D.E. Pelinovsky, Phys. Rev. Lett. **117**, 094101 (2016)
41. D. Xiong, D. Saadatmand, S.V. Dmitriev, Phys. Rev. E **96**, 042109 (2017)
42. D. Saadatmand, D. Xiong, V.A. Kuzkin, A.M. Krivtsov, A.V. Savin, S.V. Dmitriev, Phys. Rev. E **97**, 022217 (2018)
43. L. Delfini, S. Lepri, R. Livi, A. Politi, J. Stat. Mech. Theory Exp. P02007 (2007)
44. O. Narayan, S. Ramaswamy, Phys. Rev. Lett. **89**, 200601 (2002)
45. G.R. Lee-Dadswell, B.G. Nickel, C.G. Gray, Phys. Rev. E **72**, 031202 (2005)
46. H. Beijeren, Phys. Rev. Lett. **108**, 180601 (2012)
47. H. Spohn, J. Stat. Phys. **154**, 1191 (2014)
48. F. Piazza, S. Lepri, Phys. Rev. B **79**, 094306 (2009)
49. R. Kubo, M. Toda, N. Hashitsume, *Statistical Physics II: Nonequilirium Statistical Mechanics* (Springer, New York, 1991)
50. D. Andrieux, P. Gaspard, J. Stat. Mech. Theory Exp. P02006 (2007)
51. S. Chen, Y. Zhang, J. Wang, H. Zhao, Phys. Rev. E **89**, 022111 (2014)
52. E. Helfand, Phys. Rev. **119**, 1 (1960)
53. V. Zaburdaev, S. Denisov, J. Klafter, Rev. Mod. Phys. **87**, 483 (2015)
54. B. Li, J. Wang, L. Wang, G. Zhang, Chaos **15**, 015121 (2005)
55. H. Zhao, Phys. Rev. Lett. **96**, 140602 (2006)
56. S. Chen, Y. Zhang, J. Wang, H. Zhao, Phys. Rev. E **87**, 032153 (2013)
57. G.P. Tsironis, A.R. Bishop, A.V. Savin, A.V. Zolotaryuk, Phys. Rev. E **60**, 6610 (1999)
58. M. Eleftheriou, S. Flach, Physica D **202**, 142 (2005)
59. R. Khomeriki, S. Lepri, S. Ruffo, Phys. Rev. E **70**, 066626 (2004)
60. A. Scott, *Nonlinear Science* (Oxford University Press, New York, 1999)
61. A.A. Kistanov, R.T. Murzaev, S.V. Dmitriev, V.I. Dubinko, V.V. Khizhnyakov, JETP Lett. **99**, 353 (2014)
62. J.M. Thijssen, *Computational Physics* (Cambridge University Press, Cambridge, 2007)
63. T. Prosen, D.K. Campbell, Chaos **15**, 015117 (2005)
64. D. Xiong, F. Thiel, E. Barkai, Phys. Rev. E **96**, 022114 (2017)
65. A.J. Sievers, S. Takeno, Phys. Rev. Lett. **61**, 970 (1988)

66. M. Johansson, K.O. Rasmussen, Phys. Rev. E **70**, 066610 (2004)
67. T. Cretegny, S. Aubry, S. Flach, Physica D **119**, 73 (1998)
68. Y. Doi, A. Nakatani, Key Eng. Mat. **340–341**, 997 (2007)
69. P. Majumdar, S. Panda, Phys. Rev. E **82**(3), 036607 (2010)
70. B. Rumpf, Europhys. Lett. **78**, 26001 (2007)
71. S.-S. Lee, S. Kim, Int. J. Mod. Phys. B **14**, 1903 (2000)
72. J.L. Lebowitz, J.A. Scaramazza (2018), arXiv:1801.07153
73. D. Xiong, Y. Zhang, Phys. Rev. E **98**, 012130 (2018)
74. J.M. Speight, Nonlinearity **10**, 1615 (1997)
75. P.G. Kevrekidis, Physica D **183**, 68 (2003)
76. O.F. Oxtoby, D.E. Pelinovsky, I.V. Barashenkov, Nonlinearity **19**, 217 (2006)
77. S.V. Dmitriev, P.G. Kevrekidis, N. Yoshikawa, D.J. Frantzeskakis, Phys. Rev. E **74**, 046609 (2006)

Index

P. G. Kevrekidis and J. Cuevas-Maraver (eds.), *A Dynamical Perspective on the ϕ^4 Model*, Nonlinear Systems and Complexity 26,
https://doi.org/10.1007/978-3-030-11839-6

The manufacturer's authorised representative in the EU is Springer Nature Customer Service Centre GmbH, Europaplatz 3, 69115 Heidelberg, Germany. If you have any concerns regarding our products, please contact ProductSafety@springernature.com

Printed and bound by CPI Group (UK) Ltd, Croydon, CR0 4YY
30/09/2026
02225228-0012